Thermoelastic Models of Continua

SOLID MECHANICS AND ITS APPLICATIONS
Volume 118

Series Editor: G.M.L. GLADWELL
Department of Civil Engineering
University of Waterloo
Waterloo, Ontario, Canada N2L 3GI

Aims and Scope of the Series

The fundamental questions arising in mechanics are: *Why?, How?,* and *How much?* The aim of this series is to provide lucid accounts written by authoritative researchers giving vision and insight in answering these questions on the subject of mechanics as it relates to solids.

The scope of the series covers the entire spectrum of solid mechanics. Thus it includes the foundation of mechanics; variational formulations; computational mechanics; statics, kinematics and dynamics of rigid and elastic bodies: vibrations of solids and structures; dynamical systems and chaos; the theories of elasticity, plasticity and viscoelasticity; composite materials; rods, beams, shells and membranes; structural control and stability; soils, rocks and geomechanics; fracture; tribology; experimental mechanics; biomechanics and machine design.

The median level of presentation is the first year graduate student. Some texts are monographs defining the current state of the field; others are accessible to final year undergraduates; but essentially the emphasis is on readability and clarity.

For a list of related mechanics titles, see final pages.

Thermoelastic Models of Continua

D. IEŞAN

Department of Mathematics,
"Al. I. Cuza" University,
Iaşi, Romania

KLUWER ACADEMIC PUBLISHERS
DORDRECHT / BOSTON / LONDON

A C.I.P. Catalogue record for this book is available from the Library of Congress.

ISBN 1-4020-2309-X (HB)
ISBN 1-4020-2310-3 (e-book)

Published by Kluwer Academic Publishers,
P.O. Box 17, 3300 AA Dordrecht, The Netherlands.

Sold and distributed in North, Central and South America
by Kluwer Academic Publishers,
101 Philip Drive, Norwell, MA 02061, U.S.A.

In all other countries, sold and distributed
by Kluwer Academic Publishers,
P.O. Box 322, 3300 AH Dordrecht, The Netherlands.

Printed on acid-free paper

Printed in the Netherlands.

Contents

Introduction

In recent years new continuous models of deformable bodies have been intensively studied. It is well known that material response to external stimuli depends heavily on the motions of its inner structure. Classical elasticity ignores this effect by ascribing only translation degrees of freedom to material points of the body.

This work is concerned with the basic problems of the theory of thermoelasticity for three generalized models of continuous bodies: porous media, micropolar solids and nonsimple bodies. The theory of thermoelasticity is of interest both from a mathematical and a technical point of view.

This book consists of three main parts. In the first part (Chapters 1-4) we focus attention on the theory of materials with voids. Porous solids play an important role in nearly all fields of engineering: soil mechanics, petroleum industry, material science as well as in biomechanics. An account of the historical development of the porous media theories as well as reference to various contributions may be found in the monographs by Bowen [28], De Boer [85], Atkin and Craine [8], Bedford and Drumheller [16], Rajagopal and Tao [311].

The theory of elastic materials with voids is one of the simple extensions of the classical theory of elasticity for the treatment of porous solids in which the matrix material is elastic and the interstices are void of material. This theory seems to be an adequate tool to describe the behaviour of granular materials like rock, soils and manufactured porous bodies. The theory of elastic materials with voids has been established by Nunziato and Cowin [292], [73] and has recently recieved considerable interest.

The second part of the work (Chapters 5 and 6) contains a study of micropolar thermoelasticity. Micropolar theory, by including intrinsic rotations of the microconstituents, provides a model that can support body and surface couples. For engineering applications it can model composites with rigid chopped fibres, elastic solids with rigid granular inclusions, and other industrial materials. The origin of the rational theories of polar continua goes back to E. and F. Cosserat [71] who gave a development of the mechanics of continuous media in which each material point has the six degrees of freedom of a rigid body. Since there are already several expositions of the historical development of this field and more general theories of polar continua (cf. Eringen [105], Dyszlewicz [96]), we do not intend to trace the history here. The theory of micropolar thermoelasticity has been discussed in various review articles (see, e.g., Nowacki [289], Eringen [102] and Dyszlewicz [96]). We present some results which have not been discussed previously in monographs.

The third part of this book (Chapter 7) is concerned with the theory of nonsimple thermoelastic media. The theory of nonsimple elastic media, which is characterized by the inclusion of higher gradients of displacement in the basic postulates, was established by Toupin [364], [365], Mindlin and Tiersten [262] and Mindlin [263].

The domain of applicability of different non-classical models has been investigated by Kunin [230] and Eringen [105].

Our purpose in this work is to present a unified and systematic treatment of some results established in the theory of generalized models of thermoelastic bodies. By no means can any claim be made in regard to completeness of the coverage. We have tried to maintain the level of rigor now customary in applied mathematics. However, in order to ease the burden on the reader, many results are stated with hypotheses that are more stringent than necessary. The applications included are problems considered relevant to the purpose of the text.

Chapter 1

THERMOELASTIC MATERIALS WITH VOIDS

1.1 Preliminaries. The Laws of Thermodynamics

In [120], Goodman and Cowin established a continuum theory of granular materials with interstitial voids. The basic concept underlying this theory is that of a material for which the bulk density is written as the product of two fields, the density field of the matrix material and the volume fraction field. This representation was employed by Nunziato and Cowin [292] to develop a nonlinear theory of elastic materials with voids. The intended applications of the theory is to elastic bodies with small voids or vacuous pores which are distributed throughout the material. The concept of a distributed body proposed by Goodman and Cowin [120] has been used in a theory of mixtures by Passman [300] and Nunziato and Walsh [293]. In [37], Capriz and Podio-Guidugli have studied the elastic materials with voids as materials with spherical structure. The linear theory of elastic materials with voids has been established by Cowin and Nuziato [73]. The first investigations in the theory of thermoelastic materials with voids are due to Nunziato and Cowin [292], Jarić and Golubović [205] and Ieşan [180].

We consider a body that at time t_0 occupies the region B of Euclidean three-dimensional space E^3. In what follows,unless specified to the contrary, B will denote a bounded regular region (see, for example, Gurtin [139], Sect. 5). The configuration of the body at time t_0 is taken as the reference configuration. The motion of the body is referred to the reference configuration and a fixed system of rectangular Cartesian axes. We assume that B is at rest relative to the considered system of axes. We identify a typical particle X of the body with its position $\boldsymbol{X}$ in the reference configuration. The coordinates of a typical particle X in B are X_A $(A = 1, 2, 3)$. The coordinates of this particle in the position $\mathbf{x}$ at time t are denoted by x_i. We have

$$\boldsymbol{x} = \boldsymbol{x}(\boldsymbol{X}, t), \ (\boldsymbol{X}, t) \in \overline{B} \times I, \tag{1.1.1}$$

where I is a given interval of time. We assume the continuous differentiability of $\boldsymbol{x}$ with respect to the variables X_A and t as many times as required and

$$\det\left(\frac{\partial x_i}{\partial X_A}\right) > 0 \ \text{ on } \overline{B} \times I. \tag{1.1.2}$$

The configuration of the body at the time t is denoted by B' and is called present configuration.

We shall employ the usual summation and differentiation conventions: lower and upper case Latin subscripts (unless otherwise specified) are understood to range over the integers $(1, 2, 3)$ whereas Greek subscripts are confined to the range $(1, 2)$, summation over repeated subscripts is implied and subscripts preceded by a comma denote partial differentiation with respect to the corresponding Cartesian coordinates. Throughout this book, a superposed dot denotes the material derivative with respect to the time. Letters in boldface stand for tensors of an order $p \geq 1$, and if $\boldsymbol{v}$ has the order p, we write $v_{ij...k}$ (p subscripts) for the rectangular Cartesian components of $\boldsymbol{v}$. Let (t_0, t_1) denote a given interval of time. We say that f is of class $C^{M,N}$ on $B \times (t_0, t_1)$ if f is continuous on $B \times (t_0, t_1)$ and the functions $\dfrac{\partial^m}{\partial X_A \partial X_B ... \partial X_K}\left(\dfrac{\partial^n f}{\partial t^n}\right)$, $m \in \{0, 1, ..., M\}$, $n \in \{0, 1, ..., N\}$, $m + n \leq \max(M, N)$, exist and are continuous on $B \times (t_0, t_1)$. We write C^N for $C^{N,N}$.

The concept of a distributed body asserts that the mass density ρ in the present configuration has the decomposition

$$\rho = \nu \hat{\rho}, \tag{1.1.3}$$

where $\hat{\rho}$ is the density of the matrix material and ν is the volume fraction field. This representation introduces an additional degree of kinematic freedom. Clearly, the relation (1.1.3) also holds for the reference configuration B. Thus, we have

$$\rho_0 = \nu_0 \hat{\rho}_0, \tag{1.1.4}$$

where ρ_0 is the mass density in the reference configuration, ν_0 is the volume fraction field at time t_0, and $\hat{\rho}_0$ is the density of the matrix material at time t_0. We assume that ν is a function of class $C^{2,1}$ on $B \times (t_0, t_1)$.

We are now in position to postulate the balance of energy and the second law of thermodynamics governing the response of materials with voids. We consider an arbitrary region $\mathcal{P}$ of the continuum bounded by a surface $\partial\mathcal{P}$ at time t, and we suppose that P is the corresponding region at time t_0, bounded by the surface ∂P. Let the outward unit normal at ∂P be N_A, and let the outward unit normal at $\partial\mathcal{P}$ be n_i, referred to the rectangular frame of reference.

We postulate *the conservation of energy* for every regular region P of B and every time, in the form (cf. [120], [292])

$$\begin{aligned}\frac{d}{dt}\int_P \rho_0\left(e+\frac{1}{2}\,\dot{\boldsymbol{x}}^2+\frac{1}{2}\,\kappa\dot{\nu}^2\right)dV=&\int_P \rho_0(\boldsymbol{f}\cdot\dot{\boldsymbol{x}}+\ell\dot{\nu}+S)dV+\\&+\int_{\partial P}(\boldsymbol{T}\cdot\dot{\boldsymbol{x}}+H\dot{\nu}+Q)dA,\end{aligned} \tag{1.1.5}$$

where dV and dA are elements of volume and area in the reference configuration, e is the internal energy per unit mass, κ is the equilibrated inertia, $\boldsymbol{f}$ is the body force per unit mass, ℓ is the extrinsic equilibrated body force per unit mass, S is the heat supply per unit mass, $\boldsymbol{T}$ is the stress vector associated with the surface $\partial\mathcal{P}$, but measured per unit area of the surface ∂P, H is the equilibrated stress associated with the surface $\partial\mathcal{P}$ but measured per unit area of the surface ∂P, and Q is the heat flux across the surface $\partial\mathcal{P}$, measured per unit area of ∂P. The physical significances of the functions ℓ, H and κ are presented in the works of Goodman and Cowin [120], Nunziato and Cowin [292], Jenkins [206], Cowin and Leslie [72]. We assume that $\boldsymbol{f},\ell$ and S are continuous on $B\times(t_0,t_1)$, e is of class $C^{0,1}$ on $B_0\times(t_0,t_1)$, $\boldsymbol{T}$, H and Q are of class $C^{1,0}$ on $B\times(t_0,t_1)$ and continuous on $\overline{B}\times[t_0,t_1)$. We restrict our attention to the case when κ is a given function of $\boldsymbol{X}$ on B (see Nunziato and Cowin [292]).

Following the procedure of Green and Rivlin [125], we consider a second motion which differs from the given motion only by a constant superposed rigid translational velocity. We assume that $\dot{e},\boldsymbol{f},\ell,S,\boldsymbol{T},H$ and Q are unaltered by such superposed rigid velocity. If we denote $v_i=\dot{x}_i$, then the equation (1.1.5) is also true when v_i is replaced by v_i+a_i, where a_i are arbitrary constants, all other terms being unaltered. By subtraction we get

$$\left[\int_P \rho_0 f_i dV+\int_{\partial P} T_i dA-\int_P \rho_0\ddot{x}_i dV\right]a_i=0,$$

for all arbitrary constants a_i. Since the quantities in the square brackets are independent of a_i, it follows that

$$\int_P \rho_0 \ddot{x}_i dV = \int_P \rho_0 f_i dV + \int_{\partial P} T_i dA. \tag{1.1.6}$$

Let T_{Ai} be the first Piola-Kirchhoff stress tensor (see, for example, Truesdell and Noll [367], Sec. 43A). From (1.1.6), by the usual methods, we obtain

$$T_i = T_{Ki} N_K \tag{1.1.7}$$

and

$$T_{Ai,A} + \rho_0 f_i = \rho_0 \ddot{x}_i \quad \text{on } B\times(t_0, t_1). \tag{1.1.8}$$

In view of (1.1.7) and (1.1.8), the relation (1.1.5) reduces to

$$\begin{aligned} \int_P \rho_0(\dot{e} + \kappa\dot{\nu}\ddot{\nu})dV = \int_P [T_{Ai} v_{i,A} + \rho_0(\ell\dot{\nu} + S)]dV + \\ + \int_{\partial P} (H\dot{\nu} + Q)dA. \end{aligned} \tag{1.1.9}$$

We apply this equation to a region which in the reference state was a tetrahedron bounded by coordinate planes through the point X and by a plane whose unit normal is N_K, to obtain

$$(H - H_K N_K)\dot{\nu} + Q - Q_K N_K = 0, \tag{1.1.10}$$

where Q_K is the heat flux and H_K is the equilibrated stress associated with surfaces in B' which were originally coordinate planes perpendicular to the X_K-axes through the point X, measured per unit undeformed area. If we use (1.1.10) in (1.1.9) and apply the resulting equation to an arbitrary region, then we find the local form of the conservation of energy

$$\rho_0 \dot{e} = T_{Ki} v_{i,K} + H_K \dot{\nu}_{,K} - g\dot{\nu} + Q_{K,K} + \rho_0 S, \tag{1.1.11}$$

where g satisfies the equation

$$H_{K,K} + g + \rho_0 \ell = \rho_0 \kappa \ddot{\nu} \quad \text{on } B\times(t_0, t_1). \tag{1.1.12}$$

The function g is a dependent constitutive variable which is called the intrinsic equilibrated body force (cf. [120], [292]).

If we define the second Piola-Kirchhoff stress tensor T_{KL} by

$$T_{Ai} = x_{i,B} T_{AB}, \tag{1.1.13}$$

then the local balance energy (1.1.11) becomes

$$\rho_0 \dot{e} = T_{KL}\dot{x}_{i,K}x_{i,L} + H_K\dot{\nu}_{,K} - g\dot{\nu} + Q_{K,K} + \rho_0 S. \tag{1.1.14}$$

Clearly,

$$T_{KL}\dot{x}_{i,K}x_{i,L} = T_{KL}v_{i,j}x_{j,K}x_{i,L} = T_{KL}(d_{ij} + \omega_{ij})x_{j,K}x_{i,L},$$

where

$$d_{ij} = \frac{1}{2}\left(\frac{\partial v_i}{\partial x_j} + \frac{\partial v_j}{\partial x_i}\right), \quad \omega_{ij} = \frac{1}{2}\left(\frac{\partial v_i}{\partial x_j} - \frac{\partial v_j}{\partial x_i}\right).$$

We consider a motion of the continuum which differs from the given motion only by a superposed uniform rigid body angular velocity, and let us assume that $\rho_0, \dot{e}, T_{KL}, H_K, Q_K, S$ and $\dot{\nu}$ are unaltered by such motion. Suppose that at time t the body is rotated back into original orientation. Thus, the equation (1.1.14) is also true when ω_{ij} is replaced by $\omega_{ij} + \Omega_{ij}$, where Ω_{ij} is a constant arbitrary skew-symmetric tensor. Then we obtain

$$T_{KL} = T_{LK}. \tag{1.1.15}$$

We denote by E_{KL} the Lagrangian strain tensor,

$$E_{KL} = \frac{1}{2}\left(x_{i,K}x_{i,L} - \delta_{KL}\right) \text{ on } B\times(t_0, t_1), \tag{1.1.16}$$

where δ_{KL} is Kronecker's delta. In view of (1.1.15), the equation of energy (1.1.14) becomes

$$\rho_0 \dot{e} = T_{KL}\dot{E}_{KL} + H_K\dot{\nu}_{,K} - g\dot{\nu} + Q_{L,L} + \rho_0 S. \tag{1.1.17}$$

The *second law of thermodynamics* is the assertion that (cf. Truesdell and Noll [367], p. 295)

$$\frac{d}{dt}\int_P \rho_0 \eta dV - \int_P \frac{1}{T}\rho_0 S\, dV - \int_{\partial P} \frac{1}{T} Q\, dA \geq 0, \tag{1.1.18}$$

for every part P of B and every time. Here η is the entropy per unit mass, and T is the absolute temperature, which is assumed to be positive. We assume that η is of class $C^{0,1}$ and T is of class $C^{2,1}$ on $B\times(t_0, t_1)$. If we introduce the Helmholtz free-energy,

$$\psi = e - T\eta, \tag{1.1.19}$$

then the equation of energy can be written in the form

$$\rho_0(\dot{\psi} + \dot{T}\eta + T\dot{\eta}) = T_{KL}\dot{E}_{KL} + H_K\dot{\nu}_{,K} - g\dot{\nu} + Q_{L,L} + \rho_0 S. \tag{1.1.20}$$

1.2 Constitutive Equations. Thermodynamic Restrictions

A thermoelastic material with voids is defined as one for which the following constitutive equations hold

$$
\begin{aligned}
\psi &= \widetilde{\psi}(x_{i,A}, T, T_{,B}, \nu, \nu_{,M}, X_N),\\
T_{KL} &= \widetilde{T}_{KL}(x_{i,A}, T, T_{,B}, \nu, \nu_{,M}, X_N),\\
\eta &= \widetilde{\eta}(x_{i,A}, T, T_{,B}, \nu, \nu_{,M}, X_N),\\
Q_K &= \widetilde{Q}_K(x_{i,A}, T, T_{,B}, \nu, \nu_{,M}, X_N),\\
H_K &= \widetilde{H}_K(x_{i,A}, T, T_{,B}, \nu, \nu_{,M}, X_N),\\
g &= \widetilde{g}(x_{i,A}, T, T_{,B}, \nu, \nu_{,M}, X_N),\\
Q &= \widetilde{Q}(x_{i,A}, T, T_{,B}, \nu, \nu_{,M}, X_K, N_L),\\
H &= \widetilde{H}(x_{i,A}, T, T_{,B}, \nu, \nu_{,M}, X_K, N_L).
\end{aligned}
\tag{1.2.1}
$$

We assume that the response functions are of class C^2 on their domain. In the case of homogeneous bodies the constitutive functions do not depend explicitly on X_A. The constitutive equations must satisfy invariance conditions under superposed rigid-body motions. Hence

$$
\widetilde{\psi}(x_{i,A}, T, T_{,B}, \nu, \nu_{,M}, X_N) = \widetilde{\psi}(Q_{ij}x_{j,A}, T, T_{,B}, \nu, \nu_{,M}, X_N), \tag{1.2.2}
$$

for all proper orthogonal tensors Q_{ij}. The other response functionals must satisfy the same invariance conditions as $\widetilde{\psi}$. In view of (1.1.2) we may write $x_{i,A}$ in the polar form $x_{i,A} = R_{iK}M_{KA}$, where M_{KA} is a positive definite symmetric tensor and R_{iK} is a rotation tensor. We may take the special value $R_{jK}\delta_{Kr}$ for Q_{rj} in (1.2.2), δ_{Kr} being Kronecker's delta. Thus, we have

$$
\widetilde{\psi}(x_{i,A}, T, T_{,B}, \nu, \nu_{,N}, X_P) = \widetilde{\psi}(M_{KL}, T, T_{,B}, \nu, \nu_{,N}, X_P).
$$

In view of (1.1.16) we find that $2E_{KL} = M_{KA}M_{AL} - \delta_{KL}$. Since M_{KL} is a positive definite symmetric tensor, a single-valued function of M_{AB} can be replaced by a single-valued function of E_{KL}. Thus, the free-energy can be expressible in a different functional form

$$
\psi = \widehat{\psi}(E_{KL}, T, T_{,B}, \nu, \nu_{,M}, X_A). \tag{1.2.3}
$$

Clearly, $\widehat{\psi}$ satisfies the condition (1.2.2) for arbitrary proper orthogonal values of Q_{rs}. For a given deformation, $\dot{\nu}$ in (1.1.10) may be chosen arbitrarily so that, on the basis of the constitutive equations, we obtain

$$Q = Q_K N_K, \quad H = H_K N_K. \tag{1.2.4}$$

If we use (1.2.4) in (1.1.18) then we obtain, by usual arguments, the following local form of the second law of thermodynamics

$$\rho_0 T\dot{\eta} - \rho_0 S - Q_{K,K} + \frac{1}{T} Q_K T_{,K} \geq 0. \tag{1.2.5}$$

It follows from (1.1.20) and (1.2.5) that

$$T_{KL}\dot{E}_{KL} + H_K \dot{\nu}_{,K} - g\dot{\nu} - \rho_0(\dot{\psi} + \dot{T}\eta) + \frac{1}{T} Q_K T_{,K} \geq 0. \tag{1.2.6}$$

We now study the thermodynamic restrictions on the constitutive functionals. In view of (1.2.3) the inequality (1.2.6) becomes

$$\begin{aligned} &\left(T_{KL} - \frac{\partial \sigma}{\partial E_{KL}}\right) \dot{E}_{KL} - \left(\rho_0 \eta + \frac{\partial \sigma}{\partial T}\right) \dot{T} - \frac{\partial \sigma}{\partial T_{,K}} \dot{T}_{,K} - \\ &- \left(g + \frac{\partial \sigma}{\partial \nu}\right) \dot{\nu} + \left(H_K - \frac{\partial \sigma}{\partial \nu_{,K}}\right) \dot{\nu}_{,K} + \frac{1}{T} Q_K T_{,K} \geq 0, \end{aligned} \tag{1.2.7}$$

where $\sigma = \rho_0 \psi$. We assume that σ in (1.2.7) is arranged as a symmetric function of E_{KL}. For a given deformation, the inequality (1.2.7) is valid for all arbitrary values of $\dot{E}_{KL}, \dot{T}, \dot{T}_{,K}, \dot{\nu}$ and $\dot{\nu}_{,L}$, subject to $E_{KL} = E_{LK}$. Thus, in absence of internal constraints, from (1.2.7) we obtain (cf. Coleman and Noll [70], Carlson [40], p. 306, Nunziato and Cowin [292])

$$T_{KL} = \frac{\partial \sigma}{\partial E_{KL}}, \; \rho_0 \eta = -\frac{\partial \sigma}{\partial T}, \; g = -\frac{\partial \sigma}{\partial \nu}, \; H_K = \frac{\partial \sigma}{\partial \nu_{,K}}, \; \frac{\partial \sigma}{\partial T_{,K}} = 0$$

and

$$Q_K T_{,K} \geq 0. \tag{1.2.8}$$

We conclude that the constitutive equations of thermoelastic materials with voids can be written in the form

$$\begin{aligned} &\sigma = \widehat{\sigma}(E_{KL}, T, \nu, \nu_{,M}, X_A), \\ &T_{KL} = \frac{\partial \sigma}{\partial E_{KL}}, \; \rho_0 \eta = -\frac{\partial \sigma}{\partial T}, \; H_K = \frac{\partial \sigma}{\partial \nu_{,K}}, \; g = -\frac{\partial \sigma}{\partial \nu}, \\ &Q_A = \widehat{Q}_A(E_{KL}, T, T_{,M}, \nu, \nu_{,N}, X_B). \end{aligned} \tag{1.2.9}$$

In view of (1.2.9), the energy equation (1.1.20) takes the form

$$\rho_0 T\dot{\eta} = Q_{A,A} + \rho_0 S \quad \text{on } B\times(t_0, t_1). \tag{1.2.10}$$

As in classical thermoelasticity (see, e.g., Carlson [40]) the inequality (1.2.8) implies that

$$\widehat{Q}_A(E_{KL}, T, 0, \nu, \nu_{,M}, X_N) = 0. \tag{1.2.11}$$

1.3 Boundary-Initial-Value Problems

The basic equations of the nonlinear theory of thermoelastic materials with voids consist of equations of motion (1.1.8), the balance of equilibrated force (1.1.12), the energy equation (1.2.10), the constitutive equations (1.2.9) and the geometrical equations (1.1.16), on $B\times(t_0, t_1)$, where t_1 is some time instant that may be infinite. The functions $\rho_0, \kappa, f_i, \ell$ and S, and the constitutive functionals $\widehat{\sigma}$ and $\widehat{Q}_K$, are prescribed. The response functionals $\widehat{Q}_K$ are subjected to the restriction (1.2.8). To the field equations we must adjoin boundary conditions and initial conditions. In the case of the mixed boundary value problem the boundary conditions are

$$\begin{aligned} &x_i = \widehat{x}_i \quad \text{on } \overline{\Sigma}_1\times(t_0, t_1), \quad \nu = \widehat{\nu} \quad \text{on } \overline{\Sigma}_3\times(t_0, t_1), \\ &T = \widehat{T} \quad \text{on } \overline{\Sigma}_5\times(t_0, t_1), \quad T_{Ki}N_K = \widehat{T}_i \quad \text{on } \Sigma_2\times(t_0, t_1), \\ &H_K N_K = \widehat{H} \quad \text{on } \Sigma_4\times(t_0, t_1), \quad Q_K N_K = \widehat{Q} \quad \text{on } \Sigma_6\times(t_0, t_1), \end{aligned} \tag{1.3.1}$$

where Σ_i, $(i = 1, 2, ..., 6)$, are subsurfaces of ∂B such that $\overline{\Sigma}_1 \cup \Sigma_2 = \overline{\Sigma}_3 \cup \Sigma_4 = \overline{\Sigma}_5 \cup \Sigma_6 = \partial B$, $\Sigma_1 \cap \Sigma_2 = \Sigma_3 \cap \Sigma_4 = \Sigma_5 \cap \Sigma_6 = \emptyset$, and $\widetilde{x}_i, \widehat{\nu}, \widehat{T}, \widehat{H}, \widehat{T}_i$ and $\widehat{Q}$ are prescribed functions. The initial conditions are

$$\begin{aligned} &\boldsymbol{x}(\boldsymbol{X}, 0) = \boldsymbol{x}^0(\boldsymbol{X}), \quad \dot{\boldsymbol{x}}(\boldsymbol{X}, 0) = \boldsymbol{v}^0(\boldsymbol{X}), \quad \nu(\boldsymbol{X}, 0) = \nu^0(\boldsymbol{X}), \\ &\dot{\nu}(\boldsymbol{X}, 0) = \zeta^0(\boldsymbol{X}), \quad \eta(\boldsymbol{X}, 0) = \eta^0(\boldsymbol{X}), \quad \boldsymbol{X} \in \overline{B}, \end{aligned} \tag{1.3.2}$$

where $\boldsymbol{x}^0, \boldsymbol{v}^0, \nu^0, \zeta^0$ and η^0 are given. We assume that

(i) ρ_0 and κ are continuous and strictly positive on $\overline{B}$;
(ii) $\boldsymbol{f}, \ell$ and S are continuous on $\overline{B}\times[t_0, t_1)$;
(iii) $\boldsymbol{x}^0, \boldsymbol{v}^0, \nu^0, \zeta^0$ and η^0 are continuous on $\overline{B}$;
(iv) $\widehat{x}_i$ are continuous on $\overline{\Sigma}_1\times[t_0, t_1)$, $\widehat{\nu}$ is continuous on$\overline{\Sigma}_3\times[t_0, t_1)$, and $\widehat{T}$ is continuous on $\overline{\Sigma}_5\times[t_0, t_1)$;
(v) $\widehat{T}_i$ are continuous in time and piecewise regular on $\Sigma_2\times[t_0, t_1)$, $\widehat{H}$ is continuous in time and piecewise regular on $\Sigma_4\times[t_0, t_1)$, and $\widetilde{Q}$ is continuous in time and piecewise regular on $\Sigma_6\times[t_0, t_1)$.

The mixed problem of thermoelastodynamics consists in finding the functions x_i and ν of class C^2 and T of class $C^{2,1}$ on $B\times(t_0,t_1)$ that satisfy the equations (1.1.8), (1.1.12), (1.1.16), (1.2.9) and (1.2.10) in $B\times(t_0,t_1)$, the boundary conditions (1.3.1) and the initial conditions (1.3.2).

It is sometimes useful to have the above relations expressed in terms of quantities referred to the present configuration. These quantities are the Cauchy stress t_{ij}, the heat flux q_i, and the equilibrated stress h_i, associated with the coordinate planes and measured per unit area of the deformed body. In this case the equations (1.1.8) are replaced by (see, e.g., Truesdell and Noll [367], Eringen [101])

$$t_{ji,j} + \rho f_i = \rho \ddot{x}_i \quad \text{on } B'\times(t_0,t_1). \tag{1.3.3}$$

The balance of equilibrated force (1.1.12) can be written in the form

$$h_{i,i} + g' + \rho \ell = \rho \kappa \ddot{\nu} \quad \text{on } B'\times(t_0,t_1), \tag{1.3.4}$$

where

$$g' = \frac{1}{J} g, \quad J = \det(x_{i,K}). \tag{1.3.5}$$

The energy equation (1.2.10) becomes

$$q_{i,i} + \rho S = \rho T \dot{\eta} \quad \text{on } B'\times(t_0,t_1). \tag{1.3.6}$$

The boundary conditions corresponding to (1.3.1) are

$$\begin{aligned} x_i = \widetilde{x}_i \quad &\text{on } \overline{S}_1\times(t_0,t_1), & \nu = \widehat{\nu} \quad &\text{on } \overline{S}_3\times(t_0,t_1),\\ T = \widetilde{T} \quad &\text{on } \overline{S}_5\times(t_0,t_1), & t_{ji}n_j = \widetilde{t}_i \quad &\text{on } S_2\times(t_0,t_1),\\ h_i n_i = \widetilde{h} \quad &\text{on } S_4\times(t_0,t_1), & q_i n_i = \widetilde{q} \quad &\text{on } S_6\times(t_0,t_1), \end{aligned} \tag{1.3.7}$$

where S_k is the image of Σ_k, $(k = 1, 2, ..., 6)$, by motion, and $\widetilde{x}_i, \widetilde{\nu}, \widetilde{T}, \widetilde{t}_i, \widetilde{h}$ and $\widetilde{q}$ are given.

We record the following relations between the stress tensors (see, e.g., Truesdell and Noll [367], p. 124, Eringen [101], p. 109)

$$J t_{ij} = x_{i,K} T_{Kj} = x_{i,K} x_{j,L} T_{KL}. \tag{1.3.8}$$

Moreover, we have (see Green and Adkins [123], p. 260 and Eringen [101], p. 163)

$$J q_i = x_{i,K} Q_K, \quad J h_i = x_{i,K} H_K. \tag{1.3.9}$$

1.4 Continuous Data Dependence and Uniqueness Results

In [80], Dafermos has considered the classical nonlinear thermoelasticity for non-conductors and established continuous data dependence of smooth thermodynamic processes upon the initial state and supply terms. The case of heat-conducting elastic solids has been studied by Chiriţă [55]. The results of Dafermos have been extended by Ciarletta and Scalia [65], [68] to the theory of thermoelastic materials with voids.

Throughout this section we assume that the constitutive equations (1.2.1) hold and that the partial derivatives of the response functions $\widetilde{\sigma}$, $\widetilde{T}_{KL}$, $\widetilde{\eta}$, $\widetilde{Q}_K$, $\widetilde{H}_K$ and $\widetilde{g}$, at any fixed state $(x_{i,A}, T, T_{,B}, \nu, \nu_{,M}, X_N)$, are bounded functions on $\overline{B}\times[0,t_1]$. Moreover, we assume that

$$\rho_0(\boldsymbol{X}) \geq \rho_1 > 0,\ \kappa(\boldsymbol{X}) \geq \kappa_1 > 0,\ \boldsymbol{X} \in B. \tag{1.4.1}$$

We shall say that $U = (\boldsymbol{x}, \nu, T)$ is a smooth thermodynamic process if $\dot{x}_i, x_{i,K}, \dot{\nu}, \nu_{,L}, T$ and $T_{,A}$ are Lipschitz-continuous, uniformly on bounded subsets of their domain and satisfy the equations of motion (1.1.8), the balance of equilibrated force (1.1.12) and the balance law of energy

$$\begin{aligned}\frac{\partial}{\partial t}\left(\frac{1}{2}\rho_0\dot{x}_i\dot{x}_i + \frac{1}{2}\rho_0\kappa\dot{\nu}^2 + \sigma + \rho_0 T\eta\right) &= \\ = \rho_0(f_i\dot{x}_i + \ell\dot{\nu} + S) + (T_{Ki}\dot{x}_i + H_K\dot{\nu} + Q_K)_{,K},&\end{aligned} \tag{1.4.2}$$

on $B\times[0,t_1]$. A smooth thermodynamic process will be called admissible if it satisfies the Clausius-Duhem inequality (1.2.5). It follows that a smooth admissible thermodynamic process satisfies the relations (1.2.8)-(1.2.11). Throughout this section it will be convenient to denote the components of the deformation gradient $\boldsymbol{F} = (x_{i,K})$ by F_{iK} and the components of gradient $\boldsymbol{G} = (\nu_{,K})$ by G_K. Moreover, the symbol $|\cdot|$ denotes a norm either in an Euclidean vector space or in a tensor space, while $\|\cdot\|$ denotes an L_2-norm.

Let $U = (\boldsymbol{x}, \nu, T)$ and $\overline{U} = (\overline{\boldsymbol{x}}, \overline{\nu}, \overline{T})$ be two smooth admissible thermodynamic processes defined on $\overline{B}\times[0,t_1]$ and corresponding to the supply terms $(\boldsymbol{f}, \ell, S)$ and $(\overline{\boldsymbol{f}}, \overline{\ell}, \overline{S})$, respectively. We introduce the function D on $[0,t_1]$ by

$$\begin{aligned}D = \int_B \{\frac{1}{2}\rho_0[(\boldsymbol{v}-\overline{\boldsymbol{v}})^2 + \kappa(\dot{\nu}-\dot{\overline{\nu}})^2] + \sigma - \overline{\sigma} - \overline{T}_{Ki}(F_{iK}-\\ -\overline{F}_{iK}) - \overline{H}_K(G_K - \overline{G}_K) + \overline{g}(\nu-\overline{\nu}) + \rho_0\eta(T-\overline{T})\}dV,\end{aligned} \tag{1.4.3}$$

where

$$\begin{aligned}
&\boldsymbol{v} = \dot{\boldsymbol{x}},\ \overline{\boldsymbol{v}} = \dot{\overline{\boldsymbol{x}}},\ \overline{F}_{iA} = \overline{x}_{i,A},\\
&\overline{\sigma}=\widetilde{\sigma}(\overline{x}_{i,A},\overline{T},\overline{\nu},\overline{\nu}_{,M},X_N),\ \overline{T}_{Kj}=\widetilde{T}_{Kj}(\overline{x}_{i,A},\overline{T},\overline{\nu},\overline{\nu}_{,M},X_N),\\
&\overline{H}_K=\widetilde{H}_K(\overline{x}_{i,A},\overline{T},\overline{\nu},\overline{\nu}_{,M},X_N),\ \overline{g}=\widetilde{g}(\overline{x}_{i,A},\overline{T},\overline{\nu},\overline{\nu}_{,M},X_N).
\end{aligned} \tag{1.4.4}$$

By using (1.2.9), we can show that D is of quadratic order in

$$\left\|(\boldsymbol{v}-\overline{\boldsymbol{v}},\dot{\nu}-\dot{\overline{\nu}},\ \boldsymbol{F}-\overline{\boldsymbol{F}},T-\overline{T},\nu-\overline{\nu},\ \boldsymbol{G}-\overline{\boldsymbol{G}})\right\|_{L^2(B)}.$$

Let us establish an evolutionary relation for D. We introduce the function $\mathcal{E}$ on $\partial B\times[0,t_1]$ defined by

$$\begin{aligned}
\mathcal{E} = [&(T_{Ki}-\overline{T}_{Ki})(v_i-\overline{v}_i)+(H_K-\overline{H}_K)(\dot{\nu}-\dot{\overline{\nu}})+\\
&+\frac{1}{T}(T-\overline{T})(Q_K-\overline{Q}_K)]N_K.
\end{aligned} \tag{1.4.5}$$

Theorem 1.4.1. *If U and $\overline{U}$ are two smooth admissible processes corresponding to the supply terms $(\boldsymbol{f},\ell,S)$ and $(\overline{\boldsymbol{f}},\overline{\ell},\overline{S})$, then*

$$\begin{aligned}
\dot{D}=&\int_B\{\rho_0[(f_i-\overline{f}_i)(v_i-\overline{v}_i)+(\ell-\overline{\ell})(\dot{\nu}-\dot{\overline{\nu}})+\frac{1}{T}(S-\overline{S})(T-\overline{T})]-\\
&-\frac{1}{T\overline{T}}(\rho_0\overline{S}+\overline{Q}_{K,K})(T-\overline{T})^2+\dot{\overline{F}}_{iK}[T_{Ki}-\overline{T}_{Ki}-\frac{\partial\overline{T}_{Ki}}{\partial\overline{F}_{jA}}(F_{jA}-\overline{F}_{jA})-\\
&-\frac{\partial\overline{T}_{Ki}}{\partial\overline{T}}(T-\overline{T})-\frac{\partial\overline{T}_{Ki}}{\partial\overline{\nu}}(\nu-\overline{\nu})-\frac{\partial\overline{T}_{Ki}}{\partial\overline{G}_L}(G_L-\overline{G}_L)]+\dot{\overline{G}}_K[H_K-\overline{H}_K-\\
&-\frac{\partial\overline{H}_K}{\partial\overline{F}_{jL}}(F_{jL}-\overline{F}_{jL})-\frac{\partial\overline{H}_K}{\partial\overline{T}}(T-\overline{T})-\frac{\partial\overline{H}_K}{\partial\overline{\nu}}(\nu-\overline{\nu})-\\
&-\frac{\partial\overline{H}_K}{\partial\overline{G}_L}(G_L-\overline{G}_L)]-\dot{\overline{\nu}}[g-\overline{g}-\frac{\partial\overline{g}}{\partial\overline{F}_{iA}}(F_{iA}-\overline{F}_{iA})-\\
&\qquad-\frac{\partial\overline{g}}{\partial\overline{T}}(T-\overline{T})-\frac{\partial\overline{g}}{\partial\overline{\nu}}(\nu-\overline{\nu})-\frac{\partial\overline{g}}{\partial\overline{G}_A}(G_A-\overline{G}_A)]-\\
&\qquad-\rho_0\dot{\overline{T}}[\eta-\overline{\eta}-\frac{\partial\overline{\eta}}{\partial\overline{F}_{iA}}(F_{iA}-\overline{F}_{iA})-\frac{\partial\overline{\eta}}{\partial\overline{T}}(T-\overline{T})-\\
&\qquad-\frac{\partial\overline{T}}{\partial\overline{\nu}}(\nu-\overline{\nu})-\frac{\partial\overline{\eta}}{\partial\overline{G}_L}(G_L-\overline{G}_L)]-\\
&\qquad-(Q_K-\overline{Q}_K)(1-\overline{T}T^{-1})_{,K}\}dV+\int_{\partial B}\mathcal{E}\,dA.
\end{aligned} \tag{1.4.6}$$

Proof. In view of (1.4.3) we get

$$\begin{aligned}\dot{D}=&\int_B\{\frac{\partial}{\partial t}[\frac{1}{2}\rho_0(\boldsymbol{v}^2+\kappa\nu^2)+\sigma+\rho_0T\eta]-\frac{\partial}{\partial t}[\frac{1}{2}\rho_0(\overline{\boldsymbol{v}}^2+\kappa\overline{\nu}^2)+\\&+\overline{\sigma}+\rho_0\overline{T}\overline{\eta}]-\rho_0[\dot{v}_i\overline{v}_i+v_i\dot{\overline{v}}_i-2(\overline{v}_i\dot{\overline{v}}_i+\kappa\overline{\nu}\dot{\overline{\nu}})+\kappa(\dot{\nu}\overline{\nu}+\dot{\nu}\overline{\nu})]-\\&-\dot{\overline{T}}_{Ki}(F_{iK}-\overline{F}_{iK})-\overline{T}_{Ki}(\dot{F}_{iK}-\dot{\overline{F}}_{iK})-\rho_0\dot{\overline{T}}(\eta-\overline{\eta})-\\&-\rho_0\overline{T}(\dot{\eta}-\dot{\overline{\eta}})-\dot{\overline{H}}_K(G_K-\overline{G}_K)+\dot{\overline{g}}(\nu-\overline{\nu})-\\&-\overline{H}_K(\dot{G}_K-\dot{\overline{G}}_K)+\overline{g}(\dot{\nu}-\dot{\overline{\nu}})\}dV.\end{aligned}\tag{1.4.7}$$

If we use the equations (1.1.8), (1.4.2) and the divergence theorem we obtain

$$\begin{aligned}\dot{D}=&\int_{\partial B}\mathcal{E}\,dA+\int_B\{\rho_0[(f_i-\overline{f}_i)(v_i-\overline{v}_i)+(\ell-\overline{\ell})(\dot{\nu}-\dot{\overline{\nu}})+\\&+S-\overline{S}-\dot{\overline{T}}(\eta-\overline{\eta})-\overline{T}(\dot{\eta}-\dot{\overline{\eta}})]+\dot{\overline{F}}_{iK}(T_{Ki}-\overline{T}_{Ki})-\\&-\dot{\overline{T}}_{Ki}(F_{iK}-\overline{F}_{iK})-\dot{\overline{H}}_K(G_K-\overline{G}_K)+\dot{\overline{G}}_K(H_K-\overline{H}_K)+\\&+\dot{\overline{g}}(\nu-\overline{\nu})-\dot{\overline{\nu}}(g-\overline{g})+Q_{K,K}-\overline{Q}_{K,K}\}dV.\end{aligned}\tag{1.4.8}$$

By using (1.2.9) we find

$$\begin{aligned}\dot{\overline{T}}_{Ai}&=\frac{\partial\overline{T}_{Bj}}{\partial\overline{F}_{iA}}\dot{\overline{F}}_{jB}+\frac{\partial\overline{g}}{\partial\overline{F}_{iA}}\dot{\overline{\nu}}+\frac{\partial\overline{H}_K}{\partial\overline{F}_{iA}}\dot{\overline{G}}_K-\rho_0\frac{\partial\overline{\eta}}{\partial\overline{F}_{iA}}\dot{\overline{T}},\\\dot{\overline{H}}_K&=\frac{\partial\overline{T}_{Ai}}{\partial\overline{G}_K}\dot{\overline{F}}_{iA}+\frac{\partial\overline{g}}{\partial\overline{G}_K}\dot{\overline{\nu}}+\frac{\partial\overline{H}_A}{\partial\overline{G}_K}\dot{\overline{G}}_A-\rho_0\frac{\partial\overline{\eta}}{\partial\overline{G}_K}\dot{\overline{T}},\\\dot{\overline{g}}&=-\frac{\partial\overline{T}_{Ai}}{\partial\overline{\nu}}\dot{\overline{F}}_{iA}-\frac{\partial\overline{g}}{\partial\overline{\nu}}\dot{\overline{\nu}}-\frac{\partial\overline{H}_A}{\partial\overline{\nu}}\dot{\overline{G}}_A+\rho_0\frac{\partial\overline{\eta}}{\partial\overline{\nu}}\dot{\overline{T}},\\\rho_0\dot{\overline{\eta}}&=-\frac{\partial\overline{T}_{Kj}}{\partial\overline{T}}\dot{\overline{F}}_{jK}+\frac{\partial\overline{g}}{\partial\overline{T}}\dot{\overline{\nu}}-\frac{\partial\overline{H}_A}{\partial\overline{T}}\dot{\overline{G}}_A+\rho_0\frac{\partial\overline{\eta}}{\partial\overline{T}}\dot{\overline{T}},\end{aligned}$$

so that

$$\begin{aligned}&\dot{\overline{F}}_{iA}(T_{Ai}-\overline{T}_{Ai})-\dot{\overline{T}}_{Ai}(F_{iA}-\overline{F}_{iA})-\dot{\overline{H}}_K(G_K-\overline{G}_K)+\\&\quad+\dot{\overline{G}}_K(H_K-\overline{H}_K)+\dot{\overline{g}}(\nu-\overline{\nu})-\dot{\overline{\nu}}(g-\overline{g})-\rho_0\dot{\overline{T}}(\eta-\overline{\eta})=\\&\quad=\dot{\overline{F}}_{iA}[T_{Ai}-\overline{T}_{Ai}-\frac{\partial\overline{T}_{Ai}}{\partial\overline{F}_{jB}}(F_{jB}-\overline{F}_{jB})-\frac{\partial\overline{T}_{Ai}}{\partial\overline{T}}(T-\overline{T})-\\&\quad-\frac{\partial\overline{T}_{Ai}}{\partial\overline{G}_K}(G_K-\overline{G}_K)-\frac{\partial\overline{T}_{Ai}}{\partial\overline{\nu}}(\nu-\overline{\nu})]+\dot{\overline{G}}_A[H_A-\overline{H}_A-\end{aligned}$$

$$-\frac{\partial\overline{H}}{\partial\overline{F}_{jB}}\,(F_{jB}-\overline{F}_{jB})-\frac{\partial\overline{H}_A}{\partial\overline{G}_B}\,(G_B-\overline{G}_B)-\frac{\partial\overline{H}_A}{\partial\overline{\nu}}\,(\nu-\overline{\nu})+$$
$$+\frac{\partial\overline{H}_A}{\partial\overline{T}}\,(T-\overline{T})]-\dot{\overline{\nu}}[g-\overline{g}-\frac{\partial\overline{g}}{\partial\overline{F}_{jB}}\,(F_{jB}-\overline{F}_{jB})-$$
$$-\frac{\partial\overline{g}}{\partial\overline{\nu}}\,(\nu-\overline{\nu})-\frac{\partial\overline{g}}{\partial\overline{G}_K}\,(G_K-\overline{G}_K)-\frac{\partial\overline{g}}{\partial\overline{T}}\,(T-\overline{T})]- \tag{1.4.9}$$
$$-\rho_0\dot{\overline{T}}[\eta-\overline{\eta}-\frac{\partial\overline{\eta}}{\partial\overline{\nu}}\,(\nu-\overline{\nu})-\frac{\partial\overline{\eta}}{\partial\overline{G}_K}\,(G_K-\overline{G}_K)-$$
$$-\frac{\partial\overline{\eta}}{\partial\overline{F}_{iA}}\,(F_{iA}-\overline{F}_{iA})-\frac{\partial\overline{\eta}}{\partial\overline{T}}\,(T-\overline{T})]-\rho_0\dot{\overline{\eta}}(T-\overline{T}).$$

In view of (1.2.10) we obtain

$$\rho_0 S+Q_{A,A}-\rho_0\overline{S}-\overline{Q}_{L,L}-\rho_0\dot{\overline{\eta}}(T-\overline{T})-\rho_0\overline{T}(\dot{\eta}-\dot{\overline{\eta}})=$$
$$=\rho_0 S+Q_{A,A}-\rho_0\overline{S}-\overline{Q}_{A,A}+\frac{1}{T}\,\overline{T}(\rho_0\overline{S}+\overline{Q}_{K,K}-\rho_0 S-$$
$$-Q_{K,K})-\frac{1}{T\overline{T}}(\rho_0\overline{S}+\overline{Q}_{K,K})(T-\overline{T})^2=\frac{1}{T}\,\rho_0(S-\overline{S})(T-\overline{T})+ \tag{1.4.10}$$
$$+[T^{-1}(Q_K-\overline{Q}_K)(T-\overline{T})]_{,K}-(Q_K-\overline{Q}_K)(1-\overline{T}\,T^{-1})_{,K}-$$
$$-\frac{1}{T\,\overline{T}}(\rho_0\overline{S}+\overline{Q}_{K,K})(T-\overline{T})^2.$$

With the help of (1.4.9) and (1.4.10), from (1.4.8) we obtain the desired result. □

Following Dafermos [80], we say that a smooth admissible process $\overline{U}=(\overline{\boldsymbol{x}},\overline{\nu},\overline{T})$ resides in the convexity region of internal energy if

(i) for each $(\boldsymbol{X},t)\in\overline{B}\times[0,t_1]$, there exists a positive constant μ_0 such that

$$\frac{\partial^2\overline{\sigma}}{\partial\overline{F}_{iA}\partial\overline{F}_{jB}}\,U_{iA}U_{jB}+\frac{\partial^2\overline{\sigma}}{\partial\overline{\nu}^2}\,\zeta^2+\frac{\partial^2\overline{\sigma}}{\partial\overline{G}_A\partial\overline{G}_B}\,Z_AZ_B+$$
$$+2\,\frac{\partial^2\overline{\sigma}}{\partial\overline{F}_{iA}\partial\overline{\nu}}\,U_{iA}\zeta+2\,\frac{\partial^2\overline{\sigma}}{\partial\overline{\nu}\partial\overline{G}_A}\,Z_A\zeta+ \tag{1.4.11}$$
$$+2\,\frac{\partial^2\overline{\sigma}}{\partial\overline{F}_{iA}\partial\overline{G}_B}\,U_{iA}Z_B\geq\mu_0(|\boldsymbol{U}|^2+|\zeta|^2+|\boldsymbol{Z}|^2),$$

for all U_{iA}, ζ and Z_A;

(ii) $$\frac{\partial\overline{\eta}}{\partial\overline{T}}>0\quad\text{on } B\times[0,t_1]. \tag{1.4.12}$$

We note that from (1.2.1) we obtain

$$\begin{aligned} Q_M - \overline{Q}_M &= \overline{R}_{MiN}(F_{iN} - \overline{F}_{iN}) + \overline{D}_M(T - \overline{T}) + \\ &+ \overline{K}_{MN}(T_{,N} - \overline{T}_{,N}) + \overline{P}_M(\nu - \overline{\nu}) + \overline{J}_{MN}(G_N - \overline{G}_N) + \\ &+ o_M(|\boldsymbol{F} - \overline{\boldsymbol{F}}| + |T - \overline{T}| + |\boldsymbol{G} - \overline{\boldsymbol{G}}| + |\nu - \overline{\nu}| + |\boldsymbol{\Lambda} - \overline{\boldsymbol{\Lambda}}|), \end{aligned} \tag{1.4.13}$$

where $\boldsymbol{\Lambda} = \text{grad}_X T$ and

$$\overline{R}_{MiN} = \frac{\partial \overline{Q}_M}{\partial \overline{F}_{iN}}, \; \overline{D}_M = \frac{\partial \overline{Q}_M}{\partial \overline{T}}, \; \overline{K}_{MN} = \frac{\partial \overline{Q}}{\partial \overline{T}_{,N}},$$

$$\overline{P}_M = \frac{\partial \overline{Q}_M}{\partial \overline{\nu}}, \quad \overline{J}_{MN} = \frac{\partial \overline{Q}_M}{\partial \overline{\nu}_{,N}}.$$

We say that a thermodynamic process $(\boldsymbol{x}, \nu, T)$ resides in the region of state space where the material is a definite heat conductor if K_{MN} is positive definite.

With a view towards deriving a continuous dependence result, we record the following Gronwall-type inequality (Dafermos [80]).

Lemma 1.4.1. *Assume that the nonnegative functions $f \in L^\infty([0, t_1])$ and $g \in L^1([0, t_1])$ satisfy the inequality*

$$f^2(t) \le M^2 f^2(0) + \int_0^t [(2\gamma + 4\beta t) f^2(\tau) + 2N g(\tau) f(\tau)] d\tau, \; t \in [0, t_1],$$

where β, γ, M and N are nonnegative constants. Then

$$f(t) \le \left[M f(0) + N \int_0^t g(s) ds \right] \exp(\alpha t + \beta t^2), \; t \in [0, t_1],$$

where $\alpha = \gamma + \beta/\gamma$.

If U and $\overline{U}$ are two smooth admissible processes, then we denote

$$\begin{aligned} \Gamma(\boldsymbol{X}, t) &= |\boldsymbol{F}(\boldsymbol{X}, t) - \overline{\boldsymbol{F}}(\boldsymbol{X}, t)| + |T(\boldsymbol{X}, t) - \overline{T}(\boldsymbol{X}, t)| + \\ &+ |\boldsymbol{\Lambda}(\boldsymbol{X}, t) - \overline{\boldsymbol{\Lambda}}(\boldsymbol{X}, t)| + |\nu(\boldsymbol{X}, t) - \overline{\nu}(\boldsymbol{X}, t)| + \\ &+ |\boldsymbol{G}(\boldsymbol{X}, t) - \overline{\boldsymbol{G}}(\boldsymbol{X}, t)|, \quad (\boldsymbol{X}, t) \in B \times [0, t_1]. \end{aligned} \tag{1.4.14}$$

The next theorem establishes the continuous dependence of smooth thermodynamic processes upon the initial state and supply terms.

Theorem 1.4.2. *Let $\overline{U} = (\overline{\boldsymbol{x}}, \overline{\nu}, \overline{T})$ be a smooth admissible process on $B \times [0, t_1]$, residing in the region of state space where the internal energy* *is a convex function and material is a definite conductor of heat and corresponding to the supply terms* $(\overline{f}, \overline{\ell}, \overline{S}) \in L^\infty(B \times [0, t_1])$. *Then there exist the positive contants* δ, α_1, M_1 *and* N_1 *with the following property: if* $U = (\boldsymbol{x}, \nu, T)$ *is any smooth admissible process on* $B \times [0, t_1]$ *corresponding to the supply terms* $(\boldsymbol{f}, \ell, S) \in L^\infty(B \times [0, t_1])$ *such that*

$$\begin{aligned} &\Gamma(\boldsymbol{X}, t) < \delta, \ (\boldsymbol{X}, t) \in B \times [0, t_1], \\ &\mathcal{E} = 0 \quad on \quad \partial B \times (0, t_1), \end{aligned} \tag{1.4.15}$$

then we have

$$\begin{aligned} &\|(\boldsymbol{v} - \overline{\boldsymbol{v}}, \boldsymbol{F} - \overline{\boldsymbol{F}}, T - \overline{T}, \nu - \overline{\nu}, \dot{\nu} - \dot{\overline{\nu}}, \boldsymbol{G} - \overline{\boldsymbol{G}})(\cdot, s)\|_{L^2(B)} \leq \\ &\leq \big\{M_1\|(\boldsymbol{v} - \overline{\boldsymbol{v}}, \boldsymbol{F} - \overline{\boldsymbol{F}}, T - \overline{T}, \nu - \overline{\nu}, \dot{\nu} - \dot{\overline{\nu}}, \boldsymbol{G} - \overline{\boldsymbol{G}})(\cdot, 0)\|_{L^2(B)} + \\ &+ N_1 \int_0^s \|(\boldsymbol{f} - \overline{\boldsymbol{f}}, \ell - \overline{\ell}, S - \overline{S})(\cdot, t)\|_{L^2(B)} dt\big\} \exp(\alpha_1 s), \end{aligned}$$

for any $s \in [0, t_1]$.

Proof. We consider first the relation (1.4.13). In view of (1.2.9), (1.4.11) and (1.4.12), we conclude that there exist positive constants δ_1 and μ_1 such that whenever

$$\begin{aligned} &|\boldsymbol{F}(\boldsymbol{X}, t) - \overline{\boldsymbol{F}}(\boldsymbol{X}, t)| + |T(\boldsymbol{X}, t) - \overline{T}(\boldsymbol{X}, t)| + |\nu(\boldsymbol{X}, t) - \overline{\nu}(\boldsymbol{X}, t)| + \\ &\quad + |\boldsymbol{G}(\boldsymbol{X}, t) - \overline{\boldsymbol{G}}(\boldsymbol{X}, t)| < \delta_1, \ (\boldsymbol{X}, t) \in B \times [0, t_1], \end{aligned}$$

holds, then we have

$$\begin{aligned} &[\sigma - \overline{\sigma} - \overline{T}_{Ki}(F_{iK} - \overline{F}_{iK}) - \overline{H}_K(G_K - \overline{G}_K) + \overline{g}(\nu - \overline{\nu}) + \\ &\quad + \rho_0 \eta (T - \overline{T})](\boldsymbol{X}, t) \geq \mu_1(|\boldsymbol{F}(\boldsymbol{X}, t) - \overline{\boldsymbol{F}}(\boldsymbol{X}, t)|^2 + \\ &\quad + |T(\boldsymbol{X}, t) - \overline{T}(\boldsymbol{X}, t)|^2 + |\nu(\boldsymbol{X}, t) - \overline{\nu}(\boldsymbol{X}, t)|^2 + \\ &\quad + |\boldsymbol{G}(\boldsymbol{X}, t) - \overline{\boldsymbol{G}}(\boldsymbol{X}, t)|^2), \ (\boldsymbol{X}, t) \in B \times [0, t_1]. \end{aligned} \tag{1.4.16}$$

Since $\overline{U}$ resides in the region where the material is a definite conductor of heat, there exists a positive constant λ such that

$$\begin{aligned} &\int_B \overline{K}_{AB} T^{-1} (\Lambda_A - \overline{\Lambda}_A)(\Lambda_B - \overline{\Lambda}_B) dV \geq \\ &\qquad \geq \lambda \int_B (\Lambda_A - \overline{\Lambda}_A)(\Lambda_A - \overline{\Lambda}_A) dV. \end{aligned} \tag{1.4.17}$$

In view of (1.4.13) and (1.4.17), it follows that there exists a positive constant δ_2 such that whenever $\Gamma(\mathbf{X},t) < \delta_2$, $(\boldsymbol{X},t) \in B\times[0,t_1]$, we have

$$
\begin{aligned}
&-\int_B (Q_A - \overline{Q}_A)(1-\overline{T}\,T^{-1})_{,A} dV \leq \\
&\leq \int_B [p_1(T-\overline{T})^2 + p_2(\nu-\overline{\nu})(T-\overline{T}) + \\
&+m_K(T-\overline{T})(\Lambda_K-\overline{\Lambda}_K) + B_{iK}(F_{iK}-\overline{F}_{iK})(T-\overline{T}) + \\
&+M_{AiN}(\Lambda_A-\overline{\Lambda}_A)(F_{iN}-\overline{F}_{iN}) + R_N(G_N-\overline{G}_N)(T-\overline{T}) + \\
&+V_{AN}(\Lambda_A-\overline{\Lambda}_A)(G_N-\overline{G}_N) - \lambda(\Lambda_K-\overline{\Lambda}_K)(\Lambda_K-\overline{\Lambda}_K)]dV,
\end{aligned}
$$

where

$$
\begin{gathered}
p_1 = \overline{D}_A T^{-2} T_{,A},\; p_2 = \overline{P}_A T^{-2} T_{,A},\; m_B = \overline{K}_{BA} T_{,A} T^{-2} - \overline{D}_B T^{-1}, \\
B_{iK} = \overline{R}_{AiK} T_{,A} T^{-2},\; M_{AiN} = -\overline{R}_{AiN} T^{-1},\; R_N = \overline{J}_{AN} T_{,A} T^{-2}, \\
V_{AN} = -\overline{J}_{AN} T^{-1}.
\end{gathered}
$$

By means of Schwarz inequality, combined with the arithmetic-geometric mean inequality, we get

$$
\begin{aligned}
&-2\int_B (Q_K - \overline{Q}_K)(1-\overline{T}\,T^{-1})_{,K} dV \leq \\
&\leq (\alpha_1^2+\alpha_2^2+\alpha_3^2-2\lambda)\|(\boldsymbol{\Lambda}-\overline{\boldsymbol{\Lambda}})(\cdot,t)\|^2_{L^2(B)} + \\
&+(2+A_1^2+A_2^2+A_3^2\alpha_1^{-2})\|(T-\overline{T})(\cdot,t)\|^2_{L^2(B)} + \\
&+(A_4^2+A_5^2\alpha_2^{-2})\|(\boldsymbol{F}-\widetilde{\boldsymbol{F}})(\cdot,t)\|^2_{L^2(B)} + \\
&+\|(\nu-\overline{\nu})(\cdot,t)\|^2_{L^2(B)} + (A_6^2+A_7^2\alpha_3^{-2})\|(\boldsymbol{G}-\overline{\boldsymbol{G}})(\cdot,t)\|^2_{L^2(B)},
\end{aligned}
\tag{1.4.18}
$$

where α_j are arbitrary positive constants and

$$
\begin{gathered}
A_1^2 = 2\max|p_1|,\quad A_2 = \max|p_2|,\quad A_3 = \max|\boldsymbol{m}|,\quad A_4 = \max|\boldsymbol{B}|, \\
A_5 = \max|\boldsymbol{M}|,\quad A_6 = \max|\boldsymbol{R}|,\quad A_7 = \max|\boldsymbol{V}|,\quad \text{on } \overline{B}\times[0,t_1].
\end{gathered}
$$

We now take the arbitrary constants such that $\alpha_1^2+\alpha_2^2+\alpha_3^2 < 2\lambda$. Clearly, from (1.4.18) we can see that there exist the positive constants δ_2 and m_0 such that whenever $\Gamma(\boldsymbol{X},t)<\delta_2$, $(\boldsymbol{X},t)\in B\times[0,t_1]$, then we have

$$
\begin{aligned}
&-\int_B (Q_K - \overline{Q}_K)(1-\overline{T}T^{-1})_{,K} dV \leq \\
&\leq m_0\|(\boldsymbol{F}-\overline{\boldsymbol{F}}, T-\overline{T}, \nu-\overline{\nu}, \boldsymbol{G}-\overline{\boldsymbol{G}})(\cdot,t)\|^2_{L^2(B)},
\end{aligned}
\tag{1.4.19}
$$

for any $t \in [0, t_1]$. Thus, in view of Theorem 1.4.1 and (1.4.19) we conclude that there exist the positive constants δ_2, d_1 and d_2 such that whenever $\Gamma < \delta_2$ on $B \times [0, t_1]$, we have the following estimate

$$\begin{aligned} \dot{D}(t) \leq\, & d_1 \|(\boldsymbol{F} - \overline{\boldsymbol{F}}, T - \overline{T}, \nu - \overline{\nu}, \boldsymbol{G} - \overline{\boldsymbol{G}})(\cdot, t)\|^2_{L^2(B)} + \\ & + d_2 \|(\boldsymbol{f} - \overline{\boldsymbol{f}}, \ell - \overline{\ell}, S - \overline{S})(\cdot, t)\| \|(\boldsymbol{v} - \overline{\boldsymbol{v}}, \dot{\nu} - \dot{\overline{\nu}}, T - \overline{T})(\cdot, t)\|_{L^2(B)}, \end{aligned} \tag{1.4.20}$$

for all $t \in [0, t_1]$. Let

$$z(t) = \|(\boldsymbol{v} - \overline{\boldsymbol{v}}, \dot{\nu} - \dot{\overline{\nu}}, \boldsymbol{F} - \overline{\boldsymbol{F}}, T - \overline{T}, \nu - \overline{\nu}, \boldsymbol{G} - \overline{\boldsymbol{G}})(\cdot, t)\|_{L^2(B)},$$

$t \in [0, t_1]$. We choose $\delta = \min(\delta_1, \delta_2)$. Then we fix $\nu \in [0, t_1]$ and integrate the inequality (1.4.20) over $[0, \tau]$, $\tau \in [0, \nu]$. We obtain

$$\begin{aligned} D(\tau) \leq\, & D(0) + d_1 \int_0^\tau z^2(t) dt + \\ & + d_2 \int_0^\tau \|(\boldsymbol{f} - \overline{\boldsymbol{f}}, \ell - \overline{\ell}, S - \overline{S})(\cdot, t)\|_{L^2(B)} z(t) dt. \end{aligned} \tag{1.4.21}$$

With the aid of (1.4.3) and (1.4.16) we conclude that there exists a positive constant c_1 such that

$$D(t) \geq c_1 z^2(t), \quad t \in [0, t_1]. \tag{1.4.22}$$

If we use the estimate $D(0) \leq c_2 z^2(0)$, where c_2 is a positive constant, then from (1.4.21) and (1.4.22) we obtain the inequality

$$\begin{aligned} c_1 z^2(\tau) \leq\, & c_2 z^2(0) + \\ & + \int_0^\tau \left[d_1 z^2(t) + d_2 \|(\boldsymbol{f} - \overline{\boldsymbol{f}}, \ell - \overline{\ell}, S - \overline{S})(\cdot, t)\|_{L^2(B)} z(t)\right] dt, \; \tau \in [0, \nu]. \end{aligned}$$

An application of Lemma 1.4.1 yields the desired result. □

The next uniqueness result is a direct consequence of Theorem 1.4.2.

Theorem 1.4.3. *Let U and $\overline{U}$ be as in Theorem 1.4.2. Assume that the corresponding supply terms $(\boldsymbol{f}, \ell, S)$ and $(\overline{\boldsymbol{f}}, \overline{\ell}, \overline{S})$ coincide on $B \times [0, t_1]$ and that both processes originate from the same state. Then $\boldsymbol{x} = \overline{\boldsymbol{x}}$, $\nu = \overline{\nu}$ and $T = \overline{T}$ on $B \times [0, t_1]$.*

1.5 The Linear Theory

In this section we present the equations of the linear theory of thermoelastic materials with voids (cf. [180]). In the linear theory we shall employ the notations $X_i = \delta_{iA} X_A$, $f_{,i} = \partial f / \partial X_i$, where δ_{iA} is the Kronecker delta. The displacement vector field $\boldsymbol{u}$ is defined by

$$u_i = x_i - X_i. \tag{1.5.1}$$

Let us introduce the notations

$$\varphi = \nu - \nu_0, \ \theta = T - T_0, \tag{1.5.2}$$

where T_0 is the absolute temperature in the reference configuration. We suppose that T_0 and ν_0 are given constants. In the linear theory we assume that $\boldsymbol{u} = \varepsilon \boldsymbol{u}'$, $\varphi = \varepsilon \varphi'$ and $\theta = \varepsilon \theta'$ where ε is a constant small enough for squares and higher powers to be neglected, and $\boldsymbol{u}', \varphi'$ and θ' are independent of ε. In the linear theory the Lagrangian strain tensor reduces to

$$e_{ij} = \frac{1}{2}(u_{i,j} + u_{j,i}). \tag{1.5.3}$$

The response functionals now depend on the variables e_{ij}, θ, $\theta_{,i}$, φ, $\varphi_{,k}$ and X_m. In the linear theory, and assuming that the initial body is free from stresses and has zero intrinsic equilibrated body force and entropy, we have

$$\begin{aligned} \sigma = \frac{1}{2}(&C_{ijrs} e_{ij} e_{rs} - 2\beta_{ij} e_{ij} \theta - a\theta^2 + A_{ij} \varphi_{,i} \varphi_{,j} + 2B_{ij} \varphi e_{ij} + \\ &+ 2D_{ijk} e_{ij} \varphi_{,k} + 2d_i \varphi \varphi_{,i} + \xi \varphi^2 - 2m\theta\varphi - 2a_i \varphi_{,i} \theta), \end{aligned} \tag{1.5.4}$$

where the constitutive coefficients are prescribed functions on X_m. In the case of homogeneous bodies the constitutive coefficients are constants. The above coefficients have the following symmetries

$$\begin{aligned} &C_{ijrs} = C_{rsij} = C_{jirs}, \ \beta_{ij} = \beta_{ji}, \ D_{ijk} = D_{jik}, \\ &A_{ij} = A_{ji}, \ B_{ij} = B_{ji}. \end{aligned} \tag{1.5.5}$$

We introduce the notations

$$t_{ji} = \delta_{jK} T_{Ki}, \ h_i = \delta_{iK} H_K, \ q_i = \delta_{iK} Q_K.$$

Clearly, as in classical linear theory of thermoelasticity, the Cauchy stress tensor and the Piola-Kirchhoff stress tensors coincide. Moreover, the heat flux vectors $\boldsymbol{q}$ and $\boldsymbol{Q}$, as well as the equilibrated stresses $\boldsymbol{h}$ and $\boldsymbol{H}$, also coincide. It follows from (1.2.9) that

$$t_{ij} = \frac{\partial \sigma}{\partial e_{ij}}, \ \rho_0 \eta = -\frac{\partial \sigma}{\partial \theta}, \ h_i = \frac{\partial \sigma}{\partial \varphi_{,i}}, \ g = -\frac{\partial \sigma}{\partial \varphi} \tag{1.5.6}$$

and

$$q_i\theta_{,i} \geq 0. \tag{1.5.7}$$

In the linear theory the components of the surface traction $\boldsymbol{t}$, the equilibrated surface traction h, and the heat flux q at regular points of ∂B are given by

$$t_i = t_{ji}n_j, \ \ h = h_i n_i, \ \ q = q_i n_i, \tag{1.5.8}$$

where n_i are the components of the unit outward normal to ∂B. By (1.2.11), (1.5.4)–(1.5.6) we obtain the following constitutive equations of the linear theory

$$\begin{aligned} t_{ij} &= C_{ijrs}e_{rs} + B_{ij}\varphi + D_{ijk}\varphi_{,k} - \beta_{ij}\theta, \\ h_i &= A_{ij}\varphi_{,j} + D_{rsi}e_{rs} + d_i\varphi - a_i\theta, \\ g &= -B_{ij}e_{ij} - \xi\varphi - d_i\varphi_{,i} + m\theta, \\ \rho_0\eta &= \beta_{ij}e_{ij} + a\theta + m\varphi + a_i\varphi_{,i}, \\ q_i &= k_{ij}\theta_{,j}. \end{aligned} \tag{1.5.9}$$

We note that the inequality (1.5.7) implies that the conductivity tensor k_{ij} is positive semi-definite

$$k_{ij}\xi_i\xi_j \geq 0 \text{ for any } \xi_s.$$

In the case of an isotropic material, the constitutive equations become

$$\begin{aligned} t_{ij} &= \lambda e_{rr}\delta_{ij} + 2\mu e_{ij} + b\varphi\delta_{ij} - \beta\theta\delta_{ij}, \\ h_i &= \alpha\varphi_{,i}, \\ g &= -be_{rr} - \xi\varphi + m\theta, \\ \rho_0\eta &= \beta e_{rr} + a\theta + m\varphi, \\ q_i &= k\theta_{,i}, \end{aligned} \tag{1.5.10}$$

where δ_{ij} is Kronecker's delta, and $\lambda, \mu, b, \beta, \alpha, \xi, m, a$ and k are constitutive coefficients. The equations of motion (1.1.8) can be written in the form

$$t_{ji,j} + \rho_0 f_i = \rho_0\ddot{u}_i \tag{1.5.11}$$

and the balance of equilibrated force (1.1.12) becomes

$$h_{i,i} + g + \rho_0\ell = \rho_0\kappa\ddot{\varphi}. \tag{1.5.12}$$

In the linear theory, the energy equation (1.2.10) reduces to

$$\rho_0 T_0\dot{\eta} = q_{i,i} + \rho_0 S. \tag{1.5.13}$$

We consider the following boundary conditions

$$\begin{aligned}
&u_i = \widetilde{u}_i \quad \text{on } \overline{S}_1\times(0,t_1), \quad t_{ji}n_j = \widetilde{t}_i \quad \text{on } S_2\times(0,t_1),\\
&\varphi = \widetilde{\varphi} \quad \text{on } \overline{S}_3\times(0,t_1), \quad h_i n_i = \widetilde{h} \quad \text{on } S_4\times(0,t_1),\\
&\theta = \widetilde{\theta} \quad \text{on } \overline{S}_5\times(0,t_1), \quad q_i n_i = \widetilde{q} \quad \text{on } S_6\times(0,t_1),
\end{aligned} \tag{1.5.14}$$

where S_r, $(r = 1, 2, ..., 6)$, are subsets of ∂B so that $\overline{S}_1 \cup S_2 = \overline{S}_3 \cup S_4 = \overline{S}_5 \cup S_6 = \partial B$, $S_1 \cap S_2 = S_3 \cap S_4 = S_5 \cap S_6 = \emptyset$, $\widetilde{u}_i, \widetilde{t}_i, \widetilde{\varphi}, \widetilde{h}, \widetilde{\theta}$ and $\widetilde{q}$ are prescribed functions, and t_1 is some time instant that may be infinite.

The initial conditions are

$$\begin{aligned}
&\boldsymbol{u}(\boldsymbol{X},0) = \boldsymbol{u}^0(\boldsymbol{X}),\ \dot{\boldsymbol{u}}(\boldsymbol{X},0) = \boldsymbol{v}^0(\boldsymbol{X}),\ \varphi(\boldsymbol{X},0) = \varphi^0(\boldsymbol{X}),\\
&\dot{\varphi}(\boldsymbol{X},0) = \zeta^0(\boldsymbol{X}),\ \ \theta(\boldsymbol{X},0) = \theta^0(\boldsymbol{X}),\ \ \boldsymbol{X} \in B,
\end{aligned} \tag{1.5.15}$$

where $\boldsymbol{u}^0, \boldsymbol{v}^0, \varphi^0, \zeta^0$ and θ^0 are prescribed functions.

We suppose that C_{ijrs}, B_{ij}, D_{ijk}, β_{ij}, A_{ij}, d_i, a_i, ξ, m, a, k_{ij}, ρ_0, κ, T_0, f_i, ℓ, S, $\widetilde{u}_i, \widetilde{t}_i, \widetilde{\varphi}, \widetilde{h}, \widetilde{\theta}, \widetilde{q}, \boldsymbol{u}^0, \boldsymbol{v}^0, \varphi^0, \zeta^0$ and θ^0 are prescribed functions with the following properties: (i) $C_{ijrs}, B_{ij}, D_{ijk}, \beta_{ij}, A_{ij}, d_i, a_i$ and k_{ij} are smooth on $\overline{B}$ and satisfy (1.5.5), (1.5.7), while a, ξ and m are continuous on $\overline{B}$; (ii) ρ_0 and κ are continuous and strictly positive on $\overline{B}$; (iii) T_0 is a positive constant; (iv) f_i, ℓ and S are continuous on $\overline{B}\times[0,t_1)$; (v) $\boldsymbol{u}^0, \boldsymbol{v}^0, \varphi^0, \zeta^0$ and θ^0 are continuous on $\overline{B}$; (vi) $\widetilde{u}_i, \widetilde{\varphi}$ and $\widetilde{\theta}$ are continuous on $\overline{S}_1\times(0,t_1)$, $\overline{S}_3\times(0,t_1)$, $\overline{S}_5\times(0,t_1)$, respectively; (vii) $\widetilde{t}_i$, $\widetilde{h}$ and $\widetilde{q}$ are continuous in time and piecewise regular on $S_2\times(0,t_1)$, $S_4\times(0,t_1)$ and $S_6\times(0,t_1)$, respectively, and continuous in time. By an external data system we mean an ordered array $L = (\boldsymbol{f}, \ell, S, \widetilde{\boldsymbol{u}}, \widetilde{\boldsymbol{t}}, \widetilde{\varphi}, \widetilde{h}, \widetilde{\theta}, \widetilde{q}, \boldsymbol{u}^0, \boldsymbol{v}^0, \varphi^0, \zeta^0, \theta^0)$ with the properties (iv)-(vii).

We say that $(\boldsymbol{u}, \varphi, \theta)$ is a dynamically admissible state on $\overline{B}\times[0,t_1)$ provided: (i) $\boldsymbol{u}$ and φ are of class C^2 on $B\times(0,t_1)$, and θ is of class $C^{2,1}$ on $B\times(0,t_1)$; (ii) $\boldsymbol{u}, \dot{\boldsymbol{u}}, \ddot{\boldsymbol{u}}, \boldsymbol{e}, \dot{\boldsymbol{e}}, \varphi, \dot{\varphi}, \varphi_{,i}, \theta, \theta_{,i}$ and $\dot{\theta}$ are continuous on $\overline{B}\times[0,t_1)$.

By an admissible system of internal forces on $\overline{B}\times[0,t_1)$ we mean an ordered array of functions (t_{ij}, h_i, g) with the properties: (i) t_{ij} and h_k are of class $C^{1,0}$ on $B\times(0,t_1)$; (ii) $t_{ij} = t_{ji}$; (iii) $t_{ij}, t_{ij,j}, h_{s,s}$ and g are continuous on $\overline{B}\times[0,t_1)$. We say that $\boldsymbol{q}$ is an admissible heat flux vector on $\overline{B}\times[0,t_1)$ provided: (i) q_k is of class $C^{1,0}$ on $B\times(0,t_1)$; (ii) q_i and $q_{k,k}$ are continuous on $\overline{B}\times[0,t_1)$.

By an admissible process on $\overline{B}\times[0,t_1)$ we mean an ordered array of functions $p = (u_i, \varphi, \theta, e_{ij}, t_{ij}, h_i, g, \eta, q_j)$ with the following properties:

(i) $(\boldsymbol{u}, \varphi, \theta)$ is a dynamically admissible state on $\overline{B}\times[0,t_1)$; (ii) e_{ij} is a continuous symmetric tensor field on $\overline{B}\times[0,t_1)$; (iii) (t_{ij}, h_i, g) is an admissible system of internal forces on $\overline{B}\times[0,t_1)$; (iv) η is of class $C^{0,1}$ on $B\times(0,t_1)$; (v) $\boldsymbol{q}$ is an admissible heat flux vector on $\overline{B}\times[0,t_1)$. If we define addition of admissible processes and multiplication of an admissible process by a scalar λ through

$$
\begin{aligned}
&p + p' = (u_i + u_i', \varphi + \varphi', ..., q_i + q_i'),\\
&\lambda p = (\lambda u_i, \lambda \varphi_i, ..., \lambda q_i),
\end{aligned}
$$

then the set of admissible processes is a linear space. We say that $p = (u_i, \varphi, \theta, e_{ij}, t_{ij}, h_i, g, \eta, q_j)$ is a thermoelastic process corresponding to the supply terms $(\boldsymbol{f}, \ell, S)$ if p is an admissible process that satisfies the fundamental system of field equations (1.5.3), (1.5.9), (1.5.11)-(1.5.13) on $B\times(0,t_1)$.

The mixed problem of the dynamic theory of thermoelasticity is to find a thermoelastic process p corresponding to the supply terms $(\boldsymbol{f}, \ell, S)$ that satisfies the boundary conditions (1.5.14) and the initial conditions (1.5.15). We call such a thermoelastic process a solution of the mixed problem of thermoelastodynamics.

From (1.5.3), (1.5.9), (1.5.11)-(1.5.13) we obtain the basic field equations in terms of the displacement, volume fraction and temperature

$$
\begin{aligned}
&(C_{ijrs}u_{r,s} + B_{ij}\varphi + D_{ijk}\varphi_{,k} - \beta_{ij}\theta)_{,j} + \rho_0 f_i = \rho_0 \ddot{u}_i,\\
&(A_{ij}\varphi_{,j} + D_{rsi}u_{r,s} + d_i\varphi - a_i\theta)_{,i} - B_{ij}u_{i,j} - \xi\varphi - d_i\varphi_{,i} +\\
&\qquad\qquad\qquad\qquad + m\theta + \rho_0\ell = \rho_0\kappa\ddot{\varphi},\\
&(k_{ij}\theta_{,j})_{,i} - T_0\beta_{ij}\dot{u}_{i,j} - T_0 a\dot{\theta} - T_0 m\dot{\varphi} - T_0 a_i\dot{\varphi}_{,i} = -\rho_0 S,
\end{aligned}
\tag{1.5.16}
$$

on $B\times(0,t_1)$. In the case of homogeneous and isotropic bodies the equations (1.5.16) reduce to

$$
\begin{aligned}
&\mu\Delta u_i + (\lambda+\mu)u_{j,ji} + b\varphi_{,i} - \beta\theta_{,i} + \rho_0 f_i = \rho_0\ddot{u}_i,\\
&\alpha\Delta\varphi - bu_{j,j} - \xi\varphi + m\theta + \rho_0\ell = \rho_0\kappa\ddot{\varphi},\\
&k\Delta\theta - \beta T_0\dot{u}_{j,j} - aT_0\dot{\theta} - mT_0\dot{\varphi} = -\rho_0 S,
\end{aligned}
\tag{1.5.17}
$$

where Δ is the Laplacian. In the linear theory the Lagrangian coordinates are usually indicated by (x_1, x_2, x_3) and the Eulerian coordinates by (y_1, y_2, y_3).

Chapter 2

DYNAMIC THEORY

2.1 Uniqueness Results

In this section we establish uniqueness theorems in the dynamic linear theory of thermoelastic materials with voids. An existence result is presented in Section 4.2.

We introduce the notation

$$\begin{aligned} W = & \frac{1}{2}C_{ijrs}e_{ij}e_{rs} + B_{ij}e_{ij}\varphi + D_{ijr}e_{ij}\varphi_{,r}+ \\ & + d_i\varphi\varphi_{,i} + \frac{1}{2}\xi\varphi^2 + \frac{1}{2}A_{ij}\varphi_{,i}\varphi_{,j}. \end{aligned} \tag{2.1.1}$$

The following uniqueness result has been presented in [180].

Theorem 2.1.1. *Assume that*
(i) W *is a positive semi-definite quadratic form;*
(ii) a, ρ_0 *and* κ *are strictly positive;*
(iii) k_{ij} *is positive semi-definite;*
(iv) *the constitutive coefficients satisfy the symmetry relations* (1.5.5).
Then the mixed problem of thermoelastodynamics has at most one solution.

Proof. In view of (1.5.9) and (1.5.5) we obtain

$$t_{ij}\dot{e}_{ij} + h_i\dot{\varphi}_{,i} - g\dot{\varphi} + \rho_0\theta\dot{\eta} = \dot{W} + a\theta\dot{\theta}, \tag{2.1.2}$$

On the other hand, from (1.5.11), (1.5.3), (1.5.12) and (1.5.13), we find that

$$\begin{aligned} t_{ij}\dot{e}_{ij} + h_i\dot{\varphi}_{,i} - g\dot{\varphi} + \rho_0\theta\dot{\eta} = & (t_{ij}\dot{u}_i + h_j\dot{\varphi} + \frac{1}{T_0}q_j\theta)_{,j}+ \\ +\rho_0\left(f_i\dot{u}_i + \ell\dot{\varphi} + \frac{1}{T_0}S\theta\right) & - \frac{1}{T_0}q_i\theta_{,i} - \rho_0(\dot{u}_i\ddot{u}_i + \kappa\dot{\varphi}\ddot{\varphi}). \end{aligned} \tag{2.1.3}$$

By the divergence theorem and (2.1.3), it follows that

$$\int_B (t_{ij}\dot{e}_{ij}+h_i\dot{\varphi}_{,i}-g\dot{\varphi}+\rho_0\theta\dot{\eta})dV=\int_{\partial B}\left(t_i\dot{u}_i+h\dot{\varphi}+\frac{1}{T_0}\,q\theta\right)dA+$$
$$+\int_B \rho_0\left(f_i\dot{u}_i+\ell\dot{\varphi}+\frac{1}{T_0}\,S\theta\right)dV-\int_B \rho_0(\dot{u}_i\ddot{u}_i+\kappa\dot{\varphi}\ddot{\varphi})dV-\frac{1}{T_0}\int_B q_i\theta_{,i}dV. \tag{2.1.4}$$

If we introduce the total energy U on $[0, t_1)$ by

$$U=\frac{1}{2}\int_B(\rho_0\dot{u}_i\ddot{u}_i+\rho_0\kappa\dot{\varphi}^2+2W+a\theta^2)dV, \tag{2.1.5}$$

then, from (2.1.2), (2.1.4) and (1.5.7) we conclude that

$$\dot{U}-\int_B\rho_0(f_i\dot{u}_i+\ell\dot{\varphi}+\frac{1}{T_0}\,S\theta)dV-$$
$$-\int_{\partial B}(t_i\dot{u}_i+h\dot{\varphi}+\frac{1}{T_0}\,q\theta)dA=-\frac{1}{T_0}\int_B q_i\theta_{,i}dV\leq 0. \tag{2.1.6}$$

Suppose that there are two solutions $(u_i^{(\alpha)}, \varphi^{(\alpha)}, \theta^{(\alpha)}, e_{ij}^{(\alpha)}, t_{ij}^{(\alpha)}, h_i^{(\alpha)}, g^{(\alpha)}, \eta^{(\alpha)}, q_i^{(\alpha)})$, $(\alpha = 1, 2)$. Then their difference $\overline{p} = (\overline{u}_i, \overline{\varphi}, \overline{\theta}, ..., \overline{q}_i)$ corresponds to null data. If $\overline{U}$ is the total energy corresponding to $\overline{p}$ then from (2.1.6) we obtain

$$\overline{U}(t)\leq\overline{U}(0),\quad 0\leq t<t_1.$$

The initial conditions imply $\overline{U}(0)=0$. By hypothesis we find $\overline{U}(t)=0$, $0 \leq t < t_1$. Hence $\dot{\overline{u}}_i = 0$, $\dot{\overline{\varphi}} = 0$, $\overline{\theta} = 0$ on $\overline{B}\times[0, t_1)$. But $\overline{u}_i$ and $\overline{\varphi}$ vanish initially so that $\overline{u}_i = 0$, $\overline{\varphi} = 0$ on $\overline{B}\times[0, t_1)$. □

The uniqueness of the solution in the theory of elastic materials with voids has been presented in [177].

The function W defined by (2.1.1) is the internal energy density per unit of initial volume in the case of isothermal theory. If we assume that W is a positive definite quadratic form, then we find that the constitutive coefficients for isotropic bodies satisfy the following inequalities (Cowin and Nunziato [73])

$$\mu>0,\ \alpha>0,\ \xi>0,\ 2\mu+3\lambda>0,\ (2\mu+3\lambda)\xi>3b^2. \tag{2.1.7}$$

In [67], Ciarletta and Scalia established two uniqueness results with no definiteness assumption on the elastic constitutive coefficients. We suppose for the purpose of these results that the underlying time interval is $I = [0, \infty)$. The kinetic energy is given by

$$K(t)=\frac{1}{2}\int_B\rho_0(\boldsymbol{X})[\dot{\boldsymbol{u}}^2(\boldsymbol{X},t)+\kappa(\boldsymbol{X})\dot{\varphi}^2(\boldsymbol{X},t)]dV,$$
$$0\leq t<\infty. \tag{2.1.8}$$

In what follows, for convenience, the dependence of $\boldsymbol{X}$ is suppressed. We introduce the notations

$$
\begin{aligned}
V=&\int_B (W+\frac{1}{2}\,a\theta^2)dV,\\
2\Gamma(t)=&C_{ijrs}e_{ij}(2t)e_{rs}(0)+B_{ij}[e_{ij}(2t)\varphi(0)+e_{ij}(0)\varphi(2t)]+\\
&+D_{ijr}[e_{ij}(2t)\varphi_{,r}(0)+e_{ij}(0)\varphi_{,r}(2t)]+d_i[\varphi(2t)\varphi_{,i}(0)+\\
&+\varphi(0)\varphi_{,i}(2t)]+\zeta\varphi(2t)\varphi(0)+A_{ij}\varphi_{,i}(2t)\varphi_{,j}(0)+\\
&+a\theta(2t)\theta(0)-\rho_0\dot{u}_i(0)\dot{u}_i(2t)-\rho_0\kappa\dot{\varphi}(0)\dot{\varphi}(2t),\quad 0\le t<\infty,
\end{aligned}
\tag{2.1.9}
$$

where W is defined in (2.1.1).

Lemma 2.1.1. *Assume that the conductivity tensor k_{ij} is symmetric and the symmetry relations (1.5.5) hold. Let*

$$
\begin{aligned}
G(\alpha,\beta)=&\int_B \rho_0\left[f_i(\alpha)\dot{u}_i(\beta)+\ell(\alpha)\dot{\varphi}(\beta)-\frac{1}{T_0}\,S(\alpha)\theta(\beta)\right]dV+\\
&+\int_{\partial B}\left[t_i(\alpha)\dot{u}_i(\beta)+h(\alpha)\dot{\varphi}(\beta)-\frac{1}{T_0}\,q(\alpha)\theta(\beta)\right]dA,
\end{aligned}
\tag{2.1.10}
$$

for all $\alpha,\beta\in I$. Then

$$
\begin{aligned}
V(t)-K(t)=\frac{1}{2}\int_0^t&[G(t+s,t-s)-G(t-s,t+s)]ds+\\
&+\int_B\Gamma(t)dV,\quad t\in I.
\end{aligned}
\tag{2.1.11}
$$

Proof. We denote

$$
F(r,s)=t_{ij}(r)\dot{e}_{ij}(s)+h_i(r)\dot{\varphi}_{,i}(s)-g(r)\dot{\varphi}(s)+\rho_0\theta(r)\dot{\eta}(s),\tag{2.1.12}
$$

for all $r,s\in I$. By (1.5.9) and the symmetry of constitutive coefficients we find that

$$
\begin{aligned}
F(t-s,&t+s)-F(t+s,t-s)=\\
&=\frac{\partial}{\partial s}\{C_{ijmn}e_{ij}(t-s)e_{mn}(t+s)+B_{ij}[e_{ij}(t+s)\varphi(t-s)+\\
&+e_{ij}(t-s)\varphi(t+s)]+D_{ijr}[e_{ij}(t+s)\varphi_{,r}(t-s)+\\
&+e_{ij}(t-s)\varphi_{,r}(t+s)]+A_{ij}\varphi_{,i}(t+s)\varphi_{,j}(t-s)+\\
&+d_i[\varphi(t+s)\varphi_{,i}(t-s)+\varphi(t-s)\varphi_{,i}(t+s)]+\\
&+\xi\varphi(t+s)\varphi(t-s)+a\theta(t-s)\theta(t+s).
\end{aligned}
\tag{2.1.13}
$$

In view of (1.5.3), (1.5.11)-(1.5.13), we have

$$\begin{aligned}
F(t-s,t+s) = {} & [t_{ij}(t-s)\dot{u}_j(t+s) + h_i(t-s)\dot{\varphi}(t+s)+ \\
& +\frac{1}{T_0}\, q_i(t+s)\theta(t-s)]_{,i} + \rho_0[f_i(t-s)\dot{u}_i(t+s)+ \\
& +\ell(t-s)\dot{\varphi}(t+s) + \frac{1}{T_0}\, S(t+s)\theta(t-s)]- \\
& -\frac{1}{T_0}\, k_{ij}\theta_{,j}(t+s)\theta_{,i}(t-s) + \rho_0\, \frac{\partial}{\partial s}[\dot{u}_i(t-s)\dot{u}_i(t+s)+ \\
& +\kappa\dot{\varphi}(t-s)\dot{\varphi}(t+s)] - \rho_0\dot{u}_i(t-s)\ddot{u}_i(t+s)- \\
& -\rho_0\kappa\dot{\varphi}(t-s)\ddot{\varphi}(t+s).
\end{aligned} \tag{2.1.14}$$

Similarly,

$$\begin{aligned}
F(t+s,t-s) = {} & [t_{ij}(t+s)\dot{u}_j(t-s) + h_i(t+s)\dot{\varphi}(t-s)+ \\
& +\frac{1}{T_0}\, q_i(t-s)\theta(t+s)]_{,i} + \rho_0[f_i(t+s)\dot{u}_i(t-s)+ \\
& +\ell(t+s)\dot{\varphi}(t-s) + \frac{1}{T_0}\, S(t-s)\theta(t+s)]- \\
& -\frac{1}{T_0}\, k_{ij}\theta_{,j}(t-s)\theta_{,i}(t+s) - \rho_0\ddot{u}_i(t+s)\dot{u}_i(t-s)- \\
& -\rho_0\kappa\ddot{\varphi}(t+s)\dot{\varphi}(t-s).
\end{aligned} \tag{2.1.15}$$

Next, if we integrate (2.1.13) over B, we conclude, with the aid of (2.1.10), (2.1.14), (2.1.15), the symmetries of constitutive coefficients and the divergence theorem, that

$$\begin{aligned}
& \int_B \frac{\partial}{\partial s}\{C_{ijmn}e_{ij}(t-s)e_{mn}(t+s) + B_{ij}[e_{ij}(t+s)\varphi(t-s)+ \\
& \quad +e_{ij}(t-s)\varphi(t+s)] + D_{ijr}[e_{ij}(t+s)\varphi_{,r}(t-s)+ \\
& \quad +e_{ij}(t-s)\varphi_{,r}(t+s)] + A_{ij}\varphi_{,i}(t+s)\varphi_{,j}(t-s)+ \\
& \quad +d_i[\varphi(t+s)\varphi_{,i}(t-s) + \varphi(t-s)\varphi_{,i}(t+s)] + \xi\varphi(t+s)\varphi(t-s)+ \\
& \quad +a\theta(t-s)\theta(t+s)\}dV = G(t-s,t+s) - G(t+s,t-s)+ \\
& \quad +\int_B \rho_0\, \frac{\partial}{\partial s}[\dot{u}_i(t-s)\dot{u}_i(t+s) + \kappa\dot{\varphi}(t-s)\dot{\varphi}(t+s)]dV.
\end{aligned}$$

If we integrate this relation over $[0,t]$, then we arrive at the desired result. □

Lemma 2.1.2 is a counterpart of Brun's theorem in the isothermal theory.

Lemma 2.1.2. *Let the constitutive coefficients be as in Lemma* 2.1.1 *and let*

$$P = \int_B \rho_0 \left(f_i \dot{u}_i + \ell \dot{\varphi} + \frac{1}{T_0} S\theta \right) dV + \\ + \int_{\partial B} \left(t_i \dot{u}_i + h\dot{\varphi} + \frac{1}{T_0} q\theta \right) dA. \tag{2.1.16}$$

Then

$$\begin{aligned} 2V(t) &= V(0) + K(0) + \int_B \Gamma(t) dV + \frac{1}{2}\int_0^t [G(t+s, t-s) - \\ &\quad - G(t-s, t+s) + 2P(s)] ds - \frac{1}{T_0}\int_0^t \int_B k_{ij}\theta_{,j}\theta_{,i} dV\, dt, \\ 2K(t) &= V(0) + K(0) - \int_B \Gamma(t) dV - \frac{1}{2}\int_0^t [G(t+s, t-s) - \\ &\quad - G(t-s, t+s) - 2P(s)] ds - \frac{1}{T_0}\int_0^t \int_B k_{ij}\theta_{,i}\theta_{,j} dV\, dt, \end{aligned} \tag{2.1.17}$$

for all $t \in I$.

Proof. In view of (1.5.3), (1.5.11)-(1.5.13), we get

$$F(t,t) = \left(t_{ji}\dot{u}_i + h_j \dot{\varphi} + \frac{1}{T_0} q_j \theta \right)_{,j} + \rho_0 \left(f_i \dot{u}_i + \ell\dot{\varphi} + \frac{1}{T_0} S\theta \right) - \\ - \rho_0 \dot{u}_i \ddot{u}_i - \rho_0 \kappa \dot{\varphi}\ddot{\varphi} - \frac{1}{T_0} k_{ij}\theta_{,i}\theta_{,j}. \tag{2.1.18}$$

On the other hand, by (1.5.9) and (2.1.12) we obtain

$$F(t,t) = \dot{W} + a\theta\dot{\theta},$$

so that (2.1.18) implies that

$$\dot{V} + \dot{K} = P - \frac{1}{T_0}\int_B k_{ij}\theta_{,j}\theta_{,i} dV.$$

This relation leads to

$$V(t) + K(t) = V(0) + K(0) + \int_0^t P(s) ds - \frac{1}{T_0}\int_0^t \int_B k_{ij}\theta_{,j}\theta_{,i} dV, \quad t \in I. \tag{2.1.19}$$

From (2.1.11) and (2.1.19) we obtain (2.1.17). □

Lemmas 2.1.1 and 2.1.2 form the basis of the following uniqueness result.

Theorem 2.1.2. *Assume that*

(i) k_{ij} *is symmetric and the symmetry relations* (1.5.5) *hold;*
(ii) ρ_0, κ *and* a *are strictly positive;*
(iii) k_{ij} *is positive semi-definite.*

Then the mixed problem of thermoelastodynamics has at most one solution.

Proof. Suppose that there are two solutions. Then their difference $(u_i, \varphi, \theta, e_{ij}, t_{ij}, h_i, g, \eta, q_i)$ corresponds to null data. Thus, we conclude from $(2.1.17)_2$ that

$$\int_B \rho_0[\dot{\boldsymbol{u}}^2(t) + \kappa\dot{\varphi}^2(t)]dV + \frac{1}{T_0}\int_0^t\int_B k_{ij}\theta_{,i}\theta_{,j}dV\,dt = 0,\ t \in I. \tag{2.1.20}$$

In view of the hypotheses (ii) and (iii), this relation implies

$$\dot{\boldsymbol{u}} = \boldsymbol{0},\ \dot{\varphi} = 0 \ \text{ on } B\times I \tag{2.1.21}$$

and

$$\int_0^t\int_B k_{ij}\theta_{,i}\theta_{,j}dV\,dt = 0, \quad t \in I. \tag{2.1.22}$$

But $\boldsymbol{u}$ and φ vanish initially, so that the relation (2.1.21) reduces to

$$\boldsymbol{u} = \boldsymbol{0},\ \varphi = 0 \ \text{ on } B\times I. \tag{2.1.23}$$

In view of (2.1.22) and (2.1.23), the relation $(2.1.17)_1$ reduces to

$$\int_B a\theta^2 dV = 0.$$

By (ii), this relation yields $\theta = 0$ on $B\times I$ and the proof is complete. □

We now establish an integral identity which involves two admissible processes at different instants. This identity leads to another uniqueness result. If f is a continuous function on $B\times I$, then we denote by $\widehat{f}$ the function defined by

$$\widehat{f}(\boldsymbol{X}, t) = \int_0^t f(\boldsymbol{X}, s)ds, \quad (\boldsymbol{X}, t) \in B\times I.$$

We introduce the notation

$$M = \widehat{S} + T_0\eta^0, \tag{2.1.24}$$

where

$$\rho_0\eta^0 = \beta_{ij}u_{i,j}^0 + a\theta^0 + m\varphi^0 + a_i\varphi_{,i}^0.$$

The energy equation (1.5.13) and the initial conditions imply that

$$\rho_0 T_0 \eta = \widehat{q}_{i,i} + \rho_0 M. \tag{2.1.25}$$

We consider two external data systems $L^{(\alpha)}$, $(\alpha = 1, 2)$. Let $p^{(\alpha)} = (u_i^{(\alpha)}, \varphi^{(\alpha)}, \theta^{(\alpha)}, e_{ij}^{(\alpha)}, t_{ij}^{(\alpha)}, h_i^{(\alpha)}, g^{(\alpha)}, \eta^{(\alpha)}, q_i^{(\alpha)})$ be a solution corresponding to $L^{(\alpha)}$, $(\alpha = 1, 2)$. We introduce the notations

$$\begin{aligned} &t_i^{(\alpha)} = t_{ji}^{(\alpha)} n_j, \ h^{(\alpha)} = h_i^{(\alpha)} n_i, \ q^{(\alpha)} = q_i^{(\alpha)} n_i, \\ &M^{(\alpha)} = \widehat{S}^{(\alpha)} + T_0 \eta^{0(\alpha)}, \qquad\qquad (\alpha = 1, 2). \end{aligned} \tag{2.1.26}$$

Lemma 2.1.3. *Let the constitutive coefficients be as in Lemma* 2.1.1 *and let*

$$\begin{aligned} E_{\alpha\beta}(r,s) = &\int_{\partial B} [t_i^{(\alpha)}(r) u_i^{(\beta)}(s) + h^{(\alpha)}(r)\varphi^{(\beta)}(s) - \\ &-\frac{1}{T_0}\, \widehat{q}^{(\alpha)}(r)\theta^{(\beta)}(s)] dA - \int_B [\rho_0 \ddot{u}_i^{(\alpha)}(r) u_i^{(\beta)}(s) + \\ &+\rho_0 \kappa \ddot{\varphi}^{(\alpha)}(r)\varphi^{(\beta)}(s)] dV + \frac{1}{T_0}\int_B \widehat{q}_i^{(\alpha)}(r)\theta_{,i}^{(\beta)}(s) dV + \\ &+\int_B \rho_0 [f_i^{(\alpha)}(r) u_i^{(\beta)}(s) + \ell^{(\alpha)}(r)\varphi^{(\beta)}(s) - \\ &-\frac{1}{T_0}\, M^{(\alpha)}(r)\theta^{(\beta)}(s)] dV, \end{aligned} \tag{2.1.27}$$

for all $r, s \in I$. *Then*

$$E_{\alpha\beta}(r,s) = E_{\beta\alpha}(s,r). \tag{2.1.28}$$

Proof. We introduce the notations

$$\begin{aligned} J_{\alpha\beta}(r,s) = &\, t_{ij}^{(\alpha)}(r) e_{ij}^{(\beta)}(s) + h_i^{(\alpha)}(r)\varphi_{,i}^{(\beta)}(s) - \\ &-g^{(\alpha)}(r)\varphi^{(\beta)}(s) - \rho_0 \eta^{(\alpha)}(r)\theta^{(\beta)}(s). \end{aligned} \tag{2.1.29}$$

It follows from (1.5.9) and (2.1.29) that

$$\begin{aligned} J_{\alpha\beta}(r,s) = &\, C_{ijpq} e_{pq}^{(\alpha)}(r) e_{ij}^{(\beta)}(s) + A_{ij}\varphi_{,j}^{(\alpha)}(r)\varphi_{,i}^{(\beta)}(s) + \\ &+\xi\varphi^{(\alpha)}(r)\varphi^{(\beta)}(s) - a\theta^{(\alpha)}(r)\theta^{(\beta)}(s) + B_{ij}[\varphi^{(\alpha)}(r) e_{ij}^{(\beta)}(s) + \\ &+e_{ij}^{(\alpha)}(r)\varphi^{(\beta)}(s)] + D_{pqi}[\varphi_{,i}^{(\alpha)}(r) e_{pq}^{(\beta)}(s) + e_{pq}^{(\alpha)}(r)\varphi_{,i}^{(\beta)}(s)] - \end{aligned}$$

$$
\begin{aligned}
&-\beta_{ij}[\theta^{(\alpha)}(r)e_{ij}^{(\beta)}(s)+e_{ij}^{(\alpha)}(r)\theta^{(\beta)}(s)]+\\
&+d_i[\varphi^{(\alpha)}(r)\varphi_{,i}^{(\beta)}(s)+\varphi_{,i}^{(\alpha)}(r)\varphi^{(\beta)}(s)]-\\
&-a_i[\theta^{(\alpha)}(r)\varphi_{,i}^{(\beta)}(s)+\varphi_{,i}^{(\alpha)}(r)\theta^{(\beta)}(s)]-\\
&-m[\theta^{(\alpha)}(r)\varphi^{(\beta)}(s)+\varphi^{(\alpha)}(r)\theta^{(\beta)}(s)].
\end{aligned}
\tag{2.1.30}
$$

By (2.1.30) and (1.5.5),

$$
J_{\alpha\beta}(r,s)=J_{\beta\alpha}(s,r),\quad r,s\in I. \tag{2.1.31}
$$

On the other hand, in view of (1.5.3), (1.5.11), (1.5.12), (2.1.25) and (2.1.29) we obtain

$$
\begin{aligned}
J_{\alpha\beta}(r,s)&=t_{ij}^{(\alpha)}(r)u_{i,j}^{(\beta)}(s)+h_i^{(\alpha)}(r)\varphi_{,i}^{(\beta)}(s)-\\
&-g^{(\alpha)}(r)\varphi^{(\beta)}(s)-\frac{1}{T_0}\,[\widehat{q}_{i,i}^{(\alpha)}(r)+\rho_0M^{(\alpha)}(r)]\theta^{(\beta)}(s)=\\
&=[t_{ij}^{(\alpha)}(r)u_j^{(\beta)}(s)+h_i^{(\alpha)}(r)\varphi^{(\beta)}(s)-\frac{1}{T_0}\,\widehat{q}_i^{(\alpha)}(r)\theta^{(\beta)}(s)]_{,i}-\\
&-\rho_0[\ddot{u}_i^{(\alpha)}(r)u_i^{(\beta)}(s)+\kappa\ddot{\varphi}^{(\alpha)}(r)\varphi^{(\beta)}(s)]+\frac{1}{T_0}\,\widehat{q}_i^{(\alpha)}(r)\theta_{,i}^{(\beta)}(s)+\\
&+\rho_0[f_i^{(\alpha)}(r)u_i^{(\beta)}(s)+\ell^{(\alpha)}(r)\varphi^{(\beta)}(s)-\frac{1}{T_0}\,M^{(\alpha)}(r)\theta^{(\beta)}(s)].
\end{aligned}
$$

If we integrate the above relation over B then we obtain, with the help of (2.1.31) and the symmetry of k_{ij}, the desired result. □

The next lemma is a consequence of Lemma 2.1.3.

Lemma 2.1.4. *Assume that k_{ij} is symmetric and the relations* (1.5.5) *hold. Let*

$$
\begin{aligned}
\Pi(r,s)=&\int_B\rho_0\left[f_i(r)u_i(s)+\ell(r)\varphi(s)-\frac{1}{T_0}\,M(r)\theta(s)\right]dV+\\
&+\int_{\partial B}\left[t_i(r)u_i(s)+h(r)\varphi(s)-\frac{1}{T_0}\,\widehat{q}(r)\theta(s)\right]dA,
\end{aligned}
\tag{2.1.32}
$$

for all $r,s\in I$. Then

$$
\begin{aligned}
&\frac{d}{dt}\left\{\int_B\rho_0(u_iu_i+\kappa\varphi^2)dV+\frac{1}{T_0}\int_0^t\int_BK_{ij}\widehat{\theta}_{,i}\widehat{\theta}_{,j}dV\,dt\right\}=\\
&=\int_0^t[\Pi(t-s,t+s)-\Pi(t+s,t-s)]ds+\int_B\rho_0\{[\dot{u}_i(2t)u_i(0)+\\
&+\dot{u}_i(0)u_i(2t)]+\kappa[\dot{\varphi}(2t)\varphi(0)+\dot{\varphi}(0)\varphi(2t)]\}dV.
\end{aligned}
\tag{2.1.33}
$$

Proof. From (2.1.28) we obtain

$$\int_0^t E_{11}(t+s,t-s)ds = \int_0^t E_{11}(t-s,t+s)ds. \tag{2.1.34}$$

We apply the relation (2.1.34) to the process $p^{(1)} = p$. It follows from (2.1.27) and (2.1.32) that

$$\begin{aligned}\int_0^t E_{11}(t+s,t-s)ds = \int_0^t \Pi(t+s,t-s)ds-\\ -\int_0^t\int_B \rho_0[\ddot{u}_i(t+s)u_i(t-s)+\kappa\ddot{\varphi}(t+s)\varphi(t-s)]dV\,ds+\\ +\frac{1}{T_0}\int_0^t\int_B k_{ij}\widehat{\theta}_{,j}(t+s)\dot{\widehat{\theta}}_{,i}(t-s)dV\,ds.\end{aligned} \tag{2.1.35}$$

Similarly,

$$\begin{aligned}\int_0^t E_{11}(t-s,t+s)ds = \int_0^t \Pi(t-s,t+s)ds-\\ -\int_0^t\int_B \rho_0[\ddot{u}_i(t-s)u_i(t+s)+\kappa\ddot{\varphi}(t-s)\varphi(t+s)]dV\,ds+\\ +\frac{1}{T_0}\int_0^t\int_B k_{ji}\widehat{\theta}_{,i}(t-s)\dot{\widehat{\theta}}_{,j}(t+s)dV\,ds.\end{aligned} \tag{2.1.36}$$

If we use the relations

$$\begin{aligned}\int_0^t \ddot{f}(t+s)h(t-s)ds &= \dot{f}(2t)h(0)-\dot{f}(t)h(t)+\\ &\quad+\int_0^t \dot{h}(t-s)\dot{f}(t+s)ds,\\ \int_0^t \ddot{h}(t-s)f(t+s)ds &= \dot{h}(t)f(t)-\dot{h}(0)f(2t)+\\ &\quad+\int_0^t \dot{h}(t-s)\dot{f}(t+s)ds,\\ \int_0^t f(t+s)\dot{h}(t-s)ds &= -h(0)f(2t)+f(t)h(t)+\\ &\quad+\int_0^t \dot{f}(t+s)h(t-s)ds,\end{aligned}$$

then the relations (2.1.34)-(2.1.36) and the hypotheses imply the desired result. □

Lemma 2.1.4 forms the basis of the following uniqueness theorem.

Theorem 2.1.3. *Assume that*

(i) ρ_0 *and* κ *are strictly positive;*

(ii) k_{ij} *is positive definite;*

(iii) $a \neq 0$ *on* B*;*

(iv) k_{ij} *is symmetric and the relations* (1.5.5) *hold.*

Then the mixed problem of thermoelastodynamics has at most one solution.

Proof. Suppose that there are two solutions. Then their difference $(u_i, \varphi, \theta, e_{ij}, t_{ij}, h_i, g, \eta, q_i)$ corresponds to null data. Thus, from (2.1.33) we get

$$\frac{d}{dt}\left\{\int_B \rho_0(\boldsymbol{u}^2 + \kappa\varphi^2)dV + \frac{1}{T_0}\int_0^t\int_B k_{ij}\widehat{\theta}_{,i}\widehat{\theta}_{,j}dV\,dt\right\} = 0,\ t \in I.$$

Since $\boldsymbol{u}$ and φ vanish initially, the above relation implies

$$\int_B \rho_0(\boldsymbol{u}^2 + \kappa\varphi^2)dV + \frac{1}{T_0}\int_0^t\int_B k_{ij}\widehat{\theta}_{,i}\widehat{\theta}_{,j}dV\,dt = 0,\ t \in I. \qquad (2.1.37)$$

In view of hypotheses (i) and (ii), from (2.1.37) we obtain

$$\boldsymbol{u} = \boldsymbol{0},\ \varphi = 0,\ \widehat{\theta}_{,i} = 0 \ \text{ on } B\times I. \qquad (2.1.38)$$

By (2.1.38) we obtain $\theta_{,i} = 0$ on $B\times I$ so that $q_i = 0$ on $B\times I$ and $\theta = f(t)$, $t \in I$. It follows from the constitutive equations and (2.1.38) that $\rho_0\eta = af$. The energy equation implies that $aT_0\dot{f} = 0$ on $B\times I$. Since T_0 and a are different from zero we obtain $f = \text{const.}$ on $B\times I$. By initial conditions we obtain $\theta = 0$ on $B\times I$. The proof is complete. □

Another uniqueness theorem is given in Section 4.2. The spatial and temporal behaviour of solutions in linear thermoelastodynamics has been studied by Chiriţă and Scalia [58].

2.2 Reciprocal Theorem. Applications

In this section we use the relation (2.1.28) to establish a reciprocal theorem of Graffi type. The proof avoids both the use of the Laplace transform and the incorporation of the initial conditions into the equations of motion. The result has been presented in [67]. The method to obtain the reciprocal theorem has been established in [184] in the context of classical thermoelastodynamics.

Let u and v be functions on $B\times I$ that are continuous in time. We denote by $u * v$ the convolution of u and v,

$$u * v(\boldsymbol{X},t) = \int_0^t u(\boldsymbol{X},t-\tau)v(\boldsymbol{X},\tau)d\tau, \ (\boldsymbol{X},t) \in B\times I.$$

We introduce the functions

$$\begin{aligned} &j(t) = t, \quad 1(t) = 1, \quad t \in I, \\ &F_i = \rho_0(j * f_i + tv_i^0 + u_i^0), \ G = \rho_0[j * \ell + \kappa(t\zeta^0 + \varphi^0)]. \end{aligned} \tag{2.2.1}$$

By (2.1.24), we have

$$M = 1 * S + T_0\eta^0. \tag{2.2.2}$$

We consider two external data systems $L^{(\alpha)} = (\boldsymbol{f}^{(\alpha)}, \ell^{(\alpha)}, S^{(\alpha)}, \widetilde{\boldsymbol{u}}^{(\alpha)}, \widetilde{\boldsymbol{t}}^{(\alpha)}, \widetilde{\varphi}^{(\alpha)}, \widetilde{h}^{(\alpha)}, \widetilde{\theta}^{(\alpha)}, \widetilde{q}^{(\alpha)}, \boldsymbol{u}^{0(\alpha)}, \boldsymbol{v}^{0(\alpha)}, \varphi^{0(\alpha)}, \zeta^{0(\alpha)}, \theta^{0(\alpha)})$, $(\alpha = 1, 2)$. Let $p^{(\alpha)} = (u_i^{(\alpha)}, \varphi^{(\alpha)}, \theta^{(\alpha)}, e_{ij}^{(\alpha)}, t_{ij}^{(\alpha)}, h_i^{(\alpha)}, g^{(\alpha)}, \eta^{(\alpha)}, q_i^{(\alpha)})$ be a solution corresponding to $L^{(\alpha)}$, $(\alpha = 1, 2)$. We shall use the notations (2.1.26) and

$$\begin{aligned} F_i^{(\alpha)} &= \rho_0[j * f_i^{(\alpha)} + tv_i^{0(\alpha)} + u_i^{0(\alpha)}], \\ G^{(\alpha)} &= \rho_0[j * \ell^{(\alpha)} + \kappa(t\zeta^{0(\alpha)} + \varphi^{0(\alpha)})], \ (\alpha = 1, 2). \end{aligned} \tag{2.2.3}$$

The next reciprocal theorem is a consequence of Lemma 2.1.3.

Theorem 2.2.1. *Assume that the conductivity tensor k_{ij} is symmetric and the relations* (1.5.5) *hold. Let $p^{(\alpha)}$ be a solution corresponding to external data system $L^{(\alpha)}$, $(\alpha = 1, 2)$. Then*

$$\begin{aligned} &\int_{\partial B} j * \left[t_i^{(1)} * u_i^{(2)} + h^{(1)} * \varphi^{(2)} - \frac{1}{T_0} 1 * q^{(1)} * \theta^{(2)}\right] dA + \\ &+ \int_B \left[F_i^{(1)} * u_i^{(2)} + G^{(1)} * \varphi^{(2)} - \frac{\rho_0}{T_0} j * M^{(1)} * \theta^{(2)}\right] dV = \\ &= \int_{\partial B} j * \left[t_i^{(2)} * u_i^{(1)} + h^{(2)} * \varphi^{(1)} - \frac{1}{T_0} 1 * q^{(2)} * \theta^{(1)}\right] dA + \\ &+ \int_B \left[F_i^{(2)} * u_i^{(1)} + G^{(2)} * \varphi^{(1)} - \frac{\rho_0}{T_0} j * M^{(2)} * \theta^{(1)}\right] dV. \end{aligned} \tag{2.2.4}$$

Proof. If we take in (2.1.28), $r = \tau$ and $s = t - \tau$ and integrate with respect to τ from 0 to t, we arrive, with the aid of constitutive equations, the hypotheses of theorem and (2.1.27), at

$$\begin{aligned}
&\int_{\partial B}\left[t_i^{(1)} * u_i^{(2)} + h^{(1)} * \varphi^{(2)} - \frac{1}{T_0}\, 1 * q^{(1)} * \theta^{(2)}\right] dA+\\
&\quad -\int_B\left[\rho_0 \ddot{u}_i^{(1)} * u_i^{(2)} + \rho_0\kappa\ddot{\varphi}^{(1)} * \varphi^{(2)}\right] dV+\\
&\quad +\int_B \rho_0\left[f_i^{(1)} * u_i^{(2)} + \ell^{(1)} * \varphi^{(2)} - \frac{1}{T_0}\, M^{(1)} * \theta^{(2)}\right] dV =\\
&\quad =\int_{\partial B}\left[t_i^{(2)} * u_i^{(1)} + h^{(2)} * \varphi^{(1)} - \frac{1}{T_0}\, 1 * q^{(2)} * \theta^{(1)}\right] dA-\\
&\quad -\int_B\left[\rho_0 \ddot{u}_i^{(2)} * u_i^{(1)} + \rho_0\kappa\ddot{\varphi}^{(2)} * \varphi^{(1)}\right] dV+\\
&\quad +\int_B \rho_0\left[f_i^{(2)} * u_i^{(1)} + \ell^{(2)} * \varphi^{(1)} - \frac{1}{T_0}\, M^{(2)} * \theta^{(1)}\right] dV.
\end{aligned} \tag{2.2.5}$$

Taking the convolution of relation (2.2.5) with $j(t) = t$, $t \in I$, and using the relations

$$j * \ddot{u}_i^{(\alpha)} = u_i^{(\alpha)} - t v_i^{0(\alpha)} - u_i^{0(\alpha)}, \; j * \ddot{\varphi}^{(\alpha)} = \varphi^{(\alpha)} - t\zeta^{0(\alpha)} - \varphi^{0(\alpha)},$$

we obtain the desired result. □

Another method to derive reciprocal theorems has been presented in [158].

Various applications of the reciprocal theorem have been presented in [202], [289]. Let us assume that the initial data are zero. Then, from (2.1.26) and (2.2.3) we find that

$$F_i^{(\alpha)} = \rho_0 j * f_i^{(\alpha)}, \; G^{(\alpha)} = \rho_0 j * \ell^{(\alpha)}, \; M^{(\alpha)} = 1 * S^{(\alpha)},$$

and the reciprocity relation (2.2.4) reduces to

$$\begin{aligned}
&\int_{\partial B}\left[t_i^{(1)} * u_i^{(2)} + h^{(1)} * \varphi^{(2)} - \frac{1}{T_0}\, 1 * q^{(1)} * \theta^{(2)}\right] dA+\\
&\quad +\int_B \rho_0\left[f_i^{(1)} * u_i^{(2)} + \ell^{(1)} * \varphi^{(2)} - \frac{1}{T_0}\, 1 * S^{(1)} * \theta^{(2)}\right] dV=\\
&\quad =\int_{\partial B}\left[t_i^{(2)} * u_i^{(1)} + h^{(2)} * \varphi^{(1)} - \frac{1}{T_0}\, 1 * q^{(2)} * \theta^{(1)}\right] dA+\\
&\quad +\int_B \rho_0\left[f_i^{(2)} * u_i^{(1)} + \ell^{(2)} * \varphi^{(1)} - \frac{1}{T_0}\, 1 * S^{(2)} * \theta^{(1)}\right] dV.
\end{aligned} \tag{2.2.6}$$

In what follows we restrict our attention to the "traction problem" when $S_1 = S_3 = S_5 = \emptyset$. We consider the external data system $\text{Ł}^{(1)}$

characterized by

$$\begin{aligned}&\rho_0 f_i^{(1)} = \delta(X-Y)\delta(t)\delta_{ij},\ \ell^{(1)} = 0,\ S^{(1)} = 0,\\ &u_i^{0(1)} = v_i^{0(1)} = \varphi^{0(1)} = \zeta^{0(1)} = \theta^{0(1)} = 0,\end{aligned} \tag{2.2.7}$$

where δ is the Dirac measure. We assume that $L^{(1)}$ generates the displacements $U_i^{(j)}$, the volume fraction $\varphi^{(j)}$ and the temperature $T^{(j)}$. Let $P_i^{(j)}, H^{(j)}$ and $Q^{(j)}$ be, respectively, the traction, the equilibrated surface traction and the heat flux generated by $U_i^{(j)}, \varphi^{(j)}$ and $T^{(j)}$. Let u_i, φ and θ be the displacement, volume fraction and temperature corresponding to to external data system $L^{(2)} = (f_i, \ell, S, \widetilde{t}_i, \widetilde{h}, \widetilde{q}, u_i^0 = v_i^0 = \varphi^0 = \zeta^0 = \theta^0 = 0)$. It follows from (2.2.6) and (2.2.7) that

$$\begin{aligned}u_k(\boldsymbol{Y},t) =& \int_B \rho_0[f_i * U_i^{(k)} + \ell * \varphi^{(k)} - \frac{1}{T_0} 1 * S * T^{(k)}]dV+\\ &+\int_{\partial B} [\widetilde{t}_i * U_i^{(k)} - P_i^{(k)} * u_i + \widetilde{h} * \varphi^{(k)} - H^{(k)} * \varphi+\\ &+\frac{1}{T_0} 1 * (\theta * Q^{(k)} - T^{(k)} * \tilde{q})]dA.\end{aligned} \tag{2.2.8}$$

Let us assume that the external system $L^{(1)}$,

$$\begin{aligned}&f_i^{(1)} = 0,\ \rho_0\ell^{(1)} = \delta(X-Y)\delta(t),\ S^{(1)} = 0,\\ &u_i^{0(1)} = v_i^{0(1)} = \varphi^{0(1)} = \zeta^{0(1)} = \theta^{0(1)} = 0,\end{aligned}$$

generates the displacement $U_i^{(4)}$, the volume fraction $\varphi^{(4)}$ and the temperature $T^{(4)}$. Let $P_i^{(4)}, H^{(4)}$ and $Q^{(4)}$ be, respectively, the traction, the equilibrated surface traction and the heat flux generated by $U_i^{(4)}, \varphi^{(4)}$ and $T^{(4)}$. Let $(u_i, \varphi, \theta, e_{ij}, t_{ij}, h_i, g, \eta, q_i)$ be a solution corresponding to the external data system $L^{(2)} = (f_i, \ell, S, \widetilde{t}_i, \widetilde{h}, \widetilde{q}, u_i^0 = v_i^0 = \varphi^0 = \zeta^0 = \theta^0 = 0)$. It follows from (2.2.6) that

$$\begin{aligned}\varphi(\boldsymbol{Y},t) =& \int_B \rho_0[f_i * U_i^{(4)} + \ell * \varphi^{(4)} - \frac{1}{T_0} 1 * S * T^{(4)}]dV+\\ &+\int_{\partial B} [\widetilde{t}_i * U_i^{(4)} - P_i^{(4)} * u_i + \widetilde{h} * \varphi^{(4)} - H^{(4)} * \varphi+\\ &+\frac{1}{T_0} 1 * (\theta * Q^{(4)} - T^{(4)} * \tilde{q})]dA.\end{aligned} \tag{2.2.9}$$

Similarly, we can obtain a formula for the temperature field. The integral representations (2.2.8) and (2.2.9) are relations of Somigliana type.

Let us consider the case of homogeneous and isotropic bodies. We denote by (P_1) the boundary-initial-value problem characterized by the field equations (1.5.3), (1.5.10), (1.5.11)-(1.5.13) and the conditions (1.5.14), (1.5.15). If we replace the constitutive equation for the intrinsic equilibrated body force by the equation $g = -\xi\varphi$, then we obtain a new boundary-initial-value problem. We denote this problem by (P_2). We note that (P_2) is an uncoupled problem, in the sense that the function φ is independent of u_i and θ. Let $s^{(1)}$ and $s^{(2)}$ be the solutions of the problems (P_1) and (P_2), respectively. We assume that $L^{(\alpha)}$ is the external data system corresponding to the problem (P_α), $(\alpha = 1, 2)$. By a proof analogous to that of Theorem 2.2.1 we obtain the following result: if an isotropic body is subjected to two external data systems $L^{(\alpha)}$, $(\alpha = 1, 2)$, then between the solutions $s^{(\alpha)}$ corresponding to the problems (P_α) there is the following reciprocity relation

$$\begin{aligned}
&\int_B \rho_0 \left[f_i^{(1)} * u_i^{(2)} + \ell^{(1)} * \varphi^{(2)} - \frac{1}{T_0}\, 1 * S^{(1)} * \theta^{(2)} \right] dV + \\
&+ \int_{\partial B} \left[t_i^{(1)} * u_i^{(2)} + h^{(1)} * \varphi^{(2)} - \frac{1}{T_0}\, 1 * q^{(1)} * \theta^{(2)} \right] dA = \\
&= \int_B \rho_0 \left[f_i^{(2)} * u_i^{(1)} + \ell^{(2)} * \varphi^{(1)} - \frac{1}{T_0}\, 1 * S^{(2)} * \theta^{(1)} \right] dV + \\
&+ \int_{\partial B} \left[t_i^{(2)} * u_i^{(1)} + h^{(2)} * \varphi^{(1)} - \frac{1}{T_0}\, 1 * q^{(2)} * \theta^{(1)} \right] dA + \\
&\qquad + \int_B (b e_{jj}^{(2)} - m\theta^{(2)}) * \varphi^{(1)} dV.
\end{aligned} \tag{2.2.10}$$

We now assume that the uncoupled problem (P_2) corresponds to the external data system

$$\begin{aligned}
&f_i^{(2)} = 0,\ \rho_0 \ell^{(2)} = \delta(X - Y)\delta(t),\ S^{(2)} = 0,\ \widetilde{t}^{(2)} = 0, \\
&\widetilde{h}^{(2)} = 0,\ \widetilde{q}^{(2)} = 0,\ u_i^{0(2)} = v_i^{0(2)} = \varphi^{0(2)} = \zeta^{0(2)} = \theta^{0(2)} = 0.
\end{aligned} \tag{2.2.11}$$

Let $\mathbf{U}, \Phi$ and T be the displacement, volume fraction and temperature fields generated by this system. If we use reciprocity relation (2.2.10) then we obtain

$$\begin{aligned}
&\int_B \rho_0 \left[f_i^{(1)} * U_i + \ell^{(1)} * \Phi - \frac{1}{T_0}\, 1 * S^{(1)} * T \right] dV + \\
&\quad + \int_{\partial B} \left[\widetilde{t}_i^{(1)} * U_i + \widetilde{h}^{(1)} * \Phi - \frac{1}{T_0}\, \widetilde{q}^{(1)} * T \right] dA = \\
&\quad = \varphi^{(1)}(\mathbf{Y}, t) + \int_B (b U_{j,j} - mT) * \varphi^{(1)} dV.
\end{aligned} \tag{2.2.12}$$

Thus, the solving of the coupled problem (P_1) corresponding to the external data system $L^{(1)}$ and to the homogeneous initial conditions is reduced to the solving of the associated uncoupled problem (P_2) corresponding to the external data system (2.2.11) and to the solving of the integral equation (2.2.12) for the volume fraction field.

2.3 Homogeneous and Isotropic Bodies

In this section we assume that the body is homogeneous and isotropic so that the basic field equations can be written in the form (1.5.17). We assume that the relations (2.1.7) hold and introduce the notations

$$c_1 = \sqrt{\frac{\lambda + 2\mu}{\rho_0}}, \quad c_2 = \sqrt{\frac{\mu}{\rho_0}}, \quad c_3 = \sqrt{\frac{\alpha}{\kappa\rho_0}}, \quad c = aT_0. \tag{2.3.1}$$

The equations (1.5.17) become

$$\begin{aligned} &c_2^2\Delta u_i + (c_1^2 - c_2^2)u_{j,ji} + \frac{b}{\rho_0}\,\varphi_{,i} - \frac{\beta}{\rho_0}\,\theta_{,i} + f_i = \ddot{u}_i, \\ &c_3^2\Delta\varphi - \frac{b}{\rho_0\kappa}\,u_{j,j} - \frac{\xi}{\rho_0\kappa}\,\varphi + \frac{m}{\rho_0\kappa}\,\theta + \ell = \ddot{\varphi}, \\ &k\Delta\theta - \beta T_0\dot{u}_{j,j} - mT_0\dot{\varphi} - c\dot{\theta} = -\rho_0 S. \end{aligned} \tag{2.3.2}$$

We introduce the operators $\square_s$, $(s = 1, 2, 3, 4)$, defined by

$$\square_i = c_i^2\Delta - \frac{\partial^2}{\partial t^2}, \; (i = 1, 2, 3), \quad \square_4 = k\Delta - c\,\frac{\partial}{\partial t}. \tag{2.3.3}$$

The equations (2.3.2) can be expressed in the form

$$\begin{aligned} &\square_2\boldsymbol{u} + (c_1^2 - c_2^2)\,\text{grad div}\,\boldsymbol{u} + \frac{b}{\rho_0}\,\text{grad}\,\varphi - \frac{\beta}{\rho_0}\,\text{grad}\,\theta = -\boldsymbol{f}, \\ &\square_3\varphi - \frac{b}{\rho_0\kappa}\,\text{div}\,\boldsymbol{u} - \frac{\xi}{\rho_0\kappa}\,\varphi + \frac{m}{\rho_0\kappa}\,\theta = -\frac{1}{\kappa}\,\ell, \\ &\square_4\theta - \beta T_0\,\text{div}\,\dot{\boldsymbol{u}} - mT_0\dot{\varphi} = -\rho_0 S. \end{aligned} \tag{2.3.4}$$

In this section we establish a solution of Galerkin type of the field equations. The result was presented in [66] and extends the representation given by Nowacki [286] in the classical thermoelastodynamics.

For convenience, we introduce the following operators

$$\Lambda_1 = \Box_1 - \frac{b\beta}{\rho_0 m}\,\Delta,\ \Lambda_2 = \Box_3 - \frac{\xi}{\rho_0\kappa} + \frac{mb}{\rho_0\kappa\beta},$$

$$\Lambda_3 = \Box_4 - \frac{mT_0\beta}{b}\,\frac{\partial}{\partial t},\ \cap_1 = \Box_1\Box_3\Box_4 - cc_1^2\varepsilon\Box_3\Delta\frac{\partial}{\partial t} -$$

$$-\frac{\xi}{\rho_0\kappa}\,\Box_1\Box_4 + \frac{b^2}{\rho_0^2\kappa}\Box_4\Delta + \frac{m^2T_0}{\rho_0\kappa}\,\Box_1\frac{\partial}{\partial t} + \gamma\Delta\frac{\partial}{\partial t},$$

$$\cap_2 = (c_1^2 - c_2^2)\Box_3\Box_4 - \frac{\beta^2 T_0}{\rho_0}\Box_3\frac{\partial}{\partial t} - \tag{2.3.5}$$

$$-\left[(c_1^2 - c_2^2)\frac{\xi}{\rho_0\kappa} - \frac{b^2}{\rho_0^2\kappa}\right]\Box_4 + \left[\gamma + (c_1^2 - c_2^2)\frac{m^2T_0}{\rho_0\kappa}\right]\frac{\partial}{\partial t},$$

$$\cap_3 = \Box_1\Box_3 - \frac{\xi}{\rho_0\kappa}\Box_1 + \frac{b^2}{\rho_0^2\kappa}\Delta,$$

$$\cap_4 = \Box_1\Box_4 - \frac{\beta^2 T_0}{\rho_0}\Delta\frac{\partial}{\partial t},$$

where

$$\varepsilon = T_0\beta^2/\rho_0 cc_1^2,\ \gamma = T_0\beta(\xi\beta - 2mb)/\rho_0^2\kappa. \tag{2.3.6}$$

Theorem 2.3.1. *Let*

$$\begin{aligned} \boldsymbol{u} &= \cap_1\boldsymbol{F} - \cap_2\,\text{grad div}\,\boldsymbol{F} - \frac{b}{\rho_0}\Lambda_3\,\text{grad}\,G + \frac{\beta}{\rho_0}\Lambda_2\,\text{grad}\,H,\\ \varphi &= \frac{b}{\rho_0\kappa}\,\Box_2\Lambda_3\,\text{div}\,\boldsymbol{F} + \cap_4 G - \frac{m}{\rho_0\kappa}\,\Lambda_1 H,\\ \theta &= \beta T_0\Box_2\Lambda_2\,\text{div}\,\dot{\boldsymbol{F}} + mT_0\Lambda_1\dot{G} + \cap_3 H, \end{aligned} \tag{2.3.7}$$

where the fields F_i of class C^8 on $B\times I$, and G and H of class C^6 on $B\times I$ satisfy the equations

$$\Box_2\cap_1 F_i = -f_i,\ \ \cap_1 G = -\frac{1}{\kappa}\,\ell,\ \ \cap_1 H = -\rho_0 S. \tag{2.3.8}$$

Then $\boldsymbol{u}, \varphi$ and θ satisfy the field equations (2.3.4).

Proof. A straightforward calculation yields

$$\Box_2\cap_2 + (c_1^2 - c_2^2)\Delta\cap_2 - \frac{b^2}{\rho_0^2\kappa}\,\Box_2\Lambda_3 + \frac{\beta^2}{\rho_0}\,T_0\Box_2\Lambda_2\frac{\partial}{\partial t} =$$

$$= (c_1^2 - c_2^2)\cap_1,$$

$$\begin{aligned}
&\frac{b}{\rho_0}\Lambda_3\Box_2 + \frac{b}{\rho_0}(c_1^2 - c_2^2)\Lambda_3\Delta + \frac{\beta m T_0}{\rho_0}\,\Lambda_1\,\frac{\partial}{\partial t} = \frac{b}{\rho_0}\,\boxminus_4,\\
&\frac{\beta}{\rho_0}\,\Lambda_2\Box_2 + (c_1^2 - c_2^2)\,\frac{\beta}{\rho_0}\,\Lambda_2\Delta - \frac{bm}{\rho_0^2\kappa}\,\Lambda_1 = \frac{\beta}{\rho_0}\,\boxminus_3.
\end{aligned} \tag{2.3.9}$$

If we substitute $\boldsymbol{u}, \varphi$ and θ given by (2.3.7) in the left-hand side of (2.3.4) then we obtain

$$\begin{aligned}
&\Box_2\boldsymbol{u} + (c_1^2 - c_2^2)\text{grad div }\boldsymbol{u} + \frac{b}{\rho_0}\text{ grad }\varphi - \frac{\beta}{\rho_0}\text{ grad }\theta =\\
&\quad = \Box_2\boxminus_1\boldsymbol{F} + \big[(c_1^2 - c_2^2)\boxminus_1 - \Box_2\boxminus_2 - (c_1^2 - c_2^2)\Delta\boxminus_2 +\\
&\quad + \frac{b^2}{\rho_0^2\kappa}\,\Box_2\Lambda_3 - \frac{\beta^2}{\rho_0}\,T_0\Box_2\Lambda_2\,\frac{\partial}{\partial t}\big]\text{grad div }\boldsymbol{F} +\\
&\quad + \big[\frac{b}{\rho_0}\boxminus_4 - \frac{b}{\rho_0}\Lambda_3\Box_2 - \frac{b}{\rho_0}(c_1^2 - c_2^2)\Lambda_3\Delta - \frac{\beta m T_0}{\rho_0}\Lambda_1\frac{\partial}{\partial t}\big]\text{grad}\,G +\\
&\quad + \big[\frac{\beta}{\rho_0}\Lambda_2\Box_2 + (c_1^2 - c_2^2)\frac{\beta}{\rho_0}\Lambda_2\Delta - \frac{bm}{\rho_0^2\kappa}\Lambda_1 - \frac{\beta}{\rho_0}\,\boxminus_3\big]\text{grad}\,H.
\end{aligned} \tag{2.3.10}$$

Taking into account the relations (2.3.8) and (2.3.9) we obtain

$$\Box_2\boldsymbol{u} + (c_1^2 - c_2^2)\text{grad div}\,\boldsymbol{u} + \frac{b}{\rho_0}\text{ grad}\,\varphi - \frac{\beta}{\rho_0}\text{ grad}\,\theta = \Box_2\boxminus_1\boldsymbol{F} = -\boldsymbol{f}.$$

By using the identities

$$\begin{aligned}
&\Box_2\Box_3\Lambda_3 + \boxminus_2\Delta - \frac{\xi}{\rho_0\kappa}\,\Box_2\Lambda_3 + \frac{m\beta T_0}{b}\,\Box_2\Lambda_2\frac{\partial}{\partial t} = \boxminus_1,\\
&\boxminus_4\Box_3 + \frac{b^2}{\rho_0^2\kappa}\,\Lambda_3\Delta - \frac{\xi}{\rho_0\kappa}\,\boxminus_4 + \frac{m^2 T_0}{\rho_0\kappa}\,\Lambda_1\,\frac{\partial}{\partial t} = \boxminus_1,\\
&\frac{m}{\rho_0\kappa}\,\boxminus_3 + \frac{\xi m}{\rho_0^2\kappa^2}\,\Lambda_1 - \frac{b\beta}{\rho_0^2\kappa}\,\Lambda_2\Delta - \frac{m}{\rho_0\kappa}\,\Lambda_1\Box_3 = 0,
\end{aligned}$$

$$\Box_2\Box_4\Lambda_2 - \boxminus_1 + \boxminus_2\Delta - \frac{mb}{\rho_0\kappa\beta}\,\Box_2\Lambda_3 = 0,$$

$$mT_0\Lambda_1\Box_4 + \frac{b\beta T_0}{m}\,\Lambda_3\Delta - mT_0\boxminus_4 = 0,$$

$$\boxminus_3\Box_4 - \frac{\beta^2 T_0}{\rho_0}\,\Lambda_2\Delta\,\frac{\partial}{\partial t} + \frac{m^2 T_0}{\rho_0\kappa}\,\Lambda_1\,\frac{\partial}{\partial t} = \boxminus_1,$$

we obtain

$$\begin{aligned}
&\Box_3\varphi - \frac{b}{\rho_0\kappa}\text{div}\,\boldsymbol{u} - \frac{\xi}{\rho_0\kappa}\,\varphi + \frac{m}{\rho_0\kappa}\,\theta = \boxminus_1 G = -\frac{1}{\kappa}\,\ell,\\
&\Box_4\theta - \beta T_0\,\text{div}\,\dot{\boldsymbol{u}} - mT_0\dot{\varphi} = \boxminus_1 H = -\rho_0 S.
\end{aligned}$$

This completes the proof. □

In Section 2.6 we shall use the representation (2.3.7) to obtain the basic singular solutions in the case of steady vibrations.

2.4 Acceleration Waves

In this section we study the acceleration waves in homogeneous and isotropic bodies. We derive the propagation conditions and growth equations which gouvern the propagation of waves. The couplings between discontinuities are studied. The results we give here have been established in [180]. Let Σ be a moving surface defined by

$$x_i = x_i(\theta^1, \theta^2, t), \tag{2.4.1}$$

where θ^1, θ^2 are curvilinear coordinates on the surface. In the linear theory Lagrangian coordinates are usually indicated by x_i. We assume that the functions (2.4.1) are continuously differentiable with respect to their arguments and that Σ is smooth in the sense that the matrix $(\partial x_i/\partial\theta^\alpha)$ has rank two. The metric tensor of the surface is given by $a_{a\beta} = x_{i,\alpha}x_{i,\beta}$. In this section we will denote by n_i the unit normal to Σ. We note that [363]

$$n_i n_i = 1,\ n_i x_{i;\alpha} = 0,\ x_{i;\alpha\beta} = b_{a\beta} n_i,\ n_{i;\alpha} = -a^{\lambda\rho} b_{\rho\alpha} x_{i;\lambda}, \tag{2.4.2}$$

where indices followed by a semicolon represents covariant partial differentiation based on the metric of Σ, $b_{a\beta}$ is the second fundamental form of the surface and $a^{\alpha\beta}$ are the elements of the inverse of matrix $(a_{a\beta})$. We also note that

$$a^{\alpha\beta} x_{i;\alpha} x_{j;\beta} = \delta_{ij} - n_i n_j,\ H = \frac{1}{2} a^{\alpha\beta} b_{a\beta}, \tag{2.4.3}$$

where H is the mean curvature of the surface.

The propagating surface Σ is said to be an acceleration wave if u_i, φ, θ have the following properties: (i) $u_i, \varphi, \theta, \dot{u}_i, \dot{\varphi}, \dot{\theta}, u_{i,j}, \varphi_{,i}$ and $\theta_{,i}$ are continuous functions everywhere; (ii) $\ddot{u}_i, \ddot{\varphi}, \ddot{\theta}, u_{i,jk}, \varphi_{,ij}, \theta_{,ij}, \dot{u}_{i,j}, \dot{\varphi}_{,i}, \dot{\theta}_{,i}$ and all higher-order derivatives of these functions have, at most, jump discontinuities across Σ but are continuous functions everywhere else.

In this section we study the acceleration waves in homogeneous and isotropic bodies. Let f be one of the functions u_i, φ, θ. We list for future reference the compatibility conditions satisfied by the jumps on Σ of the second and third derivatives of this function

$$\begin{aligned}
[f_{,ij}] &= Cn_in_j,\ [\dot{f}_{,i}] = -CVn_i,\ [\ddot{f}] = V^2C,\\
[f_{,kij}] &= a^{\alpha\beta}(Cn_k)_{;\alpha}(n_ix_{j;\beta}+n_jx_{i;\beta})-\\
&\quad -a^{\alpha\beta}a^{\nu\rho}b_{\alpha\nu}Cn_kx_{i;\beta}x_{j;\rho}+[f_{,kpq}]n_pn_qn_in_j,\\
[\dot{f}_{,si}] &= -a^{\alpha\beta}(VCn_s)_{;\alpha}x_{i;\beta}+\frac{\delta}{\delta t}(Cn_s)n_i-V[f_{,spq}]n_pn_qn_i,\\
[\ddot{f}_{,s}] &= -Cn_s\frac{\delta V}{\delta t}-2V\frac{\delta}{\delta t}(Cn_s)+V^2[f_{,sij}]n_in_j,\\
[f_{,jjs}] &= -2HCn_s+[f_{,sij}]n_in_j,
\end{aligned} \tag{2.4.4}$$

where

$$\frac{\delta}{\delta t}=\frac{\partial}{\partial t}+Vn_i\frac{\partial}{\partial x_i},$$

is the displacement derivative, V is the speed of propagation of Σ in the direction of the normal $\boldsymbol{n}$, and $C=[f_{,ij}n_in_j]$. A detailed account of compatibility conditions holding on singular surfaces has been given in [42], [367, pp. 491-529]. Derivation of the equations (2.4.4) can be constructed in a straightforward manner from these sources. We assume that $f_i=\ell=S=0$. On forming the jump of each term in the equation (2.3.2) on the singular surface Σ and using the compatibility conditions (2.4.4), we obtain

$$(V^2-c_2^2)\lambda_i-(c_1^2-c_2^2)\lambda_jn_jn_i=0, \tag{2.4.5}$$

$$(V^2-c_3^2)\chi=0, \tag{2.4.6}$$

$$k\zeta+\beta T_0V\lambda_in_i=0, \tag{2.4.7}$$

where $\lambda_i=[u_{i,rs}n_rn_s]$, $\chi=[\varphi_{,rs}n_rn_s]$, $\zeta=[\theta_{,ij}n_in_j]$.

Equations (2.4.5) admit a non trivial solution for λ_i if and only if

$$(c_2^2-V^2)^2(c_1^2-V^2)=0.$$

It is an immediate consequence of equation (2.4.5) that longitudinal waves of first kind (for which $\lambda_i=\Lambda n_i$) propagate with the speed c_1. The speed of propagation of transverse waves (for which $\lambda_in_i=0$) is c_2. From equation (2.4.7) we note that $\zeta=-\beta T_0c_1\Lambda/k$ for a longitudinal wave of first kind, and $\zeta=0$ for a transverse wave.

If $\chi\neq 0$, the wave is an acceleration wave of compaction or distension. This wave is called the longitudinal wave of second kind. The possible speed of propagation of this wave is c_3.

Let us study now the growth of the acceleration waves. We first apply to each term of equations (2.3.2) the operator $\partial/\partial x_s$ and form jumps across the wave Σ. Next we make use of the compatibility conditions (2.4.4), then multiply by n_s and sum on the repeated index s. We obtain the equations

$$\begin{aligned}
&(V^2 - c_2^2)\mu_i - (c_1^2 - c_2^2)\{\mu_j n_j n_i + a^{\alpha\beta} x_{i;\beta}(\lambda_{j;\alpha} n_j - \\
&\quad -a^{\nu\rho} b_{\alpha\nu} \lambda_j x_{j;\rho}) + a^{\alpha\beta}\lambda_{j;\alpha} n_i x_{j;\beta}\} - 2V\,\frac{\delta\lambda_i}{\delta t} + \\
&\quad +2Hc_2^2\lambda_i - \frac{b}{\rho_0}\chi n_i + \frac{\beta}{\rho_0}\zeta n_i = 0, \\
&(V^2 - c_3^2)\tau + 2Hc_3^2\chi - 2V\,\frac{\delta\chi}{\delta t} + \frac{b}{\rho_0\kappa}\,\lambda_j n_j = 0, \\
&\beta T_0\left(V\mu_j n_j - n_i\,\frac{\delta\lambda_i}{\delta t} + Va^{\alpha\beta}\lambda_{j;\alpha} x_{j;\beta}\right) + k\gamma + \\
&\quad +(cV - 2kH)\zeta + mT_0 V\chi = 0,
\end{aligned} \tag{2.4.8}$$

where

$$\mu_i = [u_{i,pqr}]n_p n_q n_r,\quad \tau = [\varphi_{,ijk}]n_i n_j n_k,\quad \gamma = [\theta_{,ijk}]n_i n_j n_k$$

and simplifications have been effected with the aid of equations (2.4.2) and the fact that V is constant for acceleration waves. It is known [362, pp. 43-45] that $\delta n_i/\delta t = -a^{\alpha\beta} x_{i;\alpha} V_{,\beta}$. Using this result and the fact that V is constant for all waves we find

$$\frac{\delta\Lambda}{\delta t} = n_j\,\frac{\delta\lambda_j}{\delta t}\,. \tag{2.4.9}$$

With the help of (2.4.2) and (2.4.3) we obtain

$$\begin{aligned}
&a^{\alpha\beta}\lambda_{j;\alpha} x_{j;\beta} = a^{\alpha\beta}(\lambda_j x_{j;\beta})_{;\alpha} - 2H\Lambda, \\
&\lambda_{j;\alpha} n_j - \lambda_j a^{\nu\rho} b_{\alpha\nu} x_{j;\rho} = \Lambda_{,\alpha}.
\end{aligned} \tag{2.4.10}$$

Using (2.4.9) and (2.4.10), we can write (2.4.8) in the form

$$\begin{aligned}
&(V^2 - c_2^2)\mu_i - (c_1^2 - c_2^2)\{\mu_j n_j n_i + a^{\alpha\beta}(\lambda_j x_{j;\beta})_{;\alpha} n_i + \\
&\quad +a^{\alpha\beta}\Lambda_{;\alpha} x_{i;\beta}\} - 2V\,\frac{\delta\lambda_i}{\delta t} + 2H\{c_2^2\lambda_i + \\
&\quad +(c_1^2 - c_2^2)\Lambda n_i\} - \frac{b}{\rho_0}\chi n_i + \frac{\beta}{\rho_0}\zeta n_i = 0,
\end{aligned} \tag{2.4.11}$$

$$(V^2 - c_3^2)\tau - 2V\,\frac{\delta X}{\delta t} + 2c_3^2 H X + \frac{b}{\rho_0 \kappa}\,\Lambda = 0,$$

$$\beta T_0 V \mu_j n_j + k\gamma - \beta T_0 \left\{ \frac{\delta \Lambda}{\delta t} - V a^{\alpha\beta} (\lambda_j x_{j;\beta})_{;\alpha} + 2HV\Lambda \right\} +$$
$$+ (cV - 2Hk)\zeta + m T_0 V X = 0.$$

On multiplying throughout equation $(2.4.11)_1$ by n_i, summing on the repeated index i and using equations (2.4.2) and (2.4.9), we obtain

$$\begin{aligned} &(V^2 - c_1^2)\mu_j n_j - (c_1^2 - c_2^2) a^{\alpha\beta} (\lambda_j x_{j;\beta})_{;\alpha} - \\ &-2V\,\frac{\delta \Lambda}{\delta t} + 2Hc_1^2 \Lambda - \frac{b}{\rho_0} X + \frac{\beta}{\rho_0}\zeta = 0. \end{aligned} \tag{2.4.12}$$

In what follows we assume that $c_1 \neq c_2$, $c_1 \neq c_3$, $c_2 \neq c_3$. In the case of longitudinal waves of first kind $V = c_1$, $\lambda_i = \Lambda n_i$, $\zeta = -\beta T_0 c_1 \Lambda / k$ and $X = 0$. The equation (2.4.12) yields the growth equation

$$\frac{1}{c_1}\,\frac{\delta \Lambda}{\delta t} = \frac{d\Lambda}{dn} = \Lambda \left(H - \frac{\varepsilon \omega^*}{2c_1} \right), \tag{2.4.13}$$

where n is the distance measured along the normal to the wave, measured from the wavefront at $t = t_0$, and

$$\varepsilon = \beta^2 T_0 / c\rho_0 c_1^2, \quad \omega^* = c c_1^2 / k.$$

In view of (2.4.13), equations (2.4.11) reduce to

$$\begin{aligned} &\mu_i - \mu_j n_j n_i = a^{\alpha\beta} \Lambda_{,\alpha} x_{i;\beta}, \quad (c_1^2 - c_3^2)\tau = \frac{b}{\rho_0 \kappa}\,\Lambda, \\ &\beta T_0 c_1 \mu_j n_j + k\gamma = \beta T_0 c_1 \Lambda \left\{ H + \frac{\omega^*}{c_1}\left(1 - \frac{1}{2}\varepsilon\right)\right\}. \end{aligned} \tag{2.4.14}$$

Suppose that, at $t = t_0$, the mean and Gaussian curvature of the wavefront are H_0 and K_0, respectively. Then, the mean curvature at a subsequent time t is given by [363, pp. 108-113]

$$H = (H_0 - nK_0)/(1 - 2nH_0 + n^2 K_0)$$

and the transport equation (2.4.13) can be integrated to give the decay law for Λ,

$$\Lambda = \Lambda_0 (1 - 2nH_0 + n^2 K_0)^{-1/2} \exp(-\varepsilon \omega^* n / 2c_1), \tag{2.4.15}$$

where Λ_0 is the strength of the wave at time $t = t_0$. From $(2.4.14)_2$ and (2.4.15) we obtain

$$\tau = \frac{b\Lambda_0}{\rho_0\kappa(c_1^2 - c_3^2)}\,(1 - 2nH_0 + K_0)^{-1/2}\exp(-\varepsilon\omega^* n/2c_1), \qquad (2.4.16)$$

so that a disturbance in the porosity, which is of third order, is induced by a longitudinal wave of first kind. This discontinuity propagates with the same speed as the inducing longitudinal wave. The decay law (2.4.15) contains the exponential factor $\exp(-\varepsilon\omega^* n/2c_1)$ which ensures that the strength Λ, and hence the jumps λ_i, ζ, τ tend to zero as the interval $t - t_0$ increases indefinitely. It follows that λ_i, ζ, τ are completely determined if Λ_0 is known. To determine the jumps μ_i and γ, we see from equations $(2.4.14)_{1,3}$ that it is necessary to have prior knowledge of one of these jumps.

In the case of transverse wave we have $V = c_2$, $\Lambda = 0$, $X = 0$, $\zeta = 0$ and the equation (2.4.12) reduces to

$$\mu_j n_j = -a^{\alpha\beta}(\lambda_r x_{r;\beta})_{;\alpha}.$$

Using this result in (2.4.11), we get

$$\frac{1}{c_2}\frac{\delta\lambda_i}{\delta t} = H\lambda_i, \qquad (2.4.17)$$

together with the result $\tau = \gamma = 0$.

As before, we have

$$\lambda_i = \lambda_i^0(1 - 2nH_0 + n^2K_0)^{-1/2},$$

where $\lambda_i = \lambda_i^0$ when $t = t_0$. To determine the jumps μ_j it is necessary to have prior knowledge of two of the jumps μ_i.

In the case of an acceleration wave of compaction or distension we have $V = c_3$, $\lambda_i = 0$, $\zeta = 0$, and the equation (2.4.12) reduces to

$$(c_3^2 - c_1^2)\mu_j n_j = \frac{b}{\rho_0}X. \qquad (2.4.18)$$

The equations (2.4.11) become

$$\begin{aligned} &\{(c_3^2 - c_2^2)\delta_{ij} - (c_1^2 - c_2^2)n_j n_i\}\,\mu_j = \frac{b}{\rho_0}\,X n_i, \\ &\frac{1}{c_3}\frac{\delta X}{\delta t} = HX, \qquad \beta T_0 c_3 \mu_j n_j + k\gamma + mT_0 c_3 X = 0. \end{aligned} \qquad (2.4.19)$$

The solution of the equations (2.4.18), (2.4.19) is

$$\begin{aligned}
\chi &= \chi_0(1-2nH_0+n^2K_0)^{-1/2},\\
\mu_i &= \frac{bn_i\chi_0}{\rho_0(c_3^2-c_1^2)}\,(1-2nH_0+n^2K_0)^{-1/2},\\
\gamma &= -\frac{mc_3T_0\chi_0}{k}\left[1+\frac{\beta b}{\rho_0 m(c_3^2-c_1^2)}\right](1-2nH_0+n^2K_0)^{-1/2},
\end{aligned} \tag{2.4.20}$$

where $\chi = \chi_0$ when $t = t_0$. We note that an acceleration wave of compaction or distension is accompanied by third order discontinuities in mechanical and thermal fields.

2.5 Harmonic Waves

a) **Preliminaries.** In this section we consider the problem of steady vibrations. The results we present here were given by Pompei and Scalia [303]. We assume that

$$\begin{aligned}
&\boldsymbol{f} = \mathrm{Re}[\boldsymbol{f}'(x)\exp(-i\omega t)], \quad \ell = \mathrm{Re}[\ell'(x)\exp(-i\omega t)],\\
&S = \mathrm{Re}[S'(x)\exp(-i\omega t)], \quad \boldsymbol{u} = \mathrm{Re}[\boldsymbol{u}'(x)\exp(-i\omega t)],\\
&\varphi = \mathrm{Re}[\varphi'(x)\exp(-i\omega t)], \quad \theta = \mathrm{Re}[\theta'(x)\exp(-i\omega t)],
\end{aligned} \tag{2.5.1}$$

where ω is a given frequency, $x = (x_1, x_2, x_3), i = \sqrt{-1}$, and $\mathrm{Re}[f]$ denotes the real part of the function f. In this case we are led to a differential system for the amplitudes $\boldsymbol{u}', \varphi'$ and θ'. In view of (2.5.1) we obtain

$$\begin{aligned}
&e_{rs} = \mathrm{Re}[e'_{rs}(x;\omega)\exp(-i\omega t)], \quad t_{rs} = \mathrm{Re}[t'_{rs}(x;\omega)\exp(-i\omega t)],\\
&h_j = \mathrm{Re}[h'_j(x;\omega)\exp(-i\omega t)], \quad g = \mathrm{Re}[g'(x;\omega)\exp(-i\omega t)],\\
&\eta = \mathrm{Re}[\eta'(x;\omega)\exp(-i\omega t)], \quad q_j = \mathrm{Re}[q'_j(x;\omega)\exp(-i\omega t)],
\end{aligned}$$

where

$$\begin{aligned}
&e'_{ij} = \frac{1}{2}(u'_{i,j}+u'_{j,i}),\\
&t'_{ij} = \lambda e'_{rr}\delta_{ij} + 2\mu e'_{ij} + b\varphi'\delta_{ij} - \beta\theta'\delta_{ij}, \; h'_i = \alpha\varphi'_{,i},\\
&g' = -be'_{rr} - \xi\varphi' + m\theta', \; \rho_0\eta' = \beta e'_{rr} + a\theta' + m\varphi', \; q'_i = k\theta'_{,i}.
\end{aligned} \tag{2.5.2}$$

We assume that κ and k are strictly positive and that the relations (2.1.7) hold. We introduce the notations

$$\begin{aligned}
&\mu' = \frac{\mu}{\rho_0},\ \lambda' = \frac{\lambda}{\rho_0},\ b' = \frac{b}{\rho_0},\ F_i = -f_i,\ \xi' = -\frac{\xi}{\alpha},\ \gamma = \frac{b}{\alpha},\\
&m' = \frac{m}{\alpha},\ L = -\frac{\rho_0 \ell}{\alpha},\ \chi = -\xi' + \tilde{\eta}\omega^2,\ \tilde{\eta} = \rho_0\,\frac{\kappa}{\alpha},\\
&T_0' = \rho_0\,\frac{T_0}{k},\ s = -\frac{\rho_0}{k}S,\ a' = \frac{a}{\rho_0},\ d = \frac{\alpha m'}{\rho_0},\ \beta' = \frac{\beta}{\rho_0},\\
&t_{ij}' = \rho_0 T_{ij},\ h_i' = \alpha H_i,\ q_j' = k q_j.
\end{aligned} \tag{2.5.3}$$

From (1.5.17), (2.5.1) and (2.5.3) we obtain the basic set of differential equations of steady vibrations. For convenience, we suppress the primes so that these equations can be expressed in the form

$$\begin{aligned}
&\mu\Delta u_r + (\lambda+\mu)u_{j,jr} + b\varphi_{,r} - \beta\theta_{,r} + \omega^2 u_r = F_r,\\
&\Delta\varphi - \gamma u_{r,r} + \chi\varphi + m\theta = L,\\
&\Delta\theta + i\omega T_0\beta u_{r,r} + i\omega T_0 d\varphi + i\omega T_0 a\theta = s.
\end{aligned} \tag{2.5.4}$$

The system (2.5.4) can be written in the matrix form. Following Kupradze [231], the vector $\boldsymbol{v} = (v_1, v_2, ..., v_m)$ shall be considered as a column matrix. Thus, the product of the matrix $A = \|a_{ij}\|_{m\times m}$ and the vector $\boldsymbol{v} = (v_1, v_2, ..., v_m)$ is an m-dimensional vector. The vector $\boldsymbol{v}$ multiplied by the matrix A will denote the matrix product between the row matrix $\|v_1, v_2, ..., v_m\|$ and the matrix A. We introduce the matricial differential operator

$$A\left(\frac{\partial}{\partial x},\omega\right) = \left\| A_{rs}\left(\frac{\partial}{\partial x},\omega\right)\right\|_{5\times 5}, \tag{2.5.5}$$

where

$$\begin{aligned}
&A_{rs}\left(\frac{\partial}{\partial x},\omega\right) = \mu\delta_{rs}\Delta + (\lambda+\mu)\frac{\partial^2}{\partial x_r \partial x_s} + \omega^2\delta_{rs},\\
&A_{r4} = b\frac{\partial}{\partial x_r},\ A_{r5} = -\beta\frac{\partial}{\partial x_r},\ A_{4r} = -\gamma\frac{\partial}{\partial x_r},\\
&A_{44} = \Delta + \chi,\ A_{45} = m,\ A_{5r} = i\omega\beta T_0\frac{\partial}{\partial x_r},\\
&A_{54} = i\omega T_0 d,\ A_{55} = \Delta + i\omega T_0 a.
\end{aligned} \tag{2.5.6}$$

We denote

$$U = (u_1, u_2, u_3, \varphi, \theta),\quad F = (F_1, F_2, F_3, L, s). \tag{2.5.7}$$

We can write the equation (2.5.4) in the form

$$A\left(\frac{\partial}{\partial x},\omega\right)U = F. \tag{2.5.8}$$

We have

$$A\left(\frac{\partial}{\partial x},\omega\right) = A_0(\omega) + A_1\left(\frac{\partial}{\partial x},\omega\right) + A_2\left(\frac{\partial}{\partial x}\right),$$

where

$$A_0(\omega) = \|a_{rs}\|_{5\times 5},\ A_1\left(\frac{\partial}{\partial x},\omega\right) = \left\|B_{rs}\left(\frac{\partial}{\partial x},\omega\right)\right\|_{5\times 5},$$
$$A_2\left(\frac{\partial}{\partial x}\right) = \left\|\Lambda_{rs}\left(\frac{\partial}{\partial x}\right)\right\|_{5\times 5},$$

$$a_{rs}(\omega) = \omega^2\delta_{rs},\ a_{i4} = a_{4i} = a_{i5} = a_{5i} = 0,\ a_{44} = \chi,\ a_{45} = m,$$
$$a_{54} = i\omega T_0 d,\ a_{55} = i\omega T_0 a,\ B_{44} = B_{55} = B_{54} = B_{45} = 0,$$
$$B_{rs}\left(\frac{\partial}{\partial x},\omega\right) = 0,\ B_{4i} = -\gamma\frac{\partial}{\partial x_i},\ B_{5j} = i\omega T_0\beta\frac{\partial}{\partial x_j},$$
$$B_{i4} = b\frac{\partial}{\partial x_i},\ B_{i5} = -\beta\frac{\partial}{\partial x_i},\ \Lambda_{rs}\left(\frac{\partial}{\partial x}\right) = \mu\delta_{rs}\Delta + (\lambda+\mu)\frac{\partial^2}{\partial x_r \partial x_s},$$
$$\Lambda_{44} = \Lambda_{55} = \Delta,\ \Lambda_{4i} = \Lambda_{5i} = \Lambda_{i4} = \Lambda_{i5} = \Lambda_{45} = \Lambda_{54} = 0.$$

For any real unit vector $\boldsymbol{\zeta} = (\zeta_1, \zeta_2, \zeta_3)$ we have

$$\det\|\Lambda_{rs}(\zeta)\|_{5\times 5} = (\lambda + 2\mu)\mu^2,$$

so that, in view of (2.1.7) it follows that the system (2.5.4) is elliptic.

Let $\widetilde{A}$ be the adjoint operator of the operator A. Thus

$$\widetilde{A}\left(\frac{\partial}{\partial x},\omega\right) = A_0(\omega) - A_1^*\left(\frac{\partial}{\partial x},\omega\right) + A_2\left(\frac{\partial}{\partial x}\right),$$

where A^* is the transposed matrix of A. The adjoint system is

$$\widetilde{A}\left(\frac{\partial}{\partial x},\omega\right)V = \Phi, \tag{2.5.9}$$

where $V = (v_1, v_2, v_3, v_4, v_5)$ and $\Phi = (\phi_1, \phi_2, \phi_3, \phi_4, \phi_5)$. The system (2.5.9) can be written in the form

$$\begin{aligned}
&\mu\Delta v_r + (\lambda+\mu)v_{j,jr} + \gamma v_{4,r} - i\omega T_0\beta v_{5,r} + \omega^2 v_r = \phi_r,\\
&\Delta v_4 - b v_{r,r} + \chi v_4 + i\omega T_0 d v_5 = \phi_4,\\
&\Delta v_5 + \beta v_{r,r} + m v_4 + i\omega T_0 a v_5 = \phi_5.
\end{aligned} \tag{2.5.10}$$

b) **Reciprocity relations.** Let D^+ be a bounded region, bounded by the closed Liapunov surface S, and D^- the complementary of $D^+ \cup S$ to infinite space. We present reciprocity relations for steady vibrations. Letters in boldface stand for three-dimensional vectors. We say that $U = (u_1, u_2, u_3, u_4, u_5) = (\boldsymbol{u}, u_4, u_5)$ is regular in D^+ if $u_s \in C^1(\overline{D}^+) \cap C^2(D^+)$ and $u_{s,rm}$, $(r, s, m = 1, 2, ..., 5)$, are integrable on D^+. Let $\Sigma(x; \delta)$ be the sphere with the centre in x and radius δ. We say that $U = (\boldsymbol{u}, u_4, u_5)$ is regular in D^- if $u_s \in C^1(\overline{D}^-) \cap C^2(D^-)$ and $u_{s,rm}$, $(r, s, m = 1, 2, ..., 5)$, are integrable on $D^- \cap \Sigma(0, \delta)$ for any δ.

Let $A_i\left(\dfrac{\partial}{\partial x}, \omega\right)$ be the row matrix with the elements $A_{ij}\left(\dfrac{\partial}{\partial x}, \omega\right)$, $(i, j = 1, 2, ..., 5)$, given by (2.5.6). We denote

$$L_r U = A_r\left(\frac{\partial}{\partial x}, \omega\right) U, \quad (r = 1, 2, ..., 5),$$

so that the system (2.5.8) can be written in the form

$$L_i U = F_i, \quad L_4 U = L, \quad L_5 U = s. \tag{2.5.11}$$

Let $\widetilde{A}_i\left(\dfrac{\partial}{\partial x}, \omega\right)$ be the row matrix with the elements $\widetilde{A}_{ij}\left(\dfrac{\partial}{\partial x}, \omega\right)$, $(i, j = 1, 2, ..., 5)$. If we denote

$$\widetilde{L}_r V = \widetilde{A}_r\left(\frac{\partial}{\partial x}, \omega\right) V, \quad (r = 1, 2, ..., 5),$$

where $V = (\boldsymbol{v}, v_4, v_5)$, then the system (2.5.10) becomes

$$\widetilde{L}_i V = \phi_i, \; \widetilde{L}_4 V = \phi_4, \quad \widetilde{L}_5 V = \phi_5. \tag{2.5.12}$$

Let $U = (\boldsymbol{u}, u_4, u_5)$ and $V = (\boldsymbol{v}, v_4, v_5)$ be regular vectors in D^+ which satisfy the equations (2.5.8) and (2.5.9), respectively. We denote

$$\begin{aligned}
&T_{ij} = \lambda u_{r,r}\delta_{ij} + \mu(u_{i,j} + u_{j,i}) + b u_4 \delta_{ij} - \beta u_5 \delta_{ij},\\
&H_i = u_{4,i}, \; G = \chi u_4 - \gamma u_{r,r} + m u_5,\\
&Q_i = u_{5,i}, \; M = i\omega T_0(\beta u_{r,r} + d u_4 + a u_5),\\
&\widetilde{T}_{kj} = \lambda v_{r,r}\delta_{kj} + \mu(v_{k,j} + v_{j,k}) + \gamma v_4 \delta_{kj} - i\omega T_0 \beta v_5 \delta_{kj},\\
&\widetilde{H}_i = v_{4,i}, \; \widetilde{G} = \chi v_4 - b v_{r,r} + i\omega T_0 d v_5,\\
&\widetilde{Q}_i = v_{5,i}, \; \widetilde{M} = \beta v_{r,r} + m v_4 + i\omega T_0 a v_5.
\end{aligned} \tag{2.5.13}$$

The equations (2.5.11) and (2.5.12) may be written in the forms

$$T_{ji,j} + \omega^2 u_i = F_i, \quad H_{i,i} + G = L, \quad Q_{i,i} + M = s \tag{2.5.14}$$

and

$$\widetilde{T}_{ji,j} + \omega^2 v_i = \phi_i, \quad \widetilde{H}_{i,i} + \widetilde{G} = \phi_4, \quad \widetilde{Q}_{i,i} + \widetilde{M} = \phi_5, \tag{2.5.15}$$

respectively. From (2.5.11) and (2.5.14) we have

$$\begin{aligned}
v_i L_i U + v_4 L_4 U + v_5 L_5 U &= v_i(T_{ji,j} + \omega^2 u_i) + v_4(H_{i,i} + G) + \\
&+ v_5(Q_{i,i} + M) = (v_i T_{ji})_{,i} + (v_4 H_i)_{,i} + (v_5 Q_i)_{,i} - v_{i,j} T_{ji} - \\
&- v_{4,i} H_i - v_{5,i} Q_i + v_4 G + v_5 M + \omega^2 u_i v_i.
\end{aligned}$$

By the divergence theorem, we find

$$\begin{aligned}
\int_{D^+} V A\left(\frac{\partial}{\partial x}, \omega\right) U \, dV = &\int_S \left(v_i T_{ji} n_j + v_4 \frac{\partial u_4}{\partial n} + v_5 \frac{\partial u_5}{\partial n} \right) dA - \\
&- \int_{D^+} [E(\boldsymbol{u}, \boldsymbol{v}) + b u_4 v_{r,r} + u_{4,i} v_{4,i} - \chi u_4 v_4 - i\omega T_0 d u_4 v_5 - \\
&- \beta u_5 v_{r,r} + u_{5,i} v_{5,i} - m u_5 v_4 - i\omega T_0 a u_5 v_5 + \gamma u_{r,r} v_4 - \\
&- i\omega T_0 \beta u_{r,r} v_5 - \omega^2 u_i v_i] dV,
\end{aligned} \tag{2.5.16}$$

where $\partial f / \partial n = f_{,i} n_i$ and

$$E(\boldsymbol{u}, \boldsymbol{v}) = \lambda u_{r,r} v_{j,j} + \frac{1}{2} \mu \, (u_{i,j} + u_{j,i})(v_{i,j} + v_{j,i}). \tag{2.5.17}$$

Similarly,

$$\begin{aligned}
\int_{D^+} U \widetilde{A}\left(\frac{\partial}{\partial x}, \omega\right) V \, dV = &\int_S \left(u_i \widetilde{T}_{ji} n_j + u_4 \frac{\partial v_4}{\partial n} + u_5 \frac{\partial v_5}{\partial n} \right) dA - \\
&- \int_{D^+} [E(\boldsymbol{u}, \boldsymbol{v}) + b u_4 v_{r,r} + u_{4,i} v_{4,i} - \chi u_4 v_4 - \\
&- i\omega T_0 d u_4 v_5 - \beta u_5 v_{r,r} + u_{5,i} v_{5,i} - m u_5 v_4 - \\
&- i\omega T_0 a u_5 v_5 + \gamma u_{r,r} v_4 - i\omega T_0 \beta u_{r,r} v_5 - \omega^2 u_i v_i] dV.
\end{aligned} \tag{2.5.18}$$

From (2.5.16) and (2.5.18) we obtain

$$\begin{aligned}
&\int_{D^+} \left[U \widetilde{A}\left(\frac{\partial}{\partial x}, \omega\right) V - V A\left(\frac{\partial}{\partial x}, \omega\right) U \right] dV = \\
&= \int_S \left[U \widetilde{R}\left(\frac{\partial}{\partial x}, n_x\right) V - V R\left(\frac{\partial}{\partial x}, n_x\right) U \right] dA,
\end{aligned} \tag{2.5.19}$$

where

$$
\begin{aligned}
&R\left(\frac{\partial}{\partial x}, n_x\right) = \left\| R_{rs}\left(\frac{\partial}{\partial x}, n_x\right)\right\|_{5\times 5},\\
&\widetilde{R}\left(\frac{\partial}{\partial x}, n_x\right) = \left\| \widetilde{R}_{rs}\left(\frac{\partial}{\partial x}, n_x\right)\right\|_{5\times 5},\\
&R_{ij}\left(\frac{\partial}{\partial x}, n_x\right) = \mu\delta_{ij}\frac{\partial}{\partial n} + \lambda n_i \frac{\partial}{\partial x_j} + \mu n_j \frac{\partial}{\partial x_i},\\
&R_{i4} = bn_i,\ R_{i5} = -\beta n_i,\ R_{4i} = R_{5i} = R_{45} = R_{54} = 0,\\
&\widetilde{R}_{ij}\left(\frac{\partial}{\partial x}, n_x\right) = R_{ij}\left(\frac{\partial}{\partial x}, n_x\right),\ \widetilde{R}_{i4} = \gamma n_i,\\
&\widetilde{R}_{4i} = \widetilde{R}_{5i} = \widetilde{R}_{45} = \widetilde{R}_{54} = 0,\ \widetilde{R}_{j5} = -i\omega T_0 \beta n_j,\\
&R_{44} = R_{55} = \widetilde{R}_{44} = \widetilde{R}_{55} = \frac{\partial}{\partial n}.
\end{aligned}
\tag{2.5.20}
$$

Let $R_i\left(\frac{\partial}{\partial x}, n_x\right)$ be row matrix with the elements $R_{ij}\left(\frac{\partial}{\partial x}, n_x\right)$, $(i, j = 1, 2, ..., 5)$. Then, from (2.5.13) and (2.5.20) we obtain

$$
T_{ij}n_j = R_i\left(\frac{\partial}{\partial x}, n_x\right)U,\ H_i n_i = R_4\left(\frac{\partial}{\partial x}, n_x\right)U,\ Q_i n_i = R_5\left(\frac{\partial}{\partial x}, n_x\right)U.
\tag{2.5.21}
$$

Next, we establish other integral identities. We denote by $\overline{f}$ the complex conjugate of f. We assume that $U = (\boldsymbol{u}, u_4, u_5)$ is a regular vector field in D^+ that satisfies the equation

$$
A\left(\frac{\partial}{\partial x}, \omega\right) U = 0.
\tag{2.5.22}
$$

Then, the vector $\overline{U} = (\overline{\mathbf{u}}, \overline{u}_4, \overline{u}_5)$ satisfies the equations

$$
\begin{aligned}
&\mu\Delta\overline{u}_r + (\lambda + \mu)\overline{u}_{j,jr} + b\overline{u}_{4,r} - \beta\overline{u}_{5,r} + \omega^2\overline{u}_r = 0,\\
&\Delta\overline{u}_4 - \gamma\overline{u}_{r,r} + \chi\overline{u}_4 + m\overline{u}_5 = 0,\\
&\Delta\overline{u}_5 - i\omega T_0\beta\overline{u}_{r,r} - i\omega T_0 d\overline{u}_4 - i\omega T_0 a\overline{u}_5 = 0.
\end{aligned}
\tag{2.5.23}
$$

The equations (2.5.23) can be written in the form

$$
\overline{T}_{ij,j} + \omega^2\overline{u}_i = 0,\quad \overline{H}_{i,i} + \overline{G} = 0,\quad \overline{Q}_{i,i} + \overline{M} = 0,
\tag{2.5.24}
$$

where

$$
\begin{aligned}
&\overline{T}_{ij} = \lambda\overline{u}_{r,r}\delta_{ij} + \mu(\overline{u}_{i,j} + \overline{u}_{j,i}) + b\overline{u}_4\delta_{ij} - \beta\overline{u}_5\delta_{ij},\ \overline{H}_i = \overline{u}_{4,i},\\
&\overline{G} = \chi\overline{u}_4 - \gamma\overline{u}_{r,r} + m\overline{u}_5,\ \overline{Q}_i = \overline{u}_{5,i},\ \overline{M} = -i\omega T_0(\beta\overline{u}_{r,r} + d\overline{u}_4 + a\overline{u}_5).
\end{aligned}
$$

For homogeneous problem we have

$$\int_{D^+}\left[(T_{ji,j}+\omega^2 u_i)\overline{u}_i+\frac{b}{\gamma}(H_{i,i}+G)\overline{u}_4+\right.$$
$$\left.+\frac{1}{i\omega T_0}(Q_{i,i}+M)\overline{u}_5\right]dV=0. \tag{2.5.25}$$

With the help of divergence theorem, by (2.5.25) we obtain

$$\begin{aligned}&\int_S\left(T_{ji}n_j\overline{u}_i+\frac{\beta}{\gamma}H_in_i\overline{u}_4+\frac{1}{i\omega T_0}Q_jn_j\overline{u}_5\right)dA+\\&+\int_{D^+}[\omega^2u_i\overline{u}_i-E(\boldsymbol{u},\overline{\boldsymbol{u}})-b(u_4\overline{u}_{j,j}+\overline{u}_4u_{i,i})+\\&+\beta(u_5\overline{u}_{s,s}+\overline{u}_5u_{r,r})+d(u_5\overline{u}_4+\overline{u}_5u_4)-\\&-\frac{b}{\gamma}u_{4,i}\overline{u}_{4,i}+\frac{b}{\gamma}\chi u_4\overline{u}_4+au_5\overline{u}_5-\frac{1}{i\omega T_0}u_{5,j}\overline{u}_{5,j}]dV=0.\end{aligned} \tag{2.5.26}$$

Similarly we find that

$$\begin{aligned}&\int_S\left(\overline{T}_{ji}n_jn_i+\frac{b}{\gamma}\overline{H}_in_iu_4-\frac{1}{i\omega T_0}\overline{Q}_kn_ku_5\right)dA+\\&+\int_{D^+}[\omega^2u_i\overline{u}_i-E(\boldsymbol{u},\overline{\boldsymbol{u}})-b(\overline{u}_4u_{j,j}+u_4\overline{u}_{i,i})+\\&+\beta(\overline{u}_5u_{s,s}+u_5\overline{u}_{r,r})+d(\overline{u}_5u_4+u_5\overline{u}_4)-\\&-\frac{b}{\gamma}u_{4,i}\overline{u}_{4,i}+\frac{b}{\gamma}\chi u_4\overline{u}_4+au_5\overline{u}_5+\frac{1}{i\omega T_0}u_{5,k}\overline{u}_{5,k}]dV=0.\end{aligned} \tag{2.5.27}$$

From (2.5.26) and (2.5.27) we get

$$\begin{aligned}&\int_S\left(T_{ji}n_j\overline{u}_i+\frac{b}{\gamma}H_in_i\overline{u}_4+\frac{1}{i\omega T_0}Q_kn_k\overline{u}_5\right)dA-\\&-\int_S\left(\overline{T}_{ji}n_ju_i+\frac{b}{\gamma}\overline{H}_in_iu_4-\frac{1}{i\omega T_0}\overline{Q}_kn_ku_5\right)dA=\\&=\frac{2}{i\omega T_0}\int_{D^+}u_{5,k}\overline{u}_{5,k}dV.\end{aligned} \tag{2.5.28}$$

2.6 Concentrated Loads

In this section we use the representation given in Theorem 2.3.1 to establish the fundamental solutions of the system (2.5.4). This problem has been studied in [66], [303]. We introduce the notations

$$D_1 = (\lambda + 2\mu)\Delta + \omega^2, \quad D_2 = \mu\Delta + \omega^2, \\ D_3 = \Delta + \kappa, \qquad\qquad D_4 = \Delta + i\omega T_0 a. \tag{2.6.1}$$

The system (2.5.4) can be written in the form

$$\begin{aligned} &D_2\boldsymbol{u} + (\lambda + \mu)\operatorname{grad}\operatorname{div}\boldsymbol{u} + b\operatorname{grad}\varphi - \beta\operatorname{grad}\theta = \boldsymbol{F}, \\ &D_3\varphi - \gamma\operatorname{div}\boldsymbol{u} + m\theta = L, \\ &D_4\theta + i\omega T_0\beta\operatorname{div}\boldsymbol{u} + i\omega T_0 d\varphi = s. \end{aligned} \tag{2.6.2}$$

We denote

$$N = D_1D_3D_4 + b\gamma\Delta D_4 + i\omega T_0[\beta^2 D_3\Delta + \beta(d\gamma + bm)\Delta - mdD_1]. \tag{2.6.3}$$

The next proposition is an immediate consequence of Theorem 2.3.1.

Theorem 2.6.1. *Let*

$$\begin{aligned} \boldsymbol{u} =& N\boldsymbol{K} - \{(\lambda + \mu)D_3D_4 + b\gamma D_4 + i\omega T_0[mb\beta + \beta d\gamma + \beta^2 D_3 - \\ &- md(\lambda + \mu)]\}\operatorname{grad}\operatorname{div}\boldsymbol{K} - (i\omega T_0\beta d + bD_4)\operatorname{grad} R + \\ &+ (mb + \beta D_3)\operatorname{grad} Y, \\ \varphi =& (i\omega T_0 m\beta + \gamma D_4)D_2\operatorname{div}\boldsymbol{K} + (i\omega T_0\beta^2\Delta + D_1D_4)R + \\ &+ (\beta\gamma\Delta - mD_1)Y, \\ \theta =& -i\omega T_0(d\gamma + \beta D_3)D_2\operatorname{div}\boldsymbol{K} + i\omega T_0(b\beta\Delta - dD_1)R + \\ &+ (b\gamma\Delta + D_1D_3)Y, \end{aligned} \tag{2.6.4}$$

where the fields K_i of class C^8, and R and Y of class C^6 on B satisfy the equations

$$D_2N\boldsymbol{K} = \boldsymbol{F}, \quad NR = L, \quad NR = s. \tag{2.6.5}$$

Then $\boldsymbol{u}, \varphi$ and θ satisfy the field equations (2.5.4).

Let us investigate the effect of concentrated body loads in an infinite space. First, we assume that

$$\boldsymbol{F} = \boldsymbol{0}, \; L = 0. \tag{2.6.6}$$

From (2.6.5) it follows that we can take $\boldsymbol{K} = \boldsymbol{0}$ and $Y = 0$. It is easy to see that if σ_j^2 are the roots of the equation

$$\begin{aligned} &(\lambda + 2\mu)\sigma^6 + [(\chi + i\omega T_0 a)(\lambda + 2\mu) + \omega^2 + \gamma b + i\omega T_0\beta^2]\sigma^4 + \\ &+ \{\chi\omega^2 + i\omega T_0[a\omega^2 + (\lambda + 2\mu)(a\chi - md) + ab\gamma + \chi\beta^2 + \\ &+ \beta(d\gamma + bm)]\}\sigma^2 + i\omega^3 T_0(a\chi - md) = 0, \end{aligned} \tag{2.6.7}$$

then the function R satisfies the equation

$$(\Delta + \sigma_1^2)(\Delta + \sigma_2^2)(\Delta + \sigma_3^2)R = \frac{1}{\lambda + 2\mu}\, s. \tag{2.6.8}$$

In what follows we assume that $\sigma_1^2 \neq \sigma_2^2 \neq \sigma_3^2 \neq \sigma_1^2$. If the functions g_j satisfy the equations

$$(\Delta + \sigma_j^2)g_j = \frac{1}{\lambda + 2\mu}\, s, \text{ (no sum; } j = 1, 2, 3), \tag{2.6.9}$$

then the function R can be expressed as

$$R = b_1 g_1 + b_2 g_2 + b_3 g_3, \tag{2.6.10}$$

where

$$\begin{aligned} &b_1^{-1} = (\sigma_1^2 - \sigma_2^2)(\sigma_1^2 - \sigma_3^2),\ b_2^{-1} = (\sigma_2^2 - \sigma_1^2)(\sigma_2^2 - \sigma_3^2), \\ &b_3^{-1} = (\sigma_3^2 - \sigma_1^2)(\sigma_3^2 - \sigma_2^2). \end{aligned} \tag{2.6.11}$$

We now assume that $s = \delta(x - y)$, where δ is the Dirac delta and y is a fixed point of the body. In this case the solution of the equation (2.6.8) is

$$P(x; y) = -\frac{1}{4\pi(\lambda + 2\mu)} \sum_{j=1}^{3} b_j \exp(i\sigma_j r), \tag{2.6.12}$$

where $r = |\mathbf{x} - \mathbf{y}|$ and σ_j are the roots of the equation (2.6.7) with positive real parts. If we take in (2.6.4), $\boldsymbol{K} = \mathbf{0}$, $Y = 0$ and $R = P$, then we obtain the solution

$$\begin{aligned} &u_j^{(5)} = (mb + \beta D_3)P_{,j},\ \varphi^{(5)} = (\beta\gamma\Delta - mD_1)P, \\ &\theta^{(5)} = (b\gamma\Delta + D_1 D_3)P. \end{aligned} \tag{2.6.13}$$

In the case of the body loads

$$\boldsymbol{F} = \mathbf{0}, \quad L = \delta(x - y), \quad s = 0,$$

we are led to the solution

$$\begin{aligned} &u_j^{(4)} = -(i\omega T_0 \beta d + bD_4)P_{,j},\ \varphi^{(4)} = (D_1 D_4 + i\omega T_0 \beta^2 \Delta)P, \\ &\theta^{(4)} = i\omega T_0 (b\beta\Delta - dD_1)P. \end{aligned} \tag{2.6.14}$$

Let us assume that

$$F_i = \delta_{ij}\delta(x - y), \quad L = 0, \quad s = 0,$$

where j is fixed. In this case we take $K_i = E\delta_{ij}$, $Y = 0$ and $R = 0$. The function E satisfies the equation

$$(\Delta + \sigma_1^2)(\Delta + \sigma_2^2)(\Delta + \sigma_3^2)(\Delta + \sigma_4^2)E = \frac{1}{\mu(\lambda + 2\mu)}\,\delta, \tag{2.6.15}$$

where $\sigma_4^2 = \omega^2/\mu$. We assume that σ_4^2 does not satisfy the equation (2.6.7). By the procedure used to obtain (2.6.11) we find that

$$E(x; y) = -\frac{1}{4\pi\mu(\lambda + 2\mu)r}\sum_{j=1}^{4} a_j \exp(i\sigma_j r), \tag{2.6.16}$$

where

$$a_s^{-1} = \prod_{j=1\,(j\neq s)}^{4} (\sigma_s^2 - \sigma_j^2), \quad (s = 1, 2, 3, 4). \tag{2.6.17}$$

From (2.6.4) we obtain the solution in the form

$$\begin{aligned} u_k^{(j)} &= \delta_{jk} NE - \{(\lambda + \mu)D_3 D_4 + b\gamma D_4 + \\ &\quad + i\omega T_0[mb\beta + \beta d\gamma + \beta^2 D_3 - md(\lambda + \mu)]\}E_{,jk}, \\ \varphi^{(j)} &= (\gamma D_4 + i\omega T_0 m\beta)D_2 E_{,j}, \\ \theta^{(j)} &= -i\omega T_0(d\gamma + \beta D_3)D_2 E_{,j}, \ (j = 1, 2, 3). \end{aligned} \tag{2.6.18}$$

The functions $u_j^{(s)}, \varphi^{(s)}, \theta^{(s)}$, $(s = 1, 2, ..., 5)$, given by (2.6.13), (2.6.14) and (2.6.18) represent the fundamental solutions of the system of field equations for steady vibrations. Let $\Gamma(x, y; \omega)$ be the matrix of fundamental solutions of the system (2.5.4)

$$\Gamma(x, y; \omega) = \|\Gamma_{rs}(x, y; \omega)\|_{5\times5}, \tag{2.6.19}$$

where

$$\Gamma_{jk}(x, y; \omega) = u_j^{(k)}, \qquad \Gamma_{4k}(x, y; \omega) = \varphi^{(k)}(x, y),$$
$$\Gamma_{5k}(x, y; \omega) = \theta^{(k)}(x, y), \quad (k = 1, 2, ..., 5).$$

If $x \neq y$, then each column $\Gamma^{(s)}$, $(s = 1, 2, ..., 5)$, of the matrix $\Gamma(x, y; \omega)$ satisfies at x the homogeneous system (2.5.8). Thus,

$$A\left(\frac{\partial}{\partial x}, \omega\right)\Gamma(x, y; \omega) = 0, \quad x \neq y. \tag{2.6.20}$$

Let $\widetilde{\Gamma}(x, y; \omega)$ be the matrix of fundamental solutions of the system (2.5.10). We have

$$\widetilde{A}\left(\frac{\partial}{\partial x}, \omega\right) \widetilde{\Gamma}(x, y; \omega) = 0, \quad x \neq y. \tag{2.6.21}$$

We note that

$$\widetilde{\Gamma}^*(y, x; \omega) = \Gamma(x, y; \omega). \tag{2.6.22}$$

2.7 Radiation Conditions

a) **A representation of solutions.** Let us consider the homogeneous equation (2.5.22).

Theorem 2.7.1. *Let $U = (\boldsymbol{u}, u_4, u_5)$ be a regular solution of the equation (2.5.22). Then*

$$U = (\boldsymbol{u}^{(1)} + \boldsymbol{u}^{(2)}, u_4, u_5), \tag{2.7.1}$$

where

$$\begin{aligned}
&(\Delta + \sigma_1^2)(\Delta + \sigma_2^2)(\Delta + \sigma_3^2)\boldsymbol{u}^{(1)} = \mathbf{0}, \ \operatorname{curl} \boldsymbol{u}^{(1)} = \mathbf{0}, \\
&(\Delta + \sigma_4^2)\boldsymbol{u}^{(2)} = \mathbf{0}, \ (\Delta + \sigma_1^2)(\Delta + \sigma_2^2)(\Delta + \sigma_3^2)u_4 = 0, \\
&\operatorname{div} \boldsymbol{u}^{(2)} = 0, \ (\Delta + \sigma_1^2)(\Delta + \sigma_2^2)(\Delta + \sigma_3^2)u_5 = 0.
\end{aligned} \tag{2.7.2}$$

Proof. The equation (2.5.22) can be expressed in the form

$$\begin{aligned}
&\mu\Delta\boldsymbol{u} + (\lambda + \mu) \operatorname{grad} \operatorname{div} \boldsymbol{u} + \omega^2\boldsymbol{u} + b \operatorname{grad} u_4 - \beta \operatorname{grad} u_5 = \mathbf{0}, \\
&\Delta u_4 + \chi u_4 + m u_5 - \gamma \operatorname{div} \boldsymbol{u} = 0, \\
&\Delta u_5 + i\omega T_0 d u_4 + i\omega T_0 a u_5 + i\omega T_0 \beta \operatorname{div} \boldsymbol{u} = 0.
\end{aligned} \tag{2.7.3}$$

The first equation from (2.7.3) can be written in the form

$$(\lambda + 2\mu) \operatorname{grad} \operatorname{div} \boldsymbol{u} + \omega^2\boldsymbol{u} - \mu \operatorname{curl} \operatorname{curl} \boldsymbol{u} + b \operatorname{grad} u_4 - \beta \operatorname{grad} u_5 = \mathbf{0},$$

so that we have

$$\boldsymbol{u} = \boldsymbol{u}^{(1)} + \boldsymbol{u}^{(2)}, \tag{2.7.4}$$

where

$$\begin{aligned}
\boldsymbol{u}^{(1)} &= -\frac{\lambda + 2\mu}{\omega^2} \operatorname{grad} \operatorname{div} \boldsymbol{u} - \frac{b}{\omega^2} \operatorname{grad} u_4 + \frac{\beta}{\omega^2} \operatorname{grad} u_5, \\
\boldsymbol{u}^{(2)} &= \frac{\mu}{\omega^2} \operatorname{curl} \operatorname{curl} \boldsymbol{u}.
\end{aligned} \tag{2.7.5}$$

We note that

$$\operatorname{curl} \boldsymbol{u}^{(1)} = \mathbf{0}, \quad \operatorname{div} \boldsymbol{u}^{(2)} = 0. \tag{2.7.6}$$

From (2.7.3) we obtain $D_2 \operatorname{curl} \boldsymbol{u} = \mathbf{0}$, so that

$$(\Delta + \sigma_4^2)\boldsymbol{u}^{(2)} = \mathbf{0}.$$

The equations (2.7.3) lead to

$$\begin{aligned}
&(\Delta + v^2) \operatorname{div} \boldsymbol{u} + \frac{b}{\lambda + 2\mu} \Delta u_4 - \frac{\beta}{\lambda + 2\mu} \Delta u_5 = 0, \\
&(\Delta + \chi) u_4 - \gamma \operatorname{div} \boldsymbol{u} + m u_5 = 0, \\
&(\Delta + i\omega T_0 a) u_5 + i\omega T_0 \beta \operatorname{div} \boldsymbol{u} + i\omega T_0 d u_4 = 0,
\end{aligned} \tag{2.7.7}$$

where $v^2 = \omega^2/(\lambda + 2\mu)$. It follows from (2.7.7) that

$$\begin{aligned}
&\{(\Delta + v^2)[(\Delta + \chi)(\Delta + i\omega T_0 a) - i\omega T_0 m d] + \\
&+ \frac{b}{\lambda + 2\mu} \Delta[\gamma(\Delta + i\omega T_0 a) + i\omega T_0 m \beta] + \\
&+ \frac{\beta}{\lambda + 2\mu} i\omega T_0 \Delta[\beta(\Delta + \chi) + d\gamma]\} \operatorname{div} \boldsymbol{u} = 0.
\end{aligned}$$

We can see that this relation can be written in the form

$$(\Delta + \sigma_1^2)(\Delta + \sigma_2^2)(\Delta + \sigma_3^2) \operatorname{div} \boldsymbol{u} = 0.$$

Similarly, we find that u_4 and u_5 satisfy the corresponding equations from (2.7.2). This completes the proof. □

b) **A uniqueness result.** We now study the regular solutions in D^- of the equation (2.5.22). We establish conditions at infinity of Sommerfeld type and show that these conditions ensure a unique solution. Let $U = (\boldsymbol{u}^{(1)} + \boldsymbol{u}^{(2)}, u_4, u_5)$ be a regular solution in D^- of the equation (2.5.22). We say that U satisfies the radiation conditions if

$$\begin{aligned}
&\boldsymbol{u}^{(1)}(x) = o(r^{-1}), \quad \boldsymbol{u}^{(1)}_{,j} = O(r^{-2}), \\
&u_K(x) = o(r^{-1}), \quad u_{K,j} = O(r^{-2}), \quad (K = 4, 5), \\
&\boldsymbol{u}^{(2)}(x) = O(r^{-1}), \quad \frac{\partial \boldsymbol{u}^{(2)}}{\partial r} - i\sigma_4 \boldsymbol{u}^{(2)} = o(r^{-1}),
\end{aligned} \tag{2.7.8}$$

where $r = (x_1^2 + x_2^2 + x_3^2)^{1/2}$. Here, $f = O(g)$ means that f/g is bounded, and $f = o(g)$ means that f/g tends to zero when g increases. We introduce the notation

$$T_i^e(\boldsymbol{u}) = [\lambda u_{r,r} \delta_{ij} + \mu(u_{i,j} + u_{j,i})] n_j. \tag{2.7.9}$$

It is known from Kupradze [231],p.44 that if $(2.7.8)_3$ hold, then we have

$$\boldsymbol{u}^{(2)}(x) = O(r^{-1}), \quad \boldsymbol{T}^e(\boldsymbol{u}^{(2)}) - i\omega\mu^{1/2}\boldsymbol{u}^{(2)} = o(r^{-1}). \tag{2.7.10}$$

Theorem 2.7.2. *Let* $U = (\boldsymbol{u}^{(1)} + \boldsymbol{u}^{(2)}, u_4, u_5)$ *be a regular solution in* D^- *of the equation* (2.5.22) *that satisfies the radiation conditions and one of the conditions*

$$\boldsymbol{u} = \boldsymbol{0}, \qquad u_4 = 0, \qquad u_5 = 0 \qquad on\ S, \tag{2.7.11}$$

$$T_{ji}n_j = 0, \quad H_i n_i = 0, \quad Q_i n_i = 0 \quad on\ S. \tag{2.7.12}$$

Then $\boldsymbol{u}^{(1)} = \boldsymbol{0},\ \boldsymbol{u}^{(2)} = \boldsymbol{0},\ u_4 = 0,\ u_5 = 0.$

Proof. Let us consider the sphere $\Sigma(0;R)$ with sufficiently large radius such that $D^+ \subset \Sigma(0;R)$. We denote by $\partial\Sigma$ the boundary of $\Sigma(0;R)$ and $\Omega = D^- \cap \Sigma(0;R)$. If $U = (\boldsymbol{u}, u_4, u_5)$ is a regular solution in D^- of the equation (2.5.22), that satisfies the radiation conditions and one of the conditions (2.7.11), (2.7.12), then by (2.5.28) we obtain

$$\frac{2i}{\omega T_0}\int_\Omega \theta_{,k}\overline{\theta}_{,k}dV = \int_{\partial\Sigma}[u_i^{(2)}\overline{T}_i^e(\boldsymbol{u}^{(2)}) - \bar{u}_i^{(2)}T_i^e(\boldsymbol{u}^{(2)})]dA + o(1) =$$
$$= \int_{\partial\Sigma}[u_k^{(2)}(\overline{T}_k^e(\boldsymbol{u}^{(2)}) + i\omega\mu^{1/2}\overline{u}_k^{(2)})-$$
$$-\overline{u}_k^{(2)}(T_k^e(\boldsymbol{u}^{(2)}) - i\omega\mu^{1/2}u_k^{(2)}) - 2i\omega\mu^{1/2}u_k^{(2)}\overline{u}_k^{(2)}]dA + o(1). \tag{2.7.13}$$

It follows from (2.7.10) that

$$\overline{T}^e(\boldsymbol{u}^{(2)}) + i\omega\mu^{1/2}\boldsymbol{u}^{(2)} = o(r^{-1}). \tag{2.7.14}$$

If we let $R \longrightarrow \infty$ and use (2.7.14), we conclude from (2.7.13) that

$$\frac{1}{\omega T_0}\int_{D^-}\theta_{,k}\overline{\theta}_{,k}dV + \omega\mu^{1/2}\lim_{R\to\infty}\int_{\partial\Sigma}|\boldsymbol{u}^{(2)}|^2 dA = 0,$$

so that

$$\int_{D^-}|u_{5,i}|^2 dV = 0, \quad \lim_{R\to\infty}\int_{\partial\Sigma}|\boldsymbol{u}^{(2)}|^2 dA = 0. \tag{2.7.15}$$

By Theorem 2.7.1,

$$(\Delta + \sigma_4^2)\boldsymbol{u}^{(2)} = \boldsymbol{0}. \tag{2.7.16}$$

In view of a well known theorem (see Kupradze [231], p. 53) from $(2.7.15)_2$ and (2.7.16) we obtain $\boldsymbol{u}^{(2)} = \boldsymbol{0}$. It follows from $(2.7.15)_1$ and $(2.7.8)_2$ that $u_5 = 0$. With the help of (2.7.3) and $(2.7.8)_2$ we find that $u_4 = 0$ and $\operatorname{div}\boldsymbol{u}^{(1)} = 0$. Moreover, from $\operatorname{div}\boldsymbol{u}^{(1)}=0$, $\operatorname{curl}\boldsymbol{u}^{(1)}=\boldsymbol{0}$ and the conditions at infinity we conclude that $\boldsymbol{u}^{(1)}=\boldsymbol{0}$ and the proof is complete.
□

2.8 Potentials

In this section we use the method of potentials to reduce boundary value problems of the theory of steady vibrations to singular integral equations for which Fredholm's theorems hold. Following Kupradze [231], if $U = (u_1, u_2, u_3, u_4, u_5)$ is a regular vector field in D^+ and $V = \widetilde{\Gamma}^{(k)}(x, y; \omega)$, $(k = 1, 2, ..., 5)$, $y \in D^+$, then (2.5.19) implies that

$$\begin{aligned} U_k(y) = \int_S \Big[& U(x) \widetilde{R}\left(\frac{\partial}{\partial x}, n_x\right) \widetilde{\Gamma}^{(k)}(x, y; \omega) - \\ & - \widetilde{\Gamma}^{(k)}(x, y; \omega) R\left(\frac{\partial}{\partial x}, n_x\right) U(x) \Big] dA_x + \\ & + \int_{D^+} \widetilde{\Gamma}^{(k)}(x, y; \omega) A\left(\frac{\partial}{\partial x}, n_x\right) U(x) dV_x, \quad (k = 1, 2, ..., 5). \end{aligned} \tag{2.8.1}$$

It follows from (2.6.22) and (2.8.1) that

$$\begin{aligned} U(x) = & \int_{D^+} \left[\Gamma(x, y; \omega) A\left(\frac{\partial}{\partial y}, \omega\right) U(y) \right] dV_y + \\ & + \int_S \left[\Lambda(x, y; \omega) U(y) - \Gamma(x, y; \omega) R\left(\frac{\partial}{\partial y}, n_y\right) U(y) \right] dA_y, \ x \in D^+, \end{aligned} \tag{2.8.2}$$

where

$$\Lambda(x, y; \omega) = \left[\widetilde{R}\left(\frac{\partial}{\partial y}, n_y\right) \Gamma^*(x, y; \omega) \right]^*. \tag{2.8.3}$$

As in [231], [232] we can prove the following result.

Theorem 2.8.1. *Each column of the matrix $\Lambda(x, y; \omega)$ satisfies at x the equation (2.5.22) for $x \neq y$.*

If U is a regular vector field in D^- that satisfies the radiation conditions, then we can establish a representation of U analogous with (2.8.2). We introduce the potential of a single-layer

$$V(x; \psi) = \int_S \Gamma(x, y; \omega) \psi(y) dA_y, \tag{2.8.4}$$

the potential of a double-layer

$$W(x; \eta) = \int_S \Lambda(x, y; \omega) \eta(y) dA_y \tag{2.8.5}$$

and the potential of mass

$$U(x;\rho) = \int_{D^+} \Gamma(x,y;\omega)\rho(y)dV_y. \tag{2.8.6}$$

We assume that ψ is Hölder continuous on S, η is Hölder continuously differentiable on S and ρ is Hölder continuous on $\overline{D}^+$. As in the classical theory we have

$$\begin{aligned} &A\left(\frac{\partial}{\partial x},\omega\right) V(x;\psi) = 0, \quad A\left(\frac{\partial}{\partial x},\omega\right) W(x;\eta) = 0, \\ &A\left(\frac{\partial}{\partial x},\omega\right) U(x;\rho) = \rho, \quad x \in D^+. \end{aligned} \tag{2.8.7}$$

For future use, we now record the following (see, for example, [231], p. 33).

Theorem 2.8.2. *The potential of a single-layer is continuous throughout.*

Theorem 2.8.3. *The potential of a double-layer has finite limits when the point x tends to $z \in S$ both from within and from without, and these limits are respectively equal to*

$$\begin{aligned} W^+(z;\eta) &= -\frac{1}{2}\eta(z) + \int_S \Lambda(z,y;\omega)\eta(y)dA_y, \\ W^-(z;\eta) &= \frac{1}{2}\eta(z) + \int_S \Lambda(z,y;\omega)\eta(y)dA_y, \end{aligned} \tag{2.8.8}$$

the integrals being conceived in the sense of Cauchy's principal value.

Theorem 2.8.4. *$R(\partial/\partial x, n_x)V(x;\psi)$ has finite limits as the point x tends to $z \in S$ from within and from without and these limits are*

$$\left[R\left(\frac{\partial}{\partial z},n_z\right) V(z;\psi)\right]^+ = \frac{1}{2}\psi(z) + \int_S \left[R\left(\frac{\partial}{\partial z},n_z\right)\Gamma(z,y;\omega)\right]\psi(y)dA_y,$$

$$\left[R\left(\frac{\partial}{\partial z},n_z\right) V(z;\psi)\right]^- = -\frac{1}{2}\psi(z) + \int_S \left[R\left(\frac{\partial}{\partial z},n_z\right)\Gamma(z,y;\omega)\right]\psi(y)dA_y,$$

respectively.

Next, we restrict our attention to the equation (2.5.22) and consider the following boundary problems

Interior problems. To find a regular solution in D^+ of the equation (2.5.22) satisfying one of the conditions

$$(I_1) \qquad U^+ = G_1 \text{ on } S,$$

$$(I_2) \qquad \left[R\left(\frac{\partial}{\partial z}, n_z\right) U(z) \right]^+ = G_2(z), \; z \in S,$$

where G_1 and G_2 are given.

Exterior problems. To find a regular solution in D^- of the equation (2.5.22) that satisfies the radiation conditions and one of the conditions

$$(E_1) \qquad U^- = G_3 \text{ on } S,$$

$$(E_2) \qquad \left[R\left(\frac{\partial}{\partial z}, n_z\right) U(z) \right]^- = G_4(z), \; z \in S,$$

where G_3 and G_4 are prescribed vector fields.

We assume that G_1 and G_3 are Hölder continuously differentiable on S, and G_2 and G_4 are Hölder continuous on S. We denote by (I_α^0) and (E_α^0) the homogeneous problems corresponding to (I_α) and (E_α), $(\alpha = 1, 2)$, respectively.

We seek the solutions of the problems (I_1) and (E_1) in the form of a double-layer potential and the solutions of the problems (I_2) and (E_2) in the form of the a single-layer potential. In view of Theorems 2.8.3 and 2.8.4 we obtain for unknown densities the following equations

$$(I_1) \quad -\frac{1}{2}\eta(z) + \int_S \Lambda(z, y; \omega)\eta(y) dA_y = G_1(z),$$

$$(I_2) \quad \frac{1}{2}\psi(z) + \int_S \left[R\left(\frac{\partial}{\partial z}, n_z\right) \Gamma(z, y; \omega) \right] \psi(y) dA_y = G_2(z),$$

$$(E_1) \quad \frac{1}{2}\eta(z) + \int_S \Lambda(z, y; \omega)\eta(y) dA_y = G_3(z),$$

$$(E_2) \quad -\frac{1}{2}\psi(z) + \int_S \left[R\left(\frac{\partial}{\partial z}, n_z\right) \Gamma(z, y; \omega) \right] \psi(y) dA_y = G_4(z), \; z \in S.$$

These equations are two-dimensional singular integral equations. They can only be understood in the sense of principal values. Next we show that the Fredholm theory is applicable. The integral operator of the problem (I_1) may be written in the form

$$T\eta = -\eta(z) + \int_S K(z, y; \omega)\eta(y) dA_y,$$

where $K(z, y) = 2\Lambda(z, y; \omega)$. From (2.6.13), (2.6.14), (2.6.18) and (2.8.3) we can write

$$K(z, y) = K_1(z, y) + K_2(z, y),$$

where

$$\begin{aligned} &K_1(z, y) = \|k_{rs}(z, y)\|_{5\times 5}, \\ &k_{rs}(z, y) = \frac{\mu}{2\pi(\lambda + 2\mu)r^3}\left[(z_s - y_s)\cos(n_y, z_r) - (z_r - y_r)\cos(n_y, z_s)\right], \\ &k_{4j} = k_{j4} = k_{44} = k_{5j} = k_{j5} = k_{45} = k_{54} = k_{55} = 0, \\ &K_2(z, y) = O(|z - y|^{\alpha - 2}), \quad \alpha \geq 0. \end{aligned}$$

We determine the characteristic matrix of the operator T introducing a local coordinate system $(\zeta_1, \zeta_2, \zeta_3)$ at each point $z \in S$ with the ζ_3-axis directed along the outside normal to S at z. Let $\mathcal{K} = \|k_{ij}\|_{5\times 5}$ be the characteristic matrix of T and $\mathcal{S} = \|\sigma_{ij}\|_{5\times 5}$ be the symbolic matrix of the operator T. We find that

$$\begin{aligned} &k_{11} = k_{22} = k_{33} = k_{44} = k_{55} = -1, \ k_{ij} = -k_{ji}, \ (i \neq j), \\ &k_{i4} = k_{i5} = k_{4i} = k_{5i} = 0, \ \sigma_{11} = \sigma_{22} = \sigma_{33} = \sigma_{44} = \sigma_{55} = -1, \\ &\sigma_{rs} = 2\pi i k_{rs} \ (r \neq s), \quad \sigma_{i4} = \sigma_{i5} = \sigma_{4i} = \sigma_{5i} = 0, \end{aligned}$$

so that

$$\det \mathcal{S} = 1 - \frac{\mu^2}{(\lambda + 2\mu)^2}.$$

We assume that $\mu/(\lambda + 2\mu) \neq \pm 1$. In this case the integral operator corresponding to the equation (I_1) is of the normal type. Since T is of normal type and the symbolic matrix is antisymmetric, then the index of this operator is zero (cf. Mikhlin [261], p. 195). We conclude that the operator T is a Fredholm operator. We can prove that the Fredholm theory is applicable to all singular integral equations (I_α) and (E_α).

Chapter 3

EQUILIBRIUM THEORY

3.1 Thermoelastic States

In this section we consider the equilibrium theory of thermoelastic materials with voids. In the absence of time dependence, the basic equations from Section 1.5 reduce to the equations of equilibrium

$$t_{ji,j} + \rho_0 f_i = 0, \tag{3.1.1}$$

the balance of equilibrated force

$$h_{i,i} + g + \rho_0 \ell = 0, \tag{3.1.2}$$

the equilibrium energy equation

$$q_{i,i} + \rho_0 S = 0, \tag{3.1.3}$$

the constitutive equations

$$\begin{aligned} t_{ij} &= C_{ijrs}e_{rs} + B_{ij}\varphi + D_{ijk}\varphi_{,k} - \beta_{ij}\theta, \\ h_i &= A_{ij}\varphi_{,j} + D_{rsi}e_{rs} + d_i\varphi - a_i\theta, \\ g &= -B_{ij}e_{ij} - \xi\varphi - d_i\varphi_{,i} + m\theta, \\ q_i &= k_{ij}\theta_{,j} \end{aligned} \tag{3.1.4}$$

and the geometrical equations (1.5.3). To the field equations (3.1.1)-(3.1.4) we adjoin the boundary conditions

$$\begin{aligned} &u_i = \widetilde{u}_i \text{ on } \overline{S}_1, \quad t_{ji}n_j = \widetilde{t}_i \text{ on } S_2, \quad \varphi = \widetilde{\varphi} \text{ on } \overline{S}_3, \\ &h_i n_i = \widetilde{h} \text{ on } S_4, \quad \theta = \widetilde{\theta} \text{ on } \overline{S}_5, \qquad q_i n_i = \widetilde{q} \text{ on } S_6, \end{aligned} \tag{3.1.5}$$

where $\widetilde{u}_i, \widetilde{t}_i, \widetilde{\varphi}, \widetilde{h}, \widetilde{\theta}$ and $\widetilde{q}$ are prescribed functions and S_i, $(i = 1, 2, ..., 6)$, are parts of ∂B such that $\overline{S}_1 \cup S_2 = \overline{S}_3 \cup S_4 = \overline{S}_5 \cup S_6 = \partial B$, $S_1 \cap S_2 = S_3 \cap S_4 = S_5 \cap S_6 = \emptyset$.

In the equilibrium theory the function θ is the solution of the boundary value problem (3.1.3), $(3.1.4)_4$, $(3.1.5)_{5,6}$. In general, we shall treat the temperature field as having already been so determined.

By an admissible state we mean an ordered array of functions $(u_i, \varphi, e_{ij}, t_{ij}, h_i, g)$ with the properties: (i) $u_i, \varphi \in C^2(B) \cap C^1(\overline{B})$; (ii) e_{ij} is a continuous symmetric tensor field on $\overline{B}$; (iii) $t_{ij}, h_k, g \in C^1(B)$, $t_{ij}, h_k, t_{ji,j}, h_{s,s} \in C^0(\overline{B})$, $t_{ij} = t_{ji}$. If we define the addition of admissible states and the multiplication of an admissible state with a scalar by

$$(u_i', \varphi', ..., g') + (u_i'', \varphi'', ..., g'') = (u_i' + u_i'', \varphi' + \varphi'', ..., g' + g''),$$
$$\lambda(u_i, \varphi, ..., g) = (\lambda u_i, \lambda\varphi, ..., \lambda g),$$

then the set of all admissible states is a vector space.

We assume that: (i) the body loads are continuous on $\overline{B}$; (ii) the constitutive coefficients are continuously differentiable on $\overline{B}$ and satisfy (1.5.5), (1.5.8); (iii) $\widetilde{u}_i$ are continuous on $\overline{S}_1$, $\widetilde{\varphi}$ is continuous on $\overline{S}_3$, $\widetilde{\theta}$ is continuous on $\overline{S}_5$, $\widetilde{t}_i$ are piecewise regular on S_2, $\widetilde{h}$ is piecewise regular on S_4 and $\widetilde{q}$ is piecewise regular on S_6.

By a thermoelastic state corresponding to the body loads f_i, ℓ and temperature field θ we mean an admissible state that satisfies the equations (3.1.1), (3.1.2), $(3.1.4)_{1-3}$ and (1.5.3) on B. It follows that the problem of thermoelastic equilibrium is reduced to the solution of two boundary value problems. The first boundary value problem is characterized by the equations (3.1.3), $(3.1.4)_4$ on B and boundary conditions $(3.1.5)_{5,6}$. After the determination of the function θ we have to solve a second boundary value problem which consists in the finding of a thermoelastic state corresponding to the body loads f_i, ℓ and temperature field θ, and to the mechanical boundary conditions from (3.1.5). If such a thermoelastic state exists, it is called a solution of the mixed problem.

The ordered array $(f_i, \ell, \widetilde{u}_i, \widetilde{\varphi}, \widetilde{t}_i, \widetilde{h}, \theta)$ is called external data system on B for the mixed problem of thermoelastostatics. We can see from (1.5.16) that the functions u_i and φ satisfy the equations of a problem of elastostatics with body forces $f_i - (\beta_{ij}\theta)_{,j}/\rho_0$ and the extrinsic equilibrated body force $\ell + [m\theta - (a_i\theta)_{,i}]/\rho_0$. In the problem of elastostatics the boundary data are $(\widetilde{u}_i, \widetilde{\varphi}, \widetilde{t}_i + \beta_{ij}\theta n_j, \widetilde{h} + a_i\theta n_i)$. This result is known as the body force analogy.

Let us assume that the body B is subjected to two external data systems $L^{(\alpha)} = (f_i^{(\alpha)}, \ell^{(\alpha)}, \widetilde{u}_i^{(\alpha)}, \widetilde{\varphi}^{(\alpha)}, \widetilde{t}_i^{(\alpha)}, \widetilde{h}^{(\alpha)}, \theta^{(\alpha)})$, $(\alpha = 1, 2)$, and let $s^{(\alpha)} = (u_i^{(\alpha)}, \varphi^{(\alpha)}, e_{ij}^{(\alpha)}, t_{ij}^{(\alpha)}, h_i^{(\alpha)}, g^{(\alpha)})$ be a solution corresponding to

$L^{(\alpha)}$. We introduce the notations

$$t_i^{(\alpha)} = t_{ji}^{(\alpha)} n_j, \quad h^{(\alpha)} = h_i^{(\alpha)} n_i. \tag{3.1.6}$$

Theorem 3.1.1. *If a thermoelastic material with voids is subjected to two systems of loadings* $L^{(\alpha)}$, $(\alpha = 1, 2)$, *then between the corresponding states* $s^{(\alpha)}$ *there is the following relation*

$$\begin{aligned}
&\int_B \left\{ \left[\rho_0 f_i^{(1)} - (\beta_{ij}\theta^{(1)})_{,j}\right] u_i^{(2)} + \left[\rho_0 \ell^{(1)} - (a_i\theta^{(1)})_{,i} + m\theta^{(1)}\right] \varphi^{(2)} \right\} dV + \\
&+ \int_{\partial B} \left[(t_i^{(1)} + \beta_{ij}\theta^{(1)} n_j) u_i^{(2)} + (h^{(1)} + a_i\theta^{(1)} n_i)\varphi^{(2)} \right] dA = \\
&= \int_B \left\{ \left[\rho_0 f_i^{(2)} - (\beta_{ij}\theta^{(2)})_{,j}\right] u_i^{(1)} + \left[\rho_0 \ell^{(2)} - (a_i\theta^{(2)})_{,i} + m\theta^{(2)}\right] \varphi^{(1)} \right\} dV + \\
&+ \int_{\partial B} \left[(t_i^{(2)} + \beta_{ij}\theta^{(2)} n_j) u_i^{(1)} + (h^{(2)} + a_i\theta^{(2)} n_i)\varphi^{(1)} \right] dA.
\end{aligned} \tag{3.1.7}$$

Proof. Let

$$\begin{aligned}
2W_{a\beta} = {} & (t_{ij}^{(\alpha)} + \beta_{ij}\theta^{(\alpha)}) e_{ij}^{(\beta)} + (h_i^{(\alpha)} + a_i\theta^{(\alpha)})\varphi_{,i}^{(\beta)} - \\
& -(g^{(\alpha)} - m\theta^{(\alpha)})\varphi^{(\beta)}.
\end{aligned} \tag{3.1.8}$$

In view of (3.1.4) we have

$$\begin{aligned}
2W_{a\beta} = {} & C_{ijrs} e_{rs}^{(\alpha)} e_{ij}^{(\beta)} + B_{ij}(\varphi^{(\alpha)} e_{ij}^{(\beta)} + \varphi^{(\beta)} e_{ij}^{(\alpha)}) + D_{ijk}(\varphi_{,k}^{(\alpha)} e_{ij}^{(\beta)} + \\
& + \varphi_{,k}^{(\beta)} e_{ij}^{(\alpha)}) + d_i(\varphi^{(\alpha)}\varphi_{,i}^{(\beta)} + \varphi^{(\beta)}\varphi_{,i}^{(\alpha)}) + A_{ij}\varphi_{,j}^{(\alpha)}\varphi_{,i}^{(\beta)} + \xi\varphi^{(\alpha)}\varphi^{(\beta)}.
\end{aligned} \tag{3.1.9}$$

With the help of (1.5.5) we get

$$W_{12} = W_{21}. \tag{3.1.10}$$

On the other hand, by (1.5.3) we obtain

$$\begin{aligned}
2W_{a\beta} = {} & \left[\left(t_{ij}^{(\alpha)} + \beta_{ij}\theta^{(\alpha)}\right) u_i^{(\beta)}\right]_{,j} - \left[t_{ij,j}^{(\alpha)} + \left(\beta_{ij}\theta^{(\alpha)}\right)_{,j}\right] u_i^{(\beta)} + \\
& + \left[\left(h_i^{(\alpha)} + a_i\theta^{(\alpha)}\right)\varphi^{(\beta)}\right]_{,i} - \left[h_{i,i}^{(\alpha)} + \left(a_i\theta^{(\alpha)}\right)_{,i}\right]\varphi^{(\beta)} - \\
& - g^{(\alpha)}\varphi^{(\beta)} + m\theta^{(\alpha)}\varphi^{(\beta)}.
\end{aligned}$$

By using (3.1.1), (3.1.2) and the divergence theorem we find

$$
\begin{aligned}
2\int_B W_{\alpha\beta} dV = \int_B \{[\rho_0 f_i^{(\alpha)} - (\beta_{ij}\theta^{(\alpha)})_{,j}]u_i^{(\beta)} + \\
+[\rho_0 \ell^{(\alpha)} - (a_i\theta^{(\alpha)})_{,i} + m\theta^{(\alpha)}]\varphi^{(\beta)}\}dV + \\
+\int_{\partial B} [(t_i^{(\alpha)} + \beta_{ij}\theta^{(\alpha)}n_j)u_i^{(\beta)} + (h^{(\alpha)} + a_i\theta^{(\alpha)}n_i)\varphi^{(\beta)}]dA.
\end{aligned} \tag{3.1.11}
$$

From (3.1.10) and (3.1.11) we obtain the desired result. □

Let $L^{(1)} = (f_i, \ell, \widetilde{u}_i, \widetilde{\varphi}, \widetilde{t}_i, \widetilde{h}, \theta)$ and $s^{(1)} = (u_i, \varphi, e_{ij}, t_{ij}, h_i, g)$. Then from (3.1.9) and (3.1.11), for $\alpha = \beta = 1$, we get

$$
\begin{aligned}
2\int_B W\, dV = \int_B \{[\rho_0 f_i - (\beta_{ij}\theta)_{,j}]u_i + [\rho_0 \ell - (a_i\theta)_{,i} + m\theta]\varphi\}dV + \\
+\int_{\partial B} [(t_i + \beta_{ij}\theta n_j)u_i + (h + a_i\theta n_i)\varphi]dA,
\end{aligned} \tag{3.1.12}
$$

where W is given by (2.1.1). This relation can be used to derive a uniqueness result.

Theorem 3.1.2. *Assume that the potential energy density W is a positive definite quadratic form. Then any two solutions of the thermoelastic problem are equal modulo a rigid displacement. Moreover, if S_1 is non-empty, then the mixed problem has at most one solution.*

Proof. Suppose that there are two solutions. Then their difference $s^0 = (u_i^0, \varphi^0, e_{ij}^0, t_{ij}^0, h_i^0, g^0)$ corresponds to null external data. We conclude from (3.1.12) that

$$
\int_B W^0 dV = 0, \tag{3.1.13}
$$

where W^0 is the potential energy density corresponding to s^0. Since the potential energy density is positive definite, from (3.1.13) we obtain $e_{ij}^0 = 0$ and $\varphi^0 = 0$ and therefore

$$
u_i^0 = a_i^0 + \varepsilon_{ijk} b_j^0 x_k, \quad \varphi^0 = 0, \tag{3.1.14}
$$

where ε_{ijk} is the alternating symbol, and a_i^0 and b_j^0 are arbitrary constants. Clearly, if S_1 is non-empty then we obtain $u_i^0 = 0$ and $\varphi^0 = 0$ on B. □

By a kinematically admissible state we mean an admissible state $s = (u_i, \varphi, e_{ij}, t_{ij}, h_i, g)$ that satisfies the constitutive equations (3.1.4), the geometrical equations (1.5.3) and the boundary conditions

$$u_i = \widetilde{u}_i \text{ on } \overline{S}_1, \quad \varphi = \widetilde{\varphi} \text{ on } \overline{S}_3. \tag{3.1.15}$$

We now present a minimum principle.

Theorem 3.1.3. *Assume that the potential energy density is positive definite. Let K denote the set of all kinematically admissible states, and let F be the functional on K defined by*

$$F(s) = \int_B W\, dV - \int_B \{[\rho_0 f_i - (\beta_{ij}\theta)_{,j}]u_i + [\rho_0 \ell + m\theta - (a_i\theta)_{,i}]\varphi\} dV -$$
$$- \int_{S_2} (\widetilde{t}_i + \beta_{ij}\theta n_j) u_i dA - \int_{S_4} (\widetilde{h} + a_i\theta n_i)\varphi\, dA, \tag{3.1.16}$$

for every $s = (u_i, \varphi, e_{ij}, t_{ij}, h_i, g) \in K$. Further, let s be a solution of the mixed problem. Then

$$F(s) \leq F(s^*), \tag{3.1.17}$$

for every $s^* \in K$, and equality holds only if $s^* = s$ modulo a rigid displacement.

Proof. Let $s = (u_i, \varphi, e_{ij}, t_{ij}, h_i, g)$ be a solution of the mixed problem, let $s^* = (u_i^*, \varphi^*, e_{ij}^*, t_{ij}^*, h_i^*, g^*) \in K$, and define $s' = s^* - s$. Then we have $s' = (u_i', \varphi', e_{ij}', t_{ij}', h_i', g')$ where $u_i' = u_i^* - u_i$, $\varphi' = \varphi^* - \varphi$, $e_{ij}' = e_{ij}^* - e_{ij}$, $t_{ij}' = t_{ij}^* - t_{ij}$, $h_i' = h_i^* - h_i$, $g' = g^* - g$. Moreover,

$$\begin{aligned}
&2e_{ij}' = u_{i,j}' + u_{j,i}', \\
&t_{ij}' = C_{ijrs}e_{rs}' + B_{ij}\varphi' + D_{ijk}\varphi_{,k}', \\
&h_i' = A_{ij}\varphi_{,j}' + D_{rsi}e_{rs}' + d_i\varphi', \\
&g' = -B_{ij}e_{ij}' - \xi\varphi' - d_i\varphi_{,i}' \text{ on } B, \\
&u_i' = 0 \text{ on } \overline{S}_1, \quad \varphi' = 0 \text{ on } \overline{S}_3.
\end{aligned} \tag{3.1.18}$$

It follows from (3.1.16) that

$$F(s^*) = F(s) + \int_B W' dV - \int_B (t_{ji,j} + \rho_0 f_i)u_i' dV -$$
$$- \int_B (h_{i,i} + g + \rho_0 \ell)\varphi' dV + \int_{S_2} (t_{ji}n_j - \widetilde{t}_i)u_i' dA + \int_{S_4} (h_i n_i - \widetilde{h})\varphi' dA, \tag{3.1.19}$$

where W' is the potential energy density corresponding to s'. Since s is a solution of the mixed problem, (3.1.19) implies

$$F(s^*) - F(s) = \int_B W' dV.$$

Thus, since W' is positive definite, we obtain (3.1.17). Moreover, $F(s^*) = F(s)$ only if $e'_{ij} = 0$ and $\varphi' = 0$. □

The body force analogy allows us to bring all of the results of isothermal elastostatics to bear on the equilibrium theory of thermoelasticity.

3.2 The Thermoelastic Plane Strain

In this section we consider the linear theory of thermoelastostatics for homogeneous and isotropic materials with voids. The constitutive equations are given by (1.5.10). We assume that the potential energy density is positive definite so that the inequalities (2.1.7) hold. In this section the region B refers to the interior of a right cylinder with the open cross-section Σ and the lateral boundary Π. The rectangular Cartesian coordinate frame is supposed to be chosen in such a way that the x_3-axis is parallel to the generators of B. We denote by L the boundary of Σ. The results presented in this section have been established by Ieşan and Nappa [190].

a) **Statement of the problem.** The thermoelastic plane strain parallel to the x_1, x_2-plane is characterized by

$$\begin{aligned} &u_\alpha = u_\alpha(x_1, x_2), \quad u_3 = 0, \quad \varphi = \varphi(x_1, x_2), \\ &\theta = \theta(x_1, x_2), \quad (x_1, x_2) \in \Sigma, \end{aligned} \tag{3.2.1}$$

where the Greek indices have the range 1, 2. The above restrictions, in conjunction with the geometrical equations (1.5.3) and the constitutive equations (1.5.10) imply that e_{ij}, t_{ij}, h_i, g and q_i are all independent of x_3. The non-zero strain measures are given by

$$e_{\alpha\beta} = \frac{1}{2}(u_{\alpha,\beta} + u_{\beta,\alpha}). \tag{3.2.2}$$

The constitutive equations show that non-zero dependent constitutive variables are $t_{\alpha\beta}, t_{33}, h_\alpha, g$ and q_α. Further

$$\begin{aligned} t_{\alpha\nu} &= \lambda e_{\rho\rho}\delta_{\alpha\nu} + 2\mu e_{\alpha\nu} + b\varphi\delta_{\alpha\nu} - \beta\theta\delta_{\alpha\nu}, \\ h_\rho &= \alpha\varphi_{,\rho}, \quad g = -be_{\rho\rho} - \xi\varphi + m\theta, \\ q_\alpha &= k\theta_{,\alpha}. \end{aligned} \tag{3.2.3}$$

We assume that the body loads are independent of x_3 and that $f_3 = 0$. The equations of equilibrium (3.1.1) and (3.1.2) reduce to

$$t_{\beta\alpha,\beta} + \rho_0 f_\alpha = 0, \quad h_{\alpha,\alpha} + g + \rho_0 \ell = 0 \text{ on } \Sigma. \tag{3.2.4}$$

The energy equation (3.1.3) becomes

$$k\Delta\theta = -\rho_0 S \text{ on } \Sigma, \tag{3.2.5}$$

where Δ is the two-dimensional Laplacian.

We suppose that on the lateral boundary are prescribed the surface traction $\widetilde{\boldsymbol{t}}$, the equilibrated surface force $\widetilde{h}$ and the heat flux $\widetilde{q}$, with $\widetilde{\boldsymbol{t}}$ $\widetilde{h}$ and $\widetilde{q}$ independent of x_3 and $\widetilde{t}_3 = 0$. Thus we obtain the following boundary conditions

$$t_{\beta\alpha} n_\beta = \widetilde{t}_\alpha, \quad h_\alpha n_\alpha = \widetilde{h}, \quad q_\alpha n_\alpha = \widetilde{q} \text{ on } L. \tag{3.2.6}$$

Thus, the temperature field θ is the solution of the boundary value problem (3.2.5), $(3.2.6)_3$. Generally, we shall treat the temperature field as having already been so determined. The determination of the thermoelastic equilibrium consists in finding of the functions u_α and φ which satisfy the equations (3.2.2)-(3.2.4) in Σ and the boundary conditions $(3.2.6)_{1,2}$ on L. From (3.2.2), (3.2.3) and (3.2.4) we obtain the equations of equilibrium expressed in terms of the functions u_α and φ

$$\begin{aligned} &\mu\Delta u_\nu + (\lambda + \mu) u_{\rho,\rho\nu} + b\varphi_{,\nu} - \beta\theta_{,\nu} + \rho_0 f_\nu = 0, \\ &\alpha\Delta\varphi - b u_{\rho,\rho} - \xi\varphi + m\theta + \rho_0 \ell = 0 \text{ on } \Sigma. \end{aligned} \tag{3.2.7}$$

Generally, in the problems of thermoelastostatics it is assumed that the mechanical loads are absent, the principal attention being devoted to the deformation due to the temperature field.

b) **Reduction of the problem to the isothermal theory.** In this section we present two methods of reduction of the thermoelastic plane strain problem to an isothermal one with zero body loads and with certain known boundary data. Throughout this section we assume that $f_\alpha = 0$ and $\ell = 0$.

In the context of the classical thermoelasticity Muskhelishvili [270] established a method of reduction of the thermoelastic problem to an isothermal one which is based on the assumption that the heat source is zero. In what follows we extend this method to the plane strain problem for thermoelastic materials with voids. We assume that $S = 0$. Since the function θ is harmonic in Σ we can define the analytic function

$$F(z) = \theta + iP, \tag{3.2.8}$$

of the complex variable $z = x_1 + ix_2$. We introduce the functions w_α defined by

$$w_1 + iw_2 = \int F(z)dz. \tag{3.2.9}$$

If the domain Σ is simply-connected, then w_α are single-valued functions. Clearly we have

$$w_{1,1} = w_{2,2} = \theta, \quad w_{1,2} = -w_{2,1}.$$

We note that from (2.1.7) it follows that

$$d = 2[(\lambda + \mu)\xi - b^2] > 0, \tag{3.2.10}$$

so that we can define the constants

$$M = \frac{1}{d}(\beta\xi - mb), \quad \tau = \frac{2}{d}[m(\lambda + \mu) - b\beta]. \tag{3.2.11}$$

We introduce the functions u_α^0 and φ^0 by

$$u_\alpha = u_\alpha^0 + Mw_\alpha, \quad \varphi = \varphi^0 + \tau\theta. \tag{3.2.12}$$

From (3.2.2) and (3.2.12) we obtain

$$e_{\alpha\beta} = e_{\alpha\beta}^0 + M\theta\delta_{\alpha\beta}, \tag{3.2.13}$$

where

$$e_{\alpha\beta}^0 = \frac{1}{2}(u_{\alpha,\beta}^0 + u_{\beta,\alpha}^0). \tag{3.2.14}$$

We introduce the notations

$$\begin{aligned} t_{\alpha\nu}^0 &= \lambda e_{\rho\rho}^0\delta_{\alpha\nu} + 2\mu e_{\alpha\nu}^0 + b\varphi^0\delta_{\alpha\nu}, \\ h_\rho^0 &= \alpha\varphi_{,\rho}^0, \quad g^0 = -be_{\rho\rho}^0 - \xi\varphi^0. \end{aligned} \tag{3.2.15}$$

We note that the relations (3.2.15) are the constitutive equations in the isothermal theory. It follows from (3.2.3), (3.2.13) and (3.2.15) that

$$t_{\alpha\beta} = t_{\alpha\beta}^0, \quad h_\rho = h_\rho^0 + \alpha\tau\theta_{,\rho}, \quad g = g^0 \quad \text{on } \Sigma. \tag{3.2.16}$$

Since the function θ is harmonic, the equations (3.2.4) become

$$t_{\beta\alpha,\beta}^0 = 0, \quad h_{\rho,\rho}^0 + g^0 = 0 \quad \text{on } \Sigma. \tag{3.2.17}$$

The boundary conditions $(3.2.6)_{1,2}$ reduce to

$$t_{\beta\alpha}^0 n_\beta = \widetilde{t}_\alpha, \quad h_\alpha^0 n_\alpha = \widetilde{h} - \alpha\tau\frac{\partial\theta}{\partial n} \quad \text{on } L. \tag{3.2.18}$$

In this way the thermoelastic problem associated to the simply-connected domain Σ is reduced to the problem of isothermal elastostatics characterized by the field equations (3.2.14), (3.2.15), (3.2.17) on Σ and the boundary conditions (3.2.18) on L. Thus, we have the result that if Σ is simply-connected and the heat source is absent, then the thermoelastic plane strain problem corresponding to null loads reduces to solving of the isothermal elastic plane strain problem ($\theta = 0$) corresponding to zero body loads, zero boundary tractions and to the superficial equilibrated force $\widetilde{h} = -\alpha\tau q/k$, where q is the heat flux in thermoelastic problem.

Consider now a bounded multiply-connected domain Σ. We assume that inside Σ there are m contours L_k $(k = 1, 2, ..., m)$ which have no common points; the outer contour is denoted by L_{m+1}. Since the temperature is single-valued, then the function F has the form

$$F(z) = \sum_{k=1}^{m} B_k \log(z - z_k) + F_0(z), \tag{3.2.19}$$

where z_k is a point inside the contour L_k, B_k are real constants and F_0 is analytic and single-valued function on Σ. It follows from (3.2.9) and (3.2.19) that

$$w_1 + iw_2 = z\sum_{k=1}^{m} B_k \log(z - z_k) + \sum_{k=1}^{m}(a_k + ib_k)\log(z - z_k) + f_0(z), \tag{3.2.20}$$

where a_k and b_k are real constants, and f_0 is single-valued and analytic on Σ. Thus, we obtain

$$[w_1 + iw_2]_{L_k} = 2\pi i(zB_k + a_k + ib_k), \tag{3.2.21}$$

where $[\ ]_{L_k}$ denotes the change in value of the function inside on passing once round the contour L_k in the conventional positive sense which keeps the area enclosed on the left, the contour lying entirely in the body. Since the functions u_α are single-valued, it follows that the functions u_α^0 are multi-valued

$$\begin{aligned} [u_1^0]_{L_k} &= 2\pi M(B_k x_2 + b_k), \\ [u_2^0]_{L_k} &= -2\pi M(B_k x_1 + a_k). \end{aligned} \tag{3.2.22}$$

Thus, in this case the thermoelastic problem reduces to a problem of mechanical dislocations defined by the field equations (3.2.14), (3.2.15), (3.2.17) on Σ, the boundary conditions (3.2.18) on L and the dislocation characteristics

$$\varepsilon_k = -2\pi M B_k, \quad \alpha_k = 2\pi M b_k, \quad \beta_k = -2\pi M a_k. \tag{3.2.23}$$

The above method shall be used in this section to solve some special problems. We now present another method for reduction of thermoelastic problem to one for an unheated body with zero body loads. This method generalizes the result established by Goodier in the context of classical thermoelasticity (see [285]) and is not restricted to harmonic temperature fields.

Let Φ be a function of class C^4 on Σ, and let Ψ be a function of class C^2 on Σ which satisfy the equations

$$(\lambda + 2\mu)\Delta\Phi + b\Psi = \beta\theta,$$
$$\alpha\Delta\Psi - \xi\Psi - b\Delta\Phi = -m\theta,$$

on Σ. It follows that Φ and Ψ satisfy

$$\begin{aligned} \Delta(\Delta - p^2)\Phi &= \rho_1\theta - \frac{\beta}{k(\lambda + 2\mu)}S, \\ (\Delta - p^2)\Psi &= \zeta\theta, \end{aligned} \tag{3.2.24}$$

where

$$p^2 = \frac{\xi(\lambda + 2\mu) - b^2}{\alpha(\lambda + 2\mu)}, \quad \rho_1 = \frac{mb - \xi\beta}{\alpha(\lambda + 2\mu)}, \quad \zeta = \frac{b\beta - m(\lambda + 2\mu)}{\alpha(\lambda + 2\mu)}. \tag{3.2.25}$$

We note that the relations (2.1.7) imply that p^2 is strictly positive. We seek the solution of thermoelastic problem in the form

$$u_\alpha = u_\alpha^* + \Phi_{,\alpha}, \quad \varphi = \varphi^* + \Psi, \tag{3.2.26}$$

where u_α^* and φ^* are unknown functions of class C^2 on Σ, and Φ and Ψ are functions which satisfy (3.2.24). From (3.2.26) we obtain

$$e_{a\beta} = e_{\alpha\beta}^* + \Phi_{,\alpha\beta},$$

where

$$e_{\alpha\beta}^* = \frac{1}{2}(u_{\alpha,\beta}^* + u_{\beta,\alpha}^*). \tag{3.2.27}$$

Thus, we get

$$t_{\alpha\nu} = t_{\alpha\nu}^* + 2\mu\Phi_{,\alpha\nu} + (\lambda\Delta\Phi + b\Psi - \beta\theta)\delta_{\alpha\nu},$$
$$h_\rho = h_\rho^* + \alpha\Psi_{,\rho}, \quad g = g^* - b\Delta\Phi - \xi\Psi + m\theta,$$

where

$$\begin{aligned} t_{\alpha\nu}^* &= \lambda e_{\rho\rho}^*\delta_{\alpha\nu} + 2\mu e_{\alpha\nu}^* + b\varphi^*\delta_{\alpha\nu}, \\ h_\rho^* &= \alpha\varphi_{,\rho}^*, \quad g' = -be_{\rho\rho}^* - \xi\varphi^*. \end{aligned} \tag{3.2.28}$$

In view of (3.2.24), the equilibrium equations reduce to

$$t^*_{\beta\alpha,\beta} = 0, \quad h^*_{\rho,\rho} + g^* = 0 \quad \text{on } \Sigma. \tag{3.2.29}$$

The boundary conditions can be expressed as

$$\begin{aligned} t^*_{\beta\alpha} n_\beta &= \widetilde{t}_\alpha - 2\mu(\Phi_{,\alpha\beta} - \delta_{\alpha\beta}\Delta\Phi) n_\beta, \\ h^*_\rho n_\rho &= \widetilde{h} - \alpha \frac{\partial \Psi}{\partial n} \quad \text{on } L. \end{aligned} \tag{3.2.30}$$

Thus, the problem of thermoelasticity is reduced to the determination of the functions Φ and Ψ which satisfy the equations (3.2.24) and to solving the isothermal plane strain problem (3.2.27)-(3.2.30).

c) **Thermal stresses in a tube.** In this section we use the above results to solve a thermoelastic problem for a hollow cylinder. We assume that the domain Σ is defined by $\Sigma = \{\boldsymbol{x} : R_1^2 < x_1^2 + x_2^2 < R_2^2,\ x_3 = 0\}$, where R_1 and R_2 are positive constants. The body is in equilibrium in the absence of mechanical loads and heat sources. We suppose that the temperature on the inner surface of the cylinder is T_1 and that the temperature on the outer surface is T_2, where T_1 and T_2 are given constants. In this case the equation (3.2.5) reduces to

$$\Delta\theta = 0 \quad \text{on } \Sigma. \tag{3.2.31}$$

The thermal boundary conditions consist of $\theta = T_\alpha$ for $r = R_\alpha$. Thus, we obtain

$$\theta = C_1 \ln r + C_2, \tag{3.2.32}$$

where

$$C_1 = (T_2 - T_1)\left[\ln \frac{R_2}{R_1}\right]^{-1}, \quad C_2 = (T_1 \ln R_2 - T_2 \ln R_1)\left[\ln \frac{R_2}{R_1}\right]^{-1}.$$

The function F defined by (3.2.8) is

$$F(z) = C_1 \log z + C_2. \tag{3.2.33}$$

From (3.2.9) and (3.2.33) we obtain

$$w_1 + i w_2 = z(C_1 \log z + C_2 - C_1). \tag{3.2.34}$$

It follows that the constants B_1, a_1 and b_1 which appear in (3.2.21) are given by

$$B_1 = C_1, \quad a_1 = b_1 = 0. \tag{3.2.35}$$

In this case the thermoelastic problem reduces to the problem of mechanical dislocations defined by the equations (3.2.14), (3.2.15), (3.2.17) on Σ, the dislocations characteristics (Fig. 1)

$$\varepsilon_1 = -2\pi M C_1, \quad \alpha_1 = 0, \quad \beta_1 = 0 \tag{3.2.36}$$

and the boundary conditions

$$\begin{aligned} &t^0_{\beta\alpha} n_\beta = 0 \text{ for } r = R_1 \text{ and } r = R_2, \\ &h^0_\rho n_\rho = \frac{\alpha\tau}{R_1} C_1 \text{ for } r = R_1, \; h^0_\rho n_\rho = -\frac{\alpha\tau}{R_2} C_1 \text{ for } r = R_2. \end{aligned} \tag{3.2.37}$$

In what follows we shall investigate this problem by using the polar coordinates (r, ϑ).

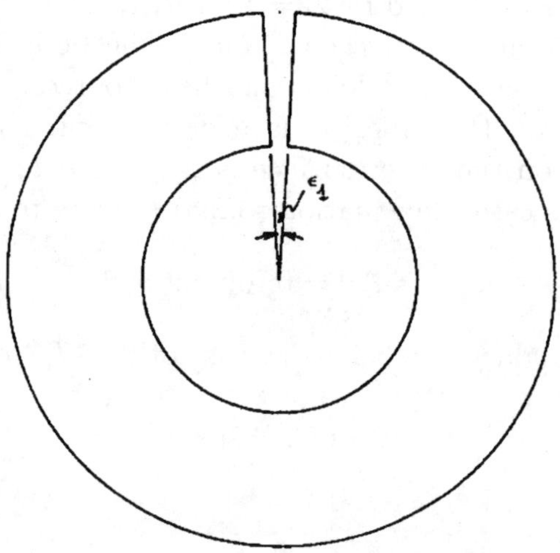

Figure 1

We denote by u and v the physical components of the displacement vector in polar coordinates,

$$u + iv = (u^0_1 + iu^0_2)e^{-i\vartheta}. \tag{3.2.38}$$

The geometrical equations (3.2.14) become

$$\varepsilon_{rr} = \frac{\partial u}{\partial r}, \quad \varepsilon_{\vartheta\vartheta} = \frac{1}{r}\left(\frac{\partial v}{\partial \vartheta} + u\right), \quad \varepsilon_{r\vartheta} = \frac{1}{2}\left(\frac{1}{r}\frac{\partial u}{\partial \vartheta} + \frac{\partial v}{\partial r} - \frac{1}{r}v\right). \tag{3.2.39}$$

The constitutive equations (3.2.15) can be expressed as

$$\begin{aligned} t_{rr} &= (\lambda + 2\mu)\varepsilon_{rr} + \lambda\varepsilon_{\vartheta\vartheta} + b\varphi^0, \\ t_{\vartheta\vartheta} &= \lambda\varepsilon_{rr} + (\lambda + 2\mu)\varepsilon_{\vartheta\vartheta} + b\varphi^0, \\ t_{r\vartheta} &= 2\mu\varepsilon_{r\vartheta}, \\ h_r &= \alpha\,\frac{\partial\varphi^0}{\partial r}, \quad h_\vartheta = \frac{1}{r}\,\alpha\,\frac{\partial\varphi^0}{\partial\vartheta}, \\ g &= -b\left[\frac{1}{r}\,\frac{\partial}{\partial r}\,(ru) + \frac{1}{r}\,\frac{\partial v}{\partial\vartheta}\right] - \xi\varphi^0. \end{aligned} \tag{3.2.40}$$

The equilibrium equations (3.2.17) reduce to

$$\begin{aligned} &\frac{\partial t_{rr}}{\partial r} + \frac{1}{r}\,\frac{\partial t_{r\vartheta}}{\partial\vartheta} + \frac{1}{r}\,(t_{rr} - t_{\vartheta\vartheta}) = 0, \\ &\frac{\partial t_{r\vartheta}}{\partial r} + \frac{1}{r}\,\frac{\partial t_{\vartheta\vartheta}}{\partial\vartheta} + \frac{2}{r}\,t_{r\vartheta} = 0, \\ &\frac{1}{r}\,\frac{\partial}{\partial r}\,(rh_r) + \frac{1}{r}\,\frac{\partial h_\vartheta}{\partial\vartheta} + g = 0. \end{aligned} \tag{3.2.41}$$

The boundary conditions (3.2.37) become

$$\begin{aligned} &t_{rr} = 0, \ t_{r\vartheta} = 0 \ \text{ for } r = R_1 \ \text{ and } \ r = R_2, \\ &\frac{\partial\varphi^0}{\partial r} = -\frac{\tau}{R_\beta}\,C_1 \ \text{ for } r = R_\beta. \end{aligned} \tag{3.2.42}$$

We seek the solution of this problem in the form

$$u = \frac{\varepsilon_1\mu}{2\pi(\lambda + 2\mu)}\,r\ln r + rF(r), \ v = \frac{\varepsilon_1}{2\pi}\,r\vartheta, \ \varphi^0 = G(r), \tag{3.2.43}$$

where F and G are unknown functions of class C^2.

It is easy to see that the functions u, v and φ^0 correspond to the dislocation characteristics (3.2.36). From (3.2.39), (3.2.40) and (3.2.43) we obtain

$$\begin{aligned} &t_{rr} = (\lambda + 2\mu)rF' + 2(\lambda + \mu)F + bG + 2\Lambda(\lambda + \mu)\ln r + \frac{\varepsilon_1}{2\pi}\,(\lambda + \mu), \\ &t_{\vartheta\vartheta} = \lambda rF' + 2(\lambda+\mu)F + bG + 2\Lambda(\lambda+\mu)\ln r + \frac{\varepsilon_1}{2\pi}\left(\frac{\lambda\mu}{\lambda+2\mu} + \lambda + 2\mu\right), \\ &t_{r\vartheta} = 0, \quad h_r = \alpha G', \quad h_\vartheta = 0, \\ &g = -b(2F + rF') - 2b\Lambda\ln r - \frac{b\varepsilon_1(\lambda + 3\mu)}{2\pi(\lambda + 2\mu)} - \xi G, \end{aligned} \tag{3.2.44}$$

where $F' = dF/dr$ and

$$\Lambda = \frac{\mu\varepsilon_1}{2\pi(\lambda + 2\mu)}. \tag{3.2.45}$$

The equilibrium equations (3.2.41) reduce to

$$\begin{aligned} \left[\frac{1}{r}(r^2F)'\right]' &= -\frac{b}{\lambda + 2\mu}\,G', \\ \frac{1}{r}\,(rG')' - \frac{\xi}{\alpha}\,G - \frac{b}{\alpha r}\,(r^2F)' &= \frac{2b}{\alpha}\,\Lambda \ln r + \Gamma, \end{aligned} \tag{3.2.46}$$

where

$$\Gamma = \frac{b\varepsilon_1(\lambda + 3\mu)}{2\pi\alpha(\lambda + 2\mu)}.$$

From $(3.2.46)_1$ we get

$$(r^2F)' = -\frac{b}{(\lambda + 2\mu)}\,rG + D_1 r, \tag{3.2.47}$$

where D_1 is an arbitrary constant. The equation $(3.2.46)_2$ becomes

$$G'' + \frac{1}{r}\,G' - p^2G = \frac{2b}{\alpha}\,\Lambda \ln r + \frac{b}{\alpha}\,D_1 + \Gamma. \tag{3.2.48}$$

The solution of (3.2.48) can be expressed in the form

$$\begin{aligned} G &= G_0(r) - \frac{2b}{\alpha p^2}\,\Lambda \ln r - \frac{1}{p^2}\left(\frac{b}{\alpha}\,D_1 + \Gamma\right), \\ G_0(r) &= A_1 I_0(pr) + A_2 K_0(pr), \end{aligned} \tag{3.2.49}$$

where A_1 and A_2 are arbitrary constants, and I_n and K_n are modified Bessel functions of order n. Equation (3.2.49) implies that

$$p^2 rG_0 = (rG_0')'.$$

It follows from (3.2.47) that

$$\begin{aligned} F = &-\frac{b}{p^2(\lambda + 2\mu)r}\,G_0' - \frac{b^2\Lambda}{2\alpha p^2(\lambda + 2\mu)}\,(1 - 2\ln r) + \\ &+ \frac{1}{2}\,\zeta D_1 + \frac{1}{r^2}\,D_2 + \frac{b\Gamma}{2p^2(\lambda + 2\mu)}, \end{aligned} \tag{3.2.50}$$

where

$$\zeta = \frac{b^2 + \alpha p^2(\lambda + 2\mu)}{\alpha p^2(\lambda + 2\mu)} = \xi/\alpha p^2.$$

Thus, we obtain

$$rF' = \frac{b}{p^2(\lambda+2\mu)}\left(\frac{2}{r}\,G_0' - p^2 G_0\right) + \frac{b^2\Lambda}{\alpha p^2(\lambda+2\mu)} - \frac{2}{r^2}\,D_2. \qquad (3.2.51)$$

In view of (3.2.49)-(3.2.51) and (3.2.44) we find that

$$\begin{aligned} t_{rr} &= \frac{d}{2\alpha p^2}\,D_1 - \frac{2\mu}{r^2}\,D_2 - f(r), \\ h_r &= \alpha p[A_1 I_1(pr) - A_2 K_1(pr)] - \frac{2b\Lambda}{p^2 r}, \end{aligned} \qquad (3.2.52)$$

where d is given by (3.2.10) and

$$f(r) = -\left(\frac{\Lambda d}{\alpha p^2}\ln r + \frac{2b\mu}{p^2(\lambda+2\mu)r}\,G_0'(r) + \frac{\varepsilon_1 d}{4\pi\alpha p^2}\right).$$

From $(3.2.37)_2$ we obtain the constants A_1 and A_2

$$\begin{aligned} A_1 &= \frac{1}{p\kappa R_1 R_2}\left(\frac{2b\Lambda}{\alpha p^2} - \tau C_1\right)[R_1 K_1(pR_1) - R_2 K_1(pR_2)], \\ A_2 &= \frac{1}{p\kappa R_1 R_2}\left(\frac{2b\Lambda}{\alpha p^2} - \tau C_1\right)[R_1 I_1(pR_1) - R_2 I_1(pR_2)], \end{aligned}$$

where

$$\kappa = I_1(pR_2)K_1(pR_1) - I_1(pR_1)K_1(pR_2).$$

As the function G_0 is known, from $(3.2.52)_1$ and the boundary conditions (3.2.42) we can determine the constants D_1 and D_2

$$\begin{aligned} D_1 &= \frac{2\mu}{\gamma R_1^2 R_2^2}\,[R_2^2 f(R_2) - R_1^2 f(R_1)], \\ D_2 &= \frac{d}{2\alpha p^2 \gamma}\,[f(R_2) - f(R_1)], \end{aligned}$$

where

$$\gamma = \frac{\mu d}{\alpha p^2 R_1^2 R_2^2}\,(R_2^2 - R_1^2).$$

Thus, the functions F and G are determined. The solution of the isothermal plane strain problem is given by (3.2.43). From (3.2.12), (3.2.34), (3.2.38) and (3.2.43) we obtain the solution of the thermoelastic problem.

d) **The deformation of a cylinder subjected to a uniform temperature gradient.** In this section we study the thermoelastic plane strain of a circular solid cylinder which is subjected to a uniform temperature gradient. We assume that the domain Σ is defined by $\Sigma = \{\boldsymbol{x} : x_1^2 + x_2^2 < a^2,\ x_3 = 0\}$, where a is a positive constant. The cylinder is in equilibrium in the absence of mechanical loads and heat sources. We assume that the temperature field θ is the solution of the equation (3.2.31) on Σ with the boundary condition

$$\frac{\partial \theta}{\partial n} = Q n_1 \quad \text{for } r = a, \tag{3.2.53}$$

where Q is a given constant. We find that the function θ can be expressed in polar coordinates in the form

$$\theta = Qr\cos\vartheta \quad \text{on } \Sigma. \tag{3.2.54}$$

Following Muskhelishvili [270], in the framework of classical thermoelasticity the temperature field (3.2.54) generates the following stresses

$$t_{\alpha i} = 0,\ t_{33} = (2\nu - 1)\beta Qr\cos\vartheta, \tag{3.2.55}$$

where ν is Poisson's ratio. In what follows we establish the solution of the problem in the context of the linear theory of thermoelastic porous materials. In this case the function F introduced in (3.2.8) becomes

$$F(z) = Qz,$$

so that from (3.2.9) we obtain

$$w_1 = \frac{1}{2}Qr^2\cos 2\vartheta,\ w_1 = \frac{1}{2}Qr^2\sin 2\vartheta. \tag{3.2.56}$$

In view of the results from Section 3.2 (b) the thermoelastic problem reduces to the solving of the isothermal elastic plane strain problem defined by the equations (3.2.14), (3.2.15), (3.2.17) on Σ and the following boundary conditions

$$t^0_{\alpha\beta} n_\beta = 0,\ h^0_\beta n_\beta = -\alpha\tau Q n_1 \quad \text{for } r = a. \tag{3.2.57}$$

It follows from (3.2.38), (3.2.40) that the conditions (3.2.57) can be expressed in the form

$$t_{rr} = 0,\ t_{r\vartheta} = 0,\ \frac{\partial \varphi^0}{\partial r} = -rQ\cos\vartheta \quad \text{for } r = a. \tag{3.2.58}$$

We seek the solution of the isothermal problem in the form

$$u = U(r)\cos\vartheta,\ v = V(r)\sin\vartheta,\ \varphi^0 = \Phi(r)\cos\vartheta, \tag{3.2.59}$$

where U, V and Φ are functions only on r. It follows from (3.2.39), (3.2.40) and (3.2.59) that

$$
\begin{aligned}
t_{rr} &= \left[(\lambda+2\mu)\frac{dU}{dr}+\frac{1}{r}\lambda(U+V)+b\Phi\right]\cos\vartheta,\\
t_{\vartheta\vartheta} &= \left[\lambda\frac{dU}{dr}+\frac{1}{r}(\lambda+2\mu)(U+V)+b\Phi\right]\cos\vartheta,\\
t_{r\vartheta} &= \mu\left[\frac{dV}{dr}-\frac{1}{r}(U+V)\right]\sin\vartheta,\\
h_r &= \alpha\frac{d\Phi}{dr}\cos\vartheta,\quad h_\theta=-\frac{1}{r}\alpha\frac{d\Phi}{dr}\sin\vartheta,\\
g &= -\left\{b\left[\frac{1}{r}\frac{d}{dr}(rU)+\frac{1}{r}V\right]+\xi\Phi\right\}\cos\vartheta.
\end{aligned}
\tag{3.2.60}
$$

If we substitute (3.2.60) into the equilibrium equations (3.2.51), then we obtain the following equations for the unknown functions U, V and Φ

$$
\begin{aligned}
&r^2\frac{d^2U}{dr^2}+r\frac{dU}{dr}-(1+\kappa_1)U+r(1-\kappa_1)\frac{dV}{dr}-(1+\kappa_1)V+r^2\kappa_2\frac{d\Phi}{dr}=0,\\
&\kappa_1\left(r^2\frac{d^2V}{dr^2}+r\frac{dV}{dr}\right)-(1+\kappa_1)V-(1-\kappa_1)\frac{dU}{dr}-(1+\kappa_1)U-\kappa_2 r\Phi=0,\\
&\alpha\left[\frac{1}{r}\frac{d}{dr}\left(r\frac{d\Phi}{dr}\right)-\frac{1}{r^2}\Phi\right]-\frac{1}{r}b\left[\frac{d}{dr}(rU)+V\right]-\xi\Phi=0,
\end{aligned}
\tag{3.2.61}
$$

where

$$\kappa_1=\frac{\mu}{\lambda+2\mu},\quad \kappa_2=\frac{b}{\lambda+2\mu}.$$

We now introduce the independent variable t through the relation

$$t=\ln r \tag{3.2.62}$$

and denote $D=\dfrac{d}{dt}\cdot$ Then, the first two equations from (3.2.61) become

$$
\begin{aligned}
&[D^2-(1+\kappa_1)]U+[(1-\kappa_1)D-(1+\kappa_1)]V=e^{-t}\kappa_2 D\Phi,\\
&[(\kappa_1-1)D-(1+\kappa_1)]U+[\kappa_1 D^2-(1+\kappa_1)]V=e^{t}\kappa_2\Phi.
\end{aligned}
\tag{3.2.63}
$$

The general solution of the homogeneous system (3.2.63) which correspond to a non-rigid displacement is given by

$$
\begin{aligned}
U_0 &= N_1 t+N_2e^{2t}+N_3e^{-2t},\\
V_0 &= -\frac{1-\kappa_1}{1+\kappa_1}N_1-N_1t-\frac{3-\kappa_1}{1-3\kappa_1}N_2e^{2t}+N_3e^{-2t},
\end{aligned}
\tag{3.2.64}
$$

where N_j are arbitrary cosntants. Particular solution of the system (3.2.63) can be seen to be

$$U^* = -\frac{1}{2}\kappa_2(J_1 + e^{-2t}J_2), \quad V^* = \frac{1}{2}\kappa_2(J_1 - e^{-2t}J_2), \tag{3.2.65}$$

where

$$J_1(t) = \int^t e^s\Phi(s)ds, \; J_2(t) = \int^t e^{3s}\Phi(s)ds. \tag{3.2.66}$$

With the help of (3.2.62), (3.2.64) and (3.2.65) we obtain

$$\begin{aligned} U &= N_1 \ln r + N_2 r^2 + N_3 r^{-2} - \frac{1}{2}\kappa_2\left(\int^r \Phi(s)ds + r^{-2}\int^r s^2\Phi(s)ds\right), \\ V &= -\frac{1-\kappa_1}{1+\kappa_1}N_1 + N_1 \ln r - \frac{3-\kappa_1}{1-3\kappa_1}N_2 r^2 + N_3 r^{-2} + \\ &+ \frac{1}{2}\kappa_2\left(\int^r \Phi(s)ds - r^{-2}\int^r s^2\Phi(s)ds\right). \end{aligned} \tag{3.2.67}$$

Since U and V must be finite for $r = 0$, we conclude from (3.2.67) that $N_1 = 0$ and $N_3 = 0$. If we substitute U and V from (3.2.67) into the third equation from (3.2.61), then we obtain the following equation

$$r^2\frac{d^2\Phi}{dr^2} + r\frac{d\Phi}{dr} - (1 + p^2r^2)\Phi = -\frac{8b}{\alpha(1-3\kappa_1)}\kappa_1 r^3 N_2, \tag{3.2.68}$$

where p has been introduced in (3.2.25). We find that

$$\Phi = PN_2 r + N_4 K_1(pr) + N_5 I_1(pr), \tag{3.2.69}$$

where

$$P = 8b\kappa_1[\alpha p^2(1-3\kappa_1)]^{-1}.$$

As Φ must be finite for $r = 0$, we must take $N_4 = 0$. The solution of the system (3.2.61) can be expressed in the form

$$\begin{aligned} U &= \left(1 - \frac{3}{8}P\kappa_2\right)N_2 r^2 - \frac{1}{2p}\kappa_2 N_5[I_0(pr) + I_2(pr)], \\ V &= \left(\frac{1}{8}P\kappa_2 + \frac{\kappa_1 - 3}{1-3\kappa_1}\right)N_2 r^2 + \frac{1}{2p}\kappa_2 N_5[I_0(pr) - I_2(pr)], \\ \Phi &= PN_2 r + N_5 I_1(pr). \end{aligned} \tag{3.2.70}$$

It follows from (3.2.39), (3.2.40) and (3.2.70) that

$$\begin{aligned} t_{rr} &= 2\mu\left[2SN_2 r + \frac{1}{pr}\,\kappa_2 N_5 I_2(pr)\right]\cos\vartheta, \\ t_{r\vartheta} &= 2\mu\left[2SN_2 r + \frac{1}{pr}\,\kappa_2 N_5 I_2(pr)\right]\sin\vartheta, \\ h_r &= \alpha[PN_2 + pN_5 I_1(pr)]\cos\vartheta, \end{aligned} \tag{3.2.71}$$

where

$$S = (\kappa_1\xi - \alpha p^2)[\alpha p^2(1 - 3\kappa_1)]^{-1}.$$

In view of (3.2.71), the boundary conditions (3.2.58) are satisfied if and only if

$$N_2 = \tau\kappa_2 I_2(pa)\Pi^{-1}Q, \quad N_5 = -2\tau pa^2 S\Pi^{-1}Q, \tag{3.2.72}$$

where

$$\Pi = p^2a^2S[I_0(pa) + I_2(pa)] - \kappa_2 P I_2(pa).$$

Thus, the solution of the isothermal problem is given by (3.2.59) where U, V and Φ have the expressions (3.2.70), and N_2 and N_5 are determined by (3.2.72). It follows from (3.2.38) and (3.2.59) that

$$u_1^0 = U\cos^2\vartheta - V\sin^2\vartheta, \quad u_2^0 = \frac{1}{2}\,(U + V)\sin 2\vartheta.$$

On the basis of (3.2.12) and (3.2.56) we obtain the solution of thermoelastic problem in the form

$$\begin{aligned} u_1 &= \left(U + \frac{1}{2}\,QMr^2\right)\cos^2\vartheta - \left(V + \frac{1}{2}\,MQr^2\right)\sin^2\vartheta, \\ u_2 &= \frac{1}{2}\,(U + V + QMr^2)\sin 2\vartheta, \\ \varphi &= (\Phi + \tau Qr)\cos\vartheta. \end{aligned}$$

We note that in the classical theory of thermoelasticity [270] the plane stresses generated by the temperature field $\theta = Qr\cos\vartheta$ are equal to zero. We can investigate the behavior of $t_{rr}, t_{r\vartheta}$ and h_r in the context of the theory of porous thermoelastic bodies. Let us consider a porous material chracterized by the constitutive coefficients (cf. Puri and Cowin [305])

$$\begin{aligned} &\lambda = 15GPa, \quad \mu = \frac{15}{2}\,GPa, \quad \alpha = 9GPa, \quad b = 10GPa, \\ &\xi = \frac{37}{3}\,GPa, \quad \beta = 3GPa, \quad m = 2GPa. \end{aligned}$$

In this case we obtain

$$\tau = \frac{6}{71},\ \kappa_1 = \frac{1}{4},\ \kappa_2 = \frac{1}{3},\ p = 1,\ P = \frac{80}{9},\ S = -\frac{71}{27}. \tag{3.2.73}$$

From (3.2.72) and (3.2.73) we find that

$$N_2 = \frac{2}{71}\,\Pi^{-1} I_0(a) Q, \quad N_5 = \frac{4}{9}\,a^2 \Pi^{-1} Q,$$

where

$$\Pi = -\frac{1}{27}[71a^2 I_0(a) + (71a^2 + 80) I_2(a)].$$

Thus, we obtain the functions

$$\begin{aligned} t_{rr} &= \frac{20}{9\Pi r}\,Q[a^2 I_2(r) - r^2 I_2(a)]\cos\vartheta, \\ t_{r\vartheta} &= \frac{20}{9\Pi r}\,Q[a^2 I_2(r) - r^2 I_2(a)]\sin\vartheta, \\ h_r &= \frac{2}{\Pi}\,Q\left\{\left[\frac{80}{71}\,I_2(a) + a^2[I_0(r) + I_2(r)\right]\right\}\cos\vartheta. \end{aligned}$$

Some diagrams have been presented in [190].

3.3 Exercises

In this section we consider the linear theory of thermoelastostatics for homogeneous and isotropic materials with voids. As in the classical theory, we assume that the mechanical loadings are absent. The equations (1.5.17) reduce to

$$\begin{aligned} \mu\Delta u_i + (\lambda+\mu) u_{j,ji} + b\varphi_{,i} - b\theta_{,i} &= 0, \\ \alpha\Delta\varphi - b u_{j,j} - \xi\varphi + m\theta &= 0 \end{aligned} \tag{3.3.1}$$

and

$$k\Delta\theta = -\rho_0 S. \tag{3.3.2}$$

We consider the boundary conditions in the form

$$t_{ji} n_j = 0, \quad \frac{\partial\varphi}{\partial n} = 0, \quad \theta = \widetilde{\theta} \ \text{ on } \partial B. \tag{3.3.3}$$

1. *Investigate the effect of a concentrated source of heat applied in an infinite space.*

Solution. We consider the case of a concentrated source of heat applied at the point $y(y_i)$ of a body occupying the entire space. Let $S = Q\delta(x -$

$y)$, where Q is a given constant and $\delta(\cdot)$ is the Dirac delta. From (3.3.2) we get

$$\theta = \frac{\rho_0 Q}{4\pi r}, \tag{3.3.4}$$

where $r = [(x_i - y_i)(x_i - y_i)]^{1/2}$. We seek the displacement vector in the form $u_i = \Phi_{,i}$ where Φ is an unknown function. The equations (3.3.1) are satisfied if the functions Φ and φ satisfy the equations

$$\Delta\Phi = \frac{1}{\lambda + 2\mu}(\beta\theta - b\varphi), \quad \Delta\varphi - p^2\varphi = \zeta\theta,$$

where p^2 and ζ are given by (3.2.25). It follows from (3.3.4) that

$$\begin{aligned} \Phi &= \frac{\rho_0 Q}{4\pi k p^2(\lambda + 2\mu)}\left[\frac{\beta\xi - mb}{2\alpha} r + \frac{\zeta b}{p^2 r}(1 - e^{-pr})\right], \\ \varphi &= -\frac{\rho_0 \zeta Q}{4\pi k p^2 r}(1 - e^{-pr}), \quad (p > 0). \end{aligned}$$

We note that in the classical theory of thermoelasticity [285, p.162] the function Φ has the form Cr where C is a constant. In the case of thermoelastic materials with voids the displacement field contains new terms characterizing the influence of the material porosity.

2. *A homogeneous and isotropic thermoelastic material with voids occupies the region $B = \{\boldsymbol{x} : a_1^2 < x_1^2 + x_2^2 + x_3^2 < a_2^2\}$, $a_1 > 0$, $a_2 > 0$. The body is in equilibrium in the absence of body forces and heat sources. The outer surface is subjected to the constant temperature T_2, and the inner surface is maintained at the constant temperature T_1. Investigate the deformation of the body when the surface of the shell is free of surface forces.*

Solution. In this case we have

$$\theta = \frac{1}{r} C_1 + C_2, \tag{3.3.5}$$

where $r = (x_1^2 + x_2^2 + x_3^2)^{1/2}$ and

$$C_1 = \frac{a_1 a_2 (T_2 - T_1)}{a_2 - a_1}, \quad C_2 = \frac{T_2 a_2 - T_1 a_1}{a_2 - a_1}.$$

We seek the solution of the system (3.3.1) in the form

$$u_i = x_i \Psi(r), \quad \varphi = \varphi(r). \tag{3.3.6}$$

The substitution from (3.3.6) in (3.3.1) yields the equations

$$
\begin{aligned}
(r^3\Psi)' &= \frac{r^2}{\lambda + 2\mu}(\beta\theta - b\varphi) + A_1 r^2, \\
\Delta\varphi - p^2\varphi &= \frac{b}{\alpha}A_1 + \zeta\theta,
\end{aligned}
\tag{3.3.7}
$$

where prime denotes the derivative with respect to r, A_1 is an unknown constant and p^2 and ζ are given by (3.2.25). From (3.3.7) we obtain

$$
\begin{aligned}
\Psi &= \frac{\xi}{3\alpha p^2}A_1 + \frac{1}{r^3}A_2 + H(r), \\
\varphi &= \varphi_0 - \frac{1}{\alpha p^2}(\alpha\zeta\theta + bA_1),
\end{aligned}
\tag{3.3.8}
$$

where

$$
\begin{aligned}
\varphi_0 &= \frac{1}{r}(B_1 e^{-pr} + B_2 e^{pr}), \\
H &= \frac{\gamma_0}{(\lambda + 2\mu)r^3}\int_0^r r^2\theta(r)dr - \frac{b}{p^2(\lambda + 2\mu)r}\varphi_0'(r), \\
\gamma_0 &= \frac{(\lambda + 2\mu)(\beta\xi - bm)}{\xi(\lambda + 2\mu) - b^2}
\end{aligned}
\tag{3.3.9}
$$

and A_2, B_1, B_2 are arbitrary constants. A simple calculation shows that the stress $P = t_{ij}n_i n_j$ in the radial direction $n_i = x_i/r$ is

$$
P = \tau A_1 - \frac{4\mu}{r^3}A_2 - 4\mu H(r), \tag{3.3.10}
$$

where

$$
\tau = \frac{1}{3\alpha p^2}[(3\lambda + 2\mu)\xi - 3b^2].
$$

For the determination of the constants B_α we have the boundary conditions

$$
\varphi'(r) = 0 \text{ for } r = a_1 \text{ and } r = a_2. \tag{3.3.11}
$$

We obtain

$$
\begin{aligned}
B_1 &= \frac{\zeta C_1}{p^2 D}[e^{pa_2}(1 + pa_2) - e^{pa_1}(1 + pa_1)], \\
B_2 &= \frac{\zeta C_1}{p^2 D}[e^{-pa_1}(1 - pa_1) - e^{-pa_2}(1 - pa_2)],
\end{aligned}
\tag{3.3.12}
$$

where

$$
D = e^{p(a_2 - a_1)}(1 - pa_1)(1 + pa_2) - e^{-p(a_2 - a_1)}(1 - pa_2)(1 + pa_1).
$$

By using (3.3.5), we get

$$
\begin{aligned}
H = {} & \frac{\gamma_0}{6(\lambda+2\mu)}\left(\frac{3}{r}C_1+2C_2\right)+ \\
& +\frac{b}{p^2(\lambda+2\mu)r^3}[B_1e^{-pr}(pr+1)-B_2e^{pr}(pr-1)].
\end{aligned}
\tag{3.3.13}
$$

The surface of the shell is free of external tractions if and only if

$$P(r)=0 \text{ for } r=a_1 \text{ and } r=a_2. \tag{3.3.14}$$

From (3.3.10) and (3.3.14) we find that

$$
\begin{aligned}
A_1 &= \frac{4\mu[a_2^3H(a_2)-a_1^3H(a_1)]}{(a_2^3-a_1^3)\tau}, \\
A_2 &= \frac{a_1^3a_2^3[H(a_2)-H(a_1)]}{a_2^3-a_1^3}.
\end{aligned}
\tag{3.3.15}
$$

It follows from (2.1.7) that $\tau>0$. With the help of (3.3.11) we obtain

$$H(a_\alpha)=\frac{\gamma_0}{6(\lambda+2\mu)}\left[3\left(\frac{1}{a_\alpha}+\frac{2b\zeta}{a_\alpha^3p^4\gamma_0}\right)C_1+2C_2\right], \quad (\alpha=1,2). \tag{3.3.16}$$

By using (3.3.16) we can express the constants A_1 and A_2 in terms of the temperatures T_1 and T_2. Thus, the functions Ψ and φ are given by (3.3.8) where φ_0, H are defined by (3.3.9), (3.3.12), (3.3.13), and the constants A_α are uniquely determined from (3.3.15), (3.3.16).

If $T_1=T_2=T^*=$ const., then $C_1=0$ and $C_2=T^*$, and (3.3.12) implies that $\varphi_0=0$. In this case we have

$$\theta=T^*,\ \Psi=\frac{\beta\xi-mb}{\xi(3\lambda+2\mu)-3b^2}T^*,\ \varphi=\frac{(3\lambda+2\mu)m-3b\beta}{\xi(3\lambda+2\mu)-3b^2}T^*.$$

The radial displacement is given by

$$u=\frac{\beta\xi-mb}{\xi(3\lambda+2\mu)-3b^2}T^*r.$$

The results established here can be used to obtain the solution in the case of the elastic space with a spherical cavity.

3. *A homogeneous and isotropic thermoelastic material occupies the region* $\{\boldsymbol{x}: R_1^2<x_1^2+x_2^2<R_2^2,\ 0<x_3<h\}$, *where* $R_2>R_1>0$. *The body is in equilibrium in the absence of body forces and heat sources. The inner surface of the tube is maintained at the constant temperature*

T_1, and the outer surface is subjected to the constant temperature T_2. Investigate the thermoelastic plane strain parallel to the x_1Ox_2 plane when the surface of the tube is free of surface forces.

Solution. This problem has been studied in Section 3.2 (c). We present here another method to solve this problem. The temperature field is given by (3.2.32). We seek the solution of the plane problem in the form

$$u_\alpha = x_\alpha \chi(r), \quad \varphi = \varphi(r), \tag{3.3.17}$$

where $r = (x_1^2 + x_2^2)^{1/2}$, and χ and φ are unknown functions. The equations (3.2.7) are satisfied if the functions χ and φ satisfy the equations

$$\begin{aligned}(r^2\chi)' &= \frac{r}{\lambda + 2\mu}(\beta\theta - b\varphi) + M_1 r,\\ \Delta\varphi - p^2\varphi &= \frac{b}{\alpha} M_1 + \zeta\theta,\end{aligned} \tag{3.3.18}$$

where M_1 is an arbitrary constant. It follows from (3.3.18) that

$$\begin{aligned}\chi &= \frac{\xi}{2\alpha p^2} M_1 + \frac{1}{r} M_2 + F(r),\\ \varphi &= \varphi_0(r) - \frac{1}{\alpha p^2}(\alpha\zeta\theta + bM_1),\end{aligned} \tag{3.3.19}$$

where

$$\begin{aligned}\varphi_0 &= E_1 I_0(pr) + E_2 K_0(pr),\\ F &= \frac{\gamma_0}{(\lambda + 2\mu)r^2}\int_0^r r\theta(r)dr - \frac{b}{(\lambda + 2\mu)p^2 r}\varphi_0'(r),\end{aligned} \tag{3.3.20}$$

M_2, E_1, E_2 are arbitrary constants, and I_n, K_n are the modified Bessel functions of order n. The radial stress is given by

$$\sigma = 2\left(\tau_0 M_1 - \frac{\mu}{r^2} M_2 - \mu F\right),$$

where

$$\tau_0 = \frac{1}{2\alpha p^2}[(\lambda + \mu)\xi - b^2].$$

The constants E_1 and E_2 are determined by the conditions $\varphi'(R_\alpha) = 0$, $(\alpha = 1, 2)$. We obtain

$$\begin{aligned}E_1 &= -\frac{\zeta C_1}{R_1 R_2 d_0}[R_2 K_1(pR_2) - R_1 K_1(pR_1)],\\ E_2 &= -\frac{\zeta C_1}{R_1 R_2 d_0}[R_2 I_1(pR_2) - R_1 I_1(pR_1)],\end{aligned}$$

where

$$d_0 = I_1(pR_2)K_1(pR_1) - I_1(pR_1)K_1(pR_2).$$

The constants M_1 and M_2 are found from the condition that $\sigma(r) = 0$ for $r = R_1$ and $r = R_2$. We find that

$$M_1 = \frac{\mu[R_1^2 F(R_1) - R_2^2 F(R_2)]}{\zeta(R_2^2 - R_1^2)},$$

$$M_2 = \frac{R_1^2 R_2^2 [F(R_2) - F(R_1)]}{R_2^2 - R_1^2}.$$

By a simple calculation we can determine the functions $t_{a\beta}$ and h_α.

If $T_1 = T_2 = T^*$, then $\theta = T^*$ and

$$\chi = \frac{\beta\xi - mb}{2[(\lambda + \mu)\xi - b^2]} T^*, \quad \varphi = \frac{(\lambda + \mu)m - b\beta}{(\lambda + \mu)\xi - b^2} T^*.$$

In this case the radial displacement is given by

$$u = \frac{\beta\xi - mb}{2[(\lambda + \mu)\xi - b^2]} T^* r.$$

In the isothermal case, the problems of the thick walled spherical and circular cylinder shells under internal and external pressure have been solved by Cowin and Puri [74].

Chapter 4

PRESTRESSED THERMOELASTIC BODIES

4.1 Equations of Perturbed Motion

In this section we derive the equations governing infinitesimal thermoelastic deformations superimposed on large deformations of materials with voids. The theory of infinitesimal deformations superposed on finite deformation has given impetus to theoretical research into the equations of elastic bodies for which little or no information is known concerning the elasticities (see Knops and Payne [220], [221]). The theory also establishes a basis for discussing stability of finite deformations (see Knops and Wilkes [222]). In what follows we present the theory of infinitesimal deformations superposed on finite deformations of thermoelastic materials with voids [183]. In the context of the classical theory, the theory of an elastic body, deformed from a state of zero stress and strain and uniform temperature, which is subsequently subjected to small displacement and temperature distribution, was initiated by Green [124].

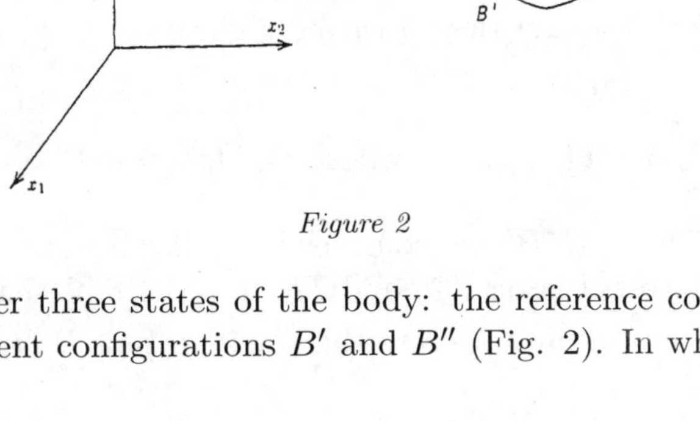

Figure 2

We consider three states of the body: the reference configuration B and two present configurations B' and B'' (Fig. 2). In what follows we wish to look at perturbations of the motion described by the equations of the Sections 1.1-1.3. We call B' the *primary* state and B'' the *secondary* state. Moreover, we shall designate as *incremental* those quantities associated with a difference of motion between the secondary and primary states. If the point X in the reference configuration B moves to x_i in the primary state and to y_i in the secondary, then $u_i = y_i - x_i$ is the incremental displacement. Thermodynamic quantities and forces associated with the secondary state will be denoted with an asterisk, and the position coordinates of the particle X at time t in B'' will be denoted by y_i. We have

$$y_i = y_i(X_1, X_2, X_3, t), \tag{4.1.1}$$

with

$$J^* = \det(y_{i,A}) \neq 0.$$

Now let

$$\theta = T^* - T, \quad \varphi = \nu^* - \nu. \tag{4.1.2}$$

We assume that

$$u_i = \varepsilon u_i', \quad \theta = \varepsilon\theta', \quad \varphi = \varepsilon\varphi', \tag{4.1.3}$$

where ε is a constant small enough for squares and higher powers to be neglected, and u_i', φ' and θ' are independent of ε. The problem consists in establishing the equations, boundary conditions, and initial conditions for u_i, θ and φ when the functions $x_i(X_A, t)$, $\nu(X_A, t)$, $T(X_A, t)$, $(X_A, t) \in B\times(t_0, t_1)$, and the loadings corresponding to primary and secondary states are known.

If we refer all quantities to the configuration B, then we have

$$\begin{gathered} u_i = u_i(X_A, t), \ \theta = \theta(X_A, t), \ \varphi = \varphi(X_A, t), \\ (X_A, t) \in B\times(t_0, t_1). \end{gathered} \tag{4.1.4}$$

When the primary state B' is a configuration of equilibrium it is convenient to refer all quantities to the configuration B'. In this case we take

$$u_i = u_i(x_j, t), \ \theta = \theta(x_j, t), \ \varphi = \varphi(x_j, t), \quad (x_j, t) \in B'\times(t_0, t_1). \tag{4.1.5}$$

In the secondary state B'' we consider the following stress tensors: t_{ij}^* is the Cauchy stress tensor; T_{Ki}^* and T_{KL}^* are the Piola-Kirchhoff stress tensors measured per unit area in the configuration B; $T_{rs}^{*(1)}$ and $T_{rs}^{*(2)}$

are, respectively, the first and the second Piola-Kirchhoff stress tensors measured per unit area in the configuration B'. We have

$$t^*_{ij} = \frac{1}{J^*} y_{i,A} T^*_{Ai} = \frac{1}{J^*} y_{i,A} y_{j,B} T^*_{AB},$$
$$t^*_{ij} = \frac{1}{J'} y_{i,s} T^{*(1)}_{sj} = \frac{1}{J'} y_{i,s} y_{j,r} T^{*(2)}_{sr}, \tag{4.1.6}$$

where

$$J' = \det(y_{i,j}) = 1 + u_{r,r}, \quad J^* = JJ'. \tag{4.1.7}$$

In the present configuration B'' we consider the following heat flux vectors: q^*_i is the heat flux across the planes y_i = const., measured per unit area of these surfaces and per unit time; Q^*_A is the heat flux across surfaces in B'' that in the reference configuration B were coordinate planes perpendicular to the X_A-axis, measured per unit area of these planes and per unit time; $\overline{Q}^*_i$ is the heat flux across surfaces in B'' that in the present configuration B' were coordinate planes perpendicular to the x_i-axis, measured per unit area of these planes and per unit time. Moreover, we consider in B'' the following equilibrated stresses: h^*_i – associated with the planes y_i = const., measured per unit area of these surfaces; H^*_A – associated with surfaces in B'' that in the reference configuration B were coordinate planes perpendicular to the X_A-axis measured per unit area of these planes; $\overline{H}^*_i$ – associated with the surfaces in B'' that in the present configuration B' were coordinate planes perpendicular to the x_i-axis, measured per unit area of these planes. We have

$$q^*_i = \frac{1}{J^*} y_{i,A} Q^*_A = \frac{1}{J'} y_{i,j} \overline{Q}^*_j,$$
$$h^*_i = \frac{1}{J^*} y_{i,A} H^*_A = \frac{1}{J'} y_{i,j} \overline{H}^*_j. \tag{4.1.8}$$

The equations of motion of the secondary state can be written in the forms

$$T^*_{Ai,A} + \rho_0 f^*_i = \rho_0 \ddot{y}_i \quad \text{on } B\times(t_0, t_1), \tag{4.1.9}$$

$$T^{*(1)}_{ji,j} + \rho f^*_i = \rho \ddot{y}_i \quad \text{on } B'\times(t_0, t_1). \tag{4.1.10}$$

The corresponding energy equation can be written in one of the forms

$$\rho_0 T^* \dot{\eta}^* = Q^*_{A,A} + \rho_0 S^* \quad \text{on } B\times(t_0, t_1), \tag{4.1.11}$$

$$\rho T^* \dot{\eta}^* = \overline{Q}^*_{i,i} + \rho S^* \quad \text{on } B'\times(t_0, t_1). \tag{4.1.12}$$

The balance of equilibrated force leads to

$$H^*_{A,A} + g^* + \rho_0 \ell^* = \rho_0 \kappa \ddot{\nu}^* \quad \text{on } B\times(t_0, t_1), \tag{4.1.13}$$

$$\overline{H}^*_{i,i} + \bar{g}^* + \rho \ell^* = \rho \kappa \ddot{\nu}^* \quad \text{on } B'\times(t_0, t_1), \tag{4.1.14}$$

where

$$\bar{g}^* = \frac{1}{J} g^*. \tag{4.1.15}$$

From the constitutive equations we get

$$\begin{aligned}
&\sigma^* = \widehat{\sigma}(E^*_{AB}, T^*, \nu^*, \nu^*_{,A}, X_M),\\
&T^*_{AB} = \frac{\partial \sigma^*}{\partial E^*_{AB}}, \quad \rho_0 \eta^* = -\frac{\partial \sigma^*}{\partial T^*}, \quad H^*_A = \frac{\partial \sigma^*}{\partial \nu^*_{,A}},\\
&g^* = -\frac{\partial \sigma^*}{\partial \nu^*}, \quad Q^*_A = \widehat{Q}_A(E^*_{MN}, T^*, T^*_{,A}, \nu^*, \nu^*_{,A}, X_K),
\end{aligned} \tag{4.1.16}$$

where

$$2E^*_{AB} = y_{i,A} y_{i,B} - \delta_{AB}. \tag{4.1.17}$$

The entropy production inequality implies

$$q_i T_{,i} \geq 0, \quad Q_A T_{,A} \geq 0, \quad Q^*_A T^*_{,A} \geq 0. \tag{4.1.18}$$

We consider the following boundary conditions for the secondary deformation

$$\begin{aligned}
&y_i = \widehat{y}_i \text{ on } \overline{\Sigma}_1\times[t_0, t_1), \ \nu^* = \widehat{\nu}^* \text{ on } \overline{\Sigma}_3\times[t_0, t_1), \ T^* = \widehat{T}^* \text{ on } \overline{\Sigma}_5\times[t_0, t_1),\\
&T^*_{Ai} N_A = \widehat{T}^*_i \text{ on } \Sigma_2\times[t_0, t_1), \ H^*_A N_A = \widehat{h}^* \text{ on } \Sigma_4\times[t_0, t_1),\\
&Q^*_A N_A = \widehat{q}^* \text{ on } \Sigma_6\times[t_0, t_1),
\end{aligned} \tag{4.1.19}$$

where $\widehat{y}_i, \widehat{\nu}^*, \widehat{T}^*, \widehat{T}^*_i, \widehat{h}^*, \widehat{q}^*$ are prescribed. We consider also the conditions

$$\begin{aligned}
&y_i = \widetilde{y}_i \text{ on } \overline{S}_1\times[t_0, t_1), \ \nu^* = \widetilde{\nu}^* \text{ on } \overline{S}_3\times[t_0, t_1), \ T^* = \widetilde{T}^* \text{ on } \overline{S}_5\times[t_0, t_1),\\
&T^{*(1)}_{ji} n_j = \widetilde{T}^*_i \text{ on } S_2\times[t_0, t_1), \ \overline{H}^*_i n_i = \widetilde{h}^* \text{ on } S_4\times[t_0, t_1),\\
&\overline{Q}^*_i n_i = \widetilde{q}^* \text{ on } S_6\times[t_0, t_1).
\end{aligned} \tag{4.1.20}$$

Let us derive the equations, boundary conditions and initial conditions for the functions u_i, θ and φ. By using the equations (1.1.16), (4.1.17) and the relation

$$y_{i,A} = x_{i,A} + u_{i,A} = x_{i,A} + u_{i,j} x_{j,A},$$

we obtain

$$E^*_{AB} = E_{AB} + \frac{1}{2}\,(x_{i,B}u_{i,A} + x_{i,A}u_{i,B}). \tag{4.1.21}$$

If we denote

$$\varepsilon_{ij} = \frac{1}{2}\,(u_{i,j} + u_{j,i}), \tag{4.1.22}$$

then the equation (4.1.21) reduces to

$$E^*_{AB} = E_{AB} + \varepsilon_{ij}x_{i,A}x_{j,B}. \tag{4.1.23}$$

If we take into account (4.1.4) and (4.1.23), then to a second-order approximation we obtain

$$\begin{aligned}
\frac{\partial\sigma^*}{\partial E^*_{AB}} &= \frac{\partial\sigma}{\partial E_{AB}} + A_{ABMN}(E^*_{MN} - E_{MN}) + D_{ABM}\varphi_{,M} + \\
&\qquad\qquad + B_{AB}\varphi - G_{AB}\theta,\\
\frac{\partial\sigma^*}{\partial T^*} &= \frac{\partial\sigma}{\partial T} - G_{AB}(E^*_{AB} - E_{AB}) - A_K\varphi_{,K} - B\varphi - A\theta,\\
\frac{\partial\sigma^*}{\partial\nu^*} &= \frac{\partial\sigma}{\partial\nu} + B_{KL}(E^*_{KL} - E_{KL}) + B_K\varphi_{,K} + \xi\varphi - B_0\theta,
\end{aligned} \tag{4.1.24}$$

where

$$\begin{aligned}
&A_{KLMN} = \frac{\partial^2\sigma}{\partial E_{KL}\partial E_{MN}},\ D_{ABC} = \frac{\partial^2\sigma}{\partial E_{AB}\partial\nu_{,C}}, B_{AB} = \frac{\partial^2\sigma}{\partial E_{AB}\partial\nu},\\
&G_{AB} = -\frac{\partial^2\sigma}{\partial E_{AB}\partial T},\ A_C = -\frac{\partial^2\sigma}{\partial T\partial\nu_{,C}},\ B_0 = -\frac{\partial^2\sigma}{\partial T\partial\nu},\\
&A = -\frac{\partial^2\sigma}{\partial T^2},\ A_{KL} = \frac{\partial^2\sigma}{\partial\nu_{,K}\partial\nu_{,L}}, B_C = \frac{\partial^2\sigma}{\partial\nu_{,C}\partial\nu},\ \xi = \frac{\partial^2\sigma}{\partial\nu^2}.
\end{aligned} \tag{4.1.25}$$

Obviously

$$\begin{aligned}
&A_{KLMN} = A_{MNKL} = A_{LKMN},\ B_{KL} = B_{LK},\\
&D_{ABC} = D_{BAC},\ G_{AB} = G_{BA},\ A_{KL} = A_{LK}.
\end{aligned} \tag{4.1.26}$$

From (1.2.9), (4.1.16), (4.1.21) and (4.1.24) we get

$$\begin{aligned}
&T^*_{AB} = T_{AB} + A_{ABMN}x_{i,M}u_{i,N} + D_{ABK}\varphi_{,K} + B_{AB}\varphi - G_{AB}\theta,\\
&\rho_0\eta^* = \rho_0\eta + G_{AB}x_{i,A}u_{i,B} + A_K\varphi_{,K} + B_0\varphi + A\theta,\\
&H^*_A = H_A + D_{KLA}x_{i,K}u_{i,L} + A_{AK}\varphi_{,K} + B_A\varphi - A_A\theta,\\
&g^* = g - B_{KL}x_{i,K}u_{i,L} - B_K\varphi_{,K} - \xi\varphi + B_0\theta,
\end{aligned} \tag{4.1.27}$$

or

$$\begin{aligned}
&T^*_{KL}=T_{KL}+A_{KLMN}x_{i,M}x_{j,N}\varepsilon_{ij}+D_{KLM}x_{i,M}\varphi_{,i}+B_{KL}\varphi-G_{KL}\theta,\\
&\rho_0\eta^* = \rho_0\eta + G_{AB}x_{i,A}x_{j,B}u_{i,j} + A_K x_{i,K}\varphi_{,i} + B_0\varphi + A\theta,\\
&H^*_K = H_K + D_{MNK}x_{i,M}x_{j,N}u_{i,j} + A_{KL}x_{i,L}\varphi_{,i} + B_K\varphi - A_K\theta,\\
&g^* = g - B_{KL}x_{i,K}x_{j,L}u_{i,j} - B_K x_{i,K}\varphi_{,i} - \xi\varphi + B_0\theta.
\end{aligned} \tag{4.1.28}$$

From (4.1.27) and the relation

$$T^*_{Ai} = (x_{i,B} + u_{i,B})T^*_{AB},$$

it follows that

$$T^*_{Ki} = T_{Ki}+A_{KijN}u_{j,N}+D_{KiM}\varphi_{,M}+B_{Ki}\varphi-G_{Ki}\theta+u_{i,B}T_{KB}, \tag{4.1.29}$$

where

$$\begin{aligned}
&A_{KijN} = A_{KLMN}x_{i,L}x_{j,M},\ D_{KiM} = D_{KBM}x_{i,B},\\
&B_{Ki} = B_{KM}x_{i,M},\ G_{Ki} = G_{KN}x_{i,N}.
\end{aligned} \tag{4.1.30}$$

If we denote

$$P_{Ki} = T^*_{Ki} - T_{Ki}, \tag{4.1.31}$$

then we have

$$P_{Ki} = A_{KijN}u_{j,N} + D_{KiM}\varphi_{,M} + B_{Ki}\varphi - G_{Ki}\theta + u_{i,B}T_{KB}. \tag{4.1.32}$$

This relation can be written in the form

$$P_{Ki} = E_{iKjL}u_{j,L} + D_{KiM}\varphi_{,M} + B_{Ki}\varphi - G_{Ki}\theta, \tag{4.1.33}$$

where

$$E_{iKjL} = A_{KijL} + \delta_{ij}T_{KL}. \tag{4.1.34}$$

We note that

$$E_{iKjL} = E_{jLiK}. \tag{4.1.35}$$

Let us establish the relation between $T^{*(1)}_{ij}$ and t_{ij}. From (4.1.6) it follows that

$$T^{*(1)}_{ij} = \frac{1}{J}x_{i,L}T^*_{Lj} = \frac{1}{J}x_{i,L}x_{r,K}T^*_{LK}y_{j,r}.$$

By using (1.3.8) and (4.1.28), from the above relations we obtain

$$T^{*(1)}_{ij} = t_{ij} + C_{ijrs}\varepsilon_{rs} + d_{ijk}\varphi_{,k} + b_{ij}\varphi - \beta_{ij}\theta + t_{ir}u_{j,r}, \tag{4.1.36}$$

where

$$C_{ijrs} = \frac{1}{J}\, x_{i,K}x_{j,L}x_{r,M}x_{s,N}A_{KLMN},$$
$$d_{ijk} = \frac{1}{J}\, x_{i,K}x_{j,L}x_{k,M}D_{KLM}, \tag{4.1.37}$$
$$b_{ij} = \frac{1}{J}\, x_{i,K}x_{j,L}B_{KL},\ \beta_{ij} = \frac{1}{J}\, x_{i,K}x_{j,L}G_{KL}.$$

We have

$$C_{ijrs} = C_{rsij} = C_{jirs},\ d_{ijk} = d_{jik},\ b_{ij} = b_{ji},\ \beta_{ij} = \beta_{ji}. \tag{4.1.38}$$

If we introduce the notation

$$s_{ij} = T^{*(1)}_{ij} - t_{ij}, \tag{4.1.39}$$

then we get

$$s_{ij} = C_{ijrs}\varepsilon_{rs} + d_{ijk}\varphi_{,k} + b_{ij}\varphi - \beta_{ij}\theta + t_{ir}u_{j,r}. \tag{4.1.40}$$

The relation (4.1.40) can be written in the form

$$s_{ji} = D_{ijrs}u_{r,s} + d_{ijk}\varphi_{,k} + b_{ij}\varphi - \beta_{ij}\theta, \tag{4.1.41}$$

where

$$D_{ijrs} = C_{ijrs} + t_{js}\delta_{ir}. \tag{4.1.42}$$

From (4.1.38) we obtain

$$D_{ijrs} = D_{rsij}. \tag{4.1.43}$$

It is easy to obtain the relations

$$\begin{aligned} t^*_{ij} &= t_{ij} - t_{ij}u_{r,r} + t_{sj}u_{j,s} + s_{ij}, \\ t^*_{ij} &= T^{*(1)}_{ij} - u_{r,r}T^{*(1)}_{ij} + u_{i,s}T^{*(1)}_{sj}. \end{aligned} \tag{4.1.44}$$

We introduce the notations

$$M_A{=}H^*_A{-}H_A,\ \sigma_i{=}H^*_i{-}h_i,\ \chi{=}g^*{-}g,\ m{=}\frac{1}{J}\,\chi,\ \gamma{=}\rho_0(\eta^* - \eta). \tag{4.1.45}$$

From (4.1.27) and (4.1.28) we get

$$\begin{aligned} \gamma &= G_{Ai}u_{i,A} + A_K\varphi_{,K} + B_0\varphi + A\theta, \\ \frac{1}{J}\,\gamma &= \beta_{ij}\varepsilon_{ij} + a_i\varphi_{,i} + b\varphi + a\theta, \\ \chi &= -B_{Li}u_{i,L} - B_K\varphi_{,K} - \xi\varphi + B_0\theta, \\ m &= -b_{ij}u_{i,j} - b_i\varphi_{,i} - \zeta\varphi + b\theta, \end{aligned} \tag{4.1.46}$$

where

$$a_i = \frac{1}{J}\,x_{i,K}A_K,\ a = \frac{1}{J}\,A,\ b = \frac{1}{J}\,B_0,\ b_i = \frac{1}{J}\,x_{i,K}B_K,\ \zeta = \frac{1}{J}\,\xi. \quad (4.1.47)$$

In view of (4.1.8) we obtain

$$\overline{H}_i^* = \frac{1}{J}\,x_{i,A}H_A^*. \quad (4.1.48)$$

By using (1.3.8), (1.3.10), (4.1.27), (4.1.28), (4.1.45) and (4.1.48) we arrive at

$$\begin{aligned} M_K &= D_{LiK}u_{i,L} + A_{KL}\varphi_{,L} + B_K\varphi - A_K\theta, \\ \sigma_i &= d_{sri}u_{s,r} + a_{ij}\varphi_{,j} + b_i\varphi - a_i\theta, \end{aligned} \quad (4.1.49)$$

where

$$a_{ij} = \frac{1}{J}\,x_{i,K}x_{j,L}A_{KL}. \quad (4.1.50)$$

Let us derive the relations between the heat flux vectors.From (4.1.16) we obtain

$$\begin{aligned} Q_A^* = Q_A + R_{AMN}(E_{MN}^* - E_{MN}) + D_A\theta+ \\ +K_{AM}\theta_{,M} + E_A\varphi + L_{AM}\varphi_{,M}, \end{aligned} \quad (4.1.51)$$

where

$$\begin{aligned} R_{AMN} &= \frac{\partial Q_A}{\partial E_{MN}} = R_{ANM},\ D_A = \frac{\partial Q_A}{\partial T}, \\ K_{AM} &= \frac{\partial Q_A}{\partial T_{,M}},\ \ E_A = \frac{\partial Q_A}{\partial \nu},\ \ L_{AM} = \frac{\partial Q_A}{\partial \nu_{,M}}. \end{aligned} \quad (4.1.52)$$

Let us note that (1.2.11) implies

$$R_{AMN} = 0,\ \ D_A = 0,\ \ E_A = 0,\ \ L_{AM} = 0\ \text{ if }\ T_{,L} = 0. \quad (4.1.53)$$

With the help of (4.1.21) and (4.1.23), the relation (4.1.51) reduces to

$$\begin{aligned} Q_A^* = Q_A + R_{AMN}x_{i,M}u_{i,N} + D_A\theta+ \\ +K_{AM}\theta_{,M} + E_A\varphi + L_{AM}\varphi_{,M}, \end{aligned} \quad (4.1.54)$$

or,

$$\begin{aligned} Q_A^* = Q_A + R_{AMN}x_{i,M}x_{j,N}\varepsilon_{ij} + D_A\theta+ \\ +K_{AM}x_{i,M}\theta_{,i} + E_A\varphi + L_{AM}x_{i,M}\varphi_{,i}. \end{aligned} \quad (4.1.55)$$

From (4.1.8) we get

$$\overline{Q_i^*} = \frac{1}{J}\, x_{i,A} Q_A^*,$$

and, by using (4.1.55), we obtain

$$\overline{Q_i^*} = q_i + h_{ijk}\varepsilon_{jk} + r_i\theta + k_{ij}\theta_{,j} + s_i\varphi + \ell_{ij}\varphi_{,j}, \tag{4.1.56}$$

where

$$\begin{aligned} h_{ijk} &= \frac{1}{J}\, x_{i,A}x_{j,M}x_{k,N}R_{AMN}, \quad & r_i &= \frac{1}{J}\, x_{i,A}D_A, \\ k_{ij} &= \frac{1}{J}\, x_{i,A}x_{j,M}K_{AM}, & s_i &= \frac{1}{J}\, x_{i,A}E_A, \\ \ell_{ij} &= \frac{1}{J}\, x_{i,A}x_{j,M}L_{AM}. \end{aligned} \tag{4.1.57}$$

Clearly,

$$h_{ijk} = 0,\ r_i = 0,\ s_i = 0,\ \ell_{ij} = 0 \ \text{ if } \ T_{,i} = 0. \tag{4.1.58}$$

If we denote

$$\Phi_A = Q_A^* - Q_A, \quad \psi_i = \overline{Q_i^*} - q_i, \tag{4.1.59}$$

then

$$\begin{aligned} \Phi_A &= R_{AiN}u_{i,N} + D_A\theta + K_{AM}\theta_{,M} + E_A\varphi + L_{AM}\varphi_{,M}, \\ \psi_i &= h_{ijk}\varepsilon_{jk} + r_i\theta + k_{ij}\theta + s_i\varphi + \ell_{ij}\varphi_{,j}, \end{aligned} \tag{4.1.60}$$

where $R_{AiN} = R_{AMN}x_{i,M}$. From (4.1.18) we obtain

$$\overline{Q_i^*}T_{,i}^* \geq 0,$$

so that we have

$$\begin{aligned} q_iT_{,i}+h_{ijk}\varepsilon_{jk}T_{,i}+r_iT_{,i}\theta+k_{ij}\theta_{,j}T_{,i}+s_i\varphi T_{,i}+\ell_{ij}\varphi_{,j}T_{,i}+q_i\theta_{,i}+ \\ +h_{ijk}\varepsilon_{jk}\theta_{,i} + r_i\theta\theta_{,i} + k_{ij}\theta_{,j}\theta_{,i} + s_i\varphi\theta_{,i} + \ell_{ij}\varphi_{,j}\theta_{,i} \geq 0, \end{aligned} \tag{4.1.61}$$

for any ε_{ij}, θ and φ.

a) **The equations of perturbed motion referred to the configuration** B. From (1.1.8), (4.1.9) and (4.1.31) we obtain

$$P_{Ai,A} + \rho_0 F_i = \rho_0 \ddot{u}_i \text{ on } B\times(t_0, t_1), \tag{4.1.62}$$

where $F_i = f_i^* - f_i$. Similarly, from (1.1.12), (4.1.13), (4.1.45) we find the equation

$$M_{A,A} + X + \rho_0 L = \rho_0\kappa\ddot{\varphi} \ \text{ on } \ B\times(t_0, t_1), \tag{4.1.63}$$

where $L = \ell^* - \ell$. From (1.2.10), (4.1.11), (4.1.59) we get

$$T\dot{\gamma} + \rho_0 \theta \dot{\eta} = \Phi_{A,A} + \rho_0 s \quad \text{on} \quad B \times (t_0, t_1), \tag{4.1.64}$$

where $s = S^* - S$. In the equations (4.1.62)-(4.1.64), the functions P_{Ai}, M_A, X, γ and Φ_A are given by (4.1.33), (4.1.49), (4.1.46) and (4.1.60).

We are led to the boundary conditions

$$\begin{aligned} &u_i = \widehat{u}_i \text{ on } \overline{\Sigma}_1 \times [t_0, t_1), \ \varphi = \widehat{\varphi} \text{ on } \overline{\Sigma}_3 \times [t_0, t_1), \ \theta = \widehat{\theta} \text{ on } \overline{\Sigma}_5 \times [t_0, t_1), \\ &P_{Ai} N_A = \widehat{P}_i \text{ on } \Sigma_2 \times [t_0, t_1), \ M_A N_A = \widetilde{M} \text{ on } \Sigma_4 \times [t_0, t_1), \\ &\Phi_A N_A = \widehat{\Phi} \text{ on } \Sigma_6 \times [t_0, t_1), \end{aligned} \tag{4.1.65}$$

where $\widehat{u}_i, \widehat{\varphi}, \widehat{\theta}, \widehat{P}_i, \widehat{M}$ and $\widehat{\Phi}$ are prescribed functions.

To complete the specification of the problem, we adjoin the initial conditions

$$\begin{aligned} &\boldsymbol{u}(\boldsymbol{X}, 0) = \boldsymbol{u}^0(\boldsymbol{X}), \ \dot{\boldsymbol{u}}(\boldsymbol{X}, 0) = \boldsymbol{v}^0(\boldsymbol{X}), \ \varphi(\boldsymbol{X}, 0) = \varphi^0(\boldsymbol{X}), \\ &\dot{\varphi}(\boldsymbol{X}, 0) = \psi^0(\boldsymbol{X}), \ \theta(\boldsymbol{X}, 0) = \theta^0(\boldsymbol{X}), \ \boldsymbol{X} \in \overline{B}, \end{aligned} \tag{4.1.66}$$

where $\boldsymbol{u}^0, \boldsymbol{v}^0, \varphi^0, \psi^0$ and θ^0 are given.

b) **The equations of perturbed motion referred to the primary configuration.** On subtracting (1.3.3) from (4.1.10) we obtain

$$s_{ji,j} + \rho F_i = \rho \ddot{u}_i \quad \text{on} \quad B' \times (t_0, t_1). \tag{4.1.67}$$

From (1.3.4), (4.1.15) and (4.1.45) we get

$$\sigma_{i,i} + m + \rho L = \rho \kappa \ddot{\varphi} \quad \text{on} \quad B' \times (t_0, t_1). \tag{4.1.68}$$

If we subtract (1.3.6) from (4.1.12) we find

$$\rho (T^* \dot{\eta}^* - T \dot{\eta}) = \rho s + \psi_{i,i} \quad \text{on} \quad B' \times (t_0, t_1). \tag{4.1.69}$$

By using (4.1.2), (4.1.3) and (4.1.45) we reduce the equation (4.1.69) to

$$\frac{1}{J} T \dot{\gamma} + \rho \theta \dot{\eta} = \psi_{i,i} + \rho s \quad \text{on} \quad B' \times (t_0, t_1). \tag{4.1.70}$$

If we take into account (4.1.41), (4.1.46), (4.1.49), (4.1.60), then the equations (4.1.67), (4.1.68), (4.1.70) lead to five equations for the functions u_i, θ and φ. From (1.3.7) and (4.1.20) we obtain the boundary conditions

$$\begin{aligned} &u_i = \widetilde{u}_i \text{ on } \overline{S}_1 \times [t_0, t_1), \ \varphi = \widetilde{\varphi} \text{ on } \overline{S}_3 \times [t_0, t_1), \ \theta = \widetilde{\theta} \text{ on } \overline{S}_4 \times [t_0, t_1), \\ &s_{ji} n_j = \widetilde{\lambda}_i \text{ on } S_2 \times [t_0, t_1), \ \sigma_i n_i = \widetilde{m} \text{ on } S_4 \times [t_0, t_1), \\ &\psi_i n_i = \widetilde{f} \text{ on } S_6 \times [t_0, t_1), \end{aligned} \tag{4.1.71}$$

where $\widetilde{u}_i, \widetilde{\varphi}, \widetilde{\theta}, \widetilde{\lambda}_i, \widetilde{m}, \widetilde{f}$ are prescribed functions. Clearly, we must adjoin initial conditions similar to (4.1.66).

If the primary configuration B is a configuration of equilibrium, then the basic equations are

$$s_{ji,j} + \rho F_i = \rho \ddot{u}_i, \quad \sigma_{i,i} + m + \rho L = \rho\kappa\ddot{\varphi}, \quad \frac{1}{J} T\dot{\gamma} = \psi_{i,i} + \rho s \tag{4.1.72}$$

and

$$\begin{aligned} s_{ji} &= D_{ijrs}u_{r,s} + d_{ijk}\varphi_{,k} + b_{ij}\varphi - \beta_{ij}\theta, \\ \sigma_i &= d_{sri}u_{s,r} + a_{ij}\varphi_{,j} + b_i\varphi - a_i\theta, \\ \gamma &= J(\beta_{ij}u_{i,j} + a_i\varphi_{,i} + b\varphi + a\theta), \\ m &= b\theta - b_{ij}u_{i,j} - b_i\varphi_{,i} - \zeta\varphi, \\ \psi_i &= h_{ijk}u_{j,k} + r_i\theta + k_{ij}\theta_{,j} + s_i\varphi + \ell_{ij}\varphi_{,j}, \end{aligned} \tag{4.1.73}$$

where $\rho, J, T, D_{ijrs}, d_{ijk}, a_{ij}, b_{ij}, \beta_{ij}, a_i, b_i, a, b, \zeta, h_{ijk}, r_i, k_{ij}, s_i$ and ℓ_{ij} are *time independent.* If the primary state is at constant temperature, then h_{ijk}, r_i, s_i and ℓ_{ij} vanish.

c) **Linear theory of materials with initial stresses and initial heat flux.** We assume here that the primary configuration B' is identical with the reference configuration B. We suppose that B is subjected to an initial stress, and is at nonuniform temperature $\overline{T}$. We consider that the distribution of voids in B' is non-uniform so that ν is a prescribed function of point. We then assume that B' (or B) is deformed and use the notations: u_i - the components of displacement vector, φ - the change in volume fraction field, θ - the temperature variation. In this case we have

$$\begin{aligned} &J = 1, \ \rho_0 = \rho, \ \nu_0 = \nu \equiv \overline{\nu}, \ E_{AB} = 0, \ T = \overline{T}, \\ &T_{AB} = \frac{\partial\sigma}{\partial E_{AB}}, \quad \rho_0\eta = -\frac{\partial\sigma}{\partial T}, \quad H_A = \frac{\partial\sigma}{\partial \nu_{,A}}, \quad g = -\frac{\partial\sigma}{\partial\nu}, \\ &Q_A = \widehat{Q}_A(0, T, T_{,K}, \nu, \nu_{,L}), \end{aligned} \tag{4.1.74}$$

the derivatives being evaluated at $E_{AB} = 0, T = \overline{T}$ and $\nu = \overline{\nu}$. Obviously, the tensors T_{AB}, T_{Ai} and t_{ij} coincide. Also, we have $Q_i = q_i$, $H_i = h_i$. The functions t_{ij} (or T_{Ai}, or T_{AB}) and h_i (or H_i) are the components of the initial stresses. The functions q_i are the components of the initial heat-flux vector. In this case the functions C_{ijrs}, d_{sri}, β_{ij}, a_{ij}, b_{ij}, a_i, b_i, a, b, c, h_{ijk}, r_i, k_{ij}, s_i and ℓ_{ij} defined by (4.1.25), (4.1.37), (4.1.47), (4.1.50), (4.1.52), (4.1.57) are evaluated at $E_{AB} = 0$, $\nu = \overline{\nu}$, $T = \overline{T}$. The work of the previous sections can be applied in this case, and it yields a linear theory of materials with voids in the presence of initial stress and initial heat flux. These equations consist of the equations of motion

$$s_{ji,j} + \rho F_i = \rho \ddot{u}_i, \quad \sigma_{i,i} + m + \rho L = \rho \kappa \ddot{\varphi}, \tag{4.1.75}$$

the energy equation

$$T\dot{\gamma} = \psi_{i,i} + \rho s \tag{4.1.76}$$

and the constitutive equations

$$\begin{aligned}
s_{ji} &= (C_{ijrs} + t_{js}\delta_{ir})u_{r,s} + d_{ijk}\varphi_{,k} + b_{ij}\varphi - \beta_{ij}\theta,\\
\sigma_i &= d_{sri}u_{s,r} + a_{ij}\varphi_{,j} + b_i\varphi - a_i\theta,\\
m &= -b_{ij}u_{i,j} - b_i\varphi_{,i} - \zeta\varphi + b\theta,\\
\gamma &= \beta_{ij}u_{i,j} + a_i\varphi_{,i} + b\varphi + a\theta,\\
\psi_i &= h_{ijk}u_{j,k} + r_i\theta + k_{ij}\theta_{,j} + s_i\varphi + \ell_{ij}\varphi_{,j}.
\end{aligned}$$

If $\overline{T}$ is constant, then $q_i = 0$, $h_{ijk} = 0$, $r_i = 0$, $s_i = 0$, $\ell_{ij} = 0$. To the system of equations we adjoin the boundary conditions (4.1.71) and the initial conditions

$$\begin{aligned}
&u_i(\boldsymbol{x},0) = u_i^0(\boldsymbol{x}),\ \dot{u}_i(\boldsymbol{x},0) = v_i^0(\boldsymbol{x}),\ \varphi(\boldsymbol{x},0) = \varphi^0(\boldsymbol{x}),\\
&\dot{\varphi}(\boldsymbol{x},0) = \zeta^0(\boldsymbol{x}),\ \gamma(\boldsymbol{x},0) = \gamma^0(\boldsymbol{x}),\ \boldsymbol{x} \in \overline{B},
\end{aligned}$$

where $u_i^0, v_i^0, \varphi^0, \zeta^0, \gamma^0$ are prescribed functions.

4.2 An Existence Result

In this section we use results of the semigroups theory of linear operators to obtain an existence result in the theory of infinitesimal deformations superposed on finite deformations. The results we give here are due to Martinez and Quintanilla [253]. Another existence result in linear thermoelasticity of materials with voids has been presented by Rusu [321].

In what follows we consider the equations of perturbed deformations referred to the configuration B. To the equations (4.1.62)-(4.1.64), (4.1.33), (4.1.46), (4.1.49), (4.1.60) we adjoin the initial conditions (4.1.66) and the following boundary conditions

$$\boldsymbol{u} = 0,\ \varphi = 0,\ \theta = 0 \ \text{ on } \ \partial B \times (0, t_1). \tag{4.2.1}$$

The behavior of the body is studied on the finite time interval $[0, t_1]$. We assume that: (i) E_{iKjN}, D_{KiM}, B_{Ki}, G_{Ki}, B_K, ξ, B_0, A_{KL}, A_K, R_{AiN},

D_A, K_{AM}, E_A, L_{AM}, $\dot{G}_{Ai}$, $\dot{A}_K$, $\dot{B}_0$ and $\dot{A}$ are continuously differentiable and, for each $t \in [0, t_1]$, Lebesgue measurable and essentially bounded; (ii) $0 < \rho_1 \leq \operatorname*{ess\,inf}_{X \in B} \rho_0 \leq \rho_2$, $0 < \kappa_1 \leq \operatorname*{ess\,inf}_{X \in B} \kappa \leq \kappa_2$, $\operatorname*{ess\,inf}_{X \in B} A \geq A_0 > 0$; (iii) there exists a positive constant d such that

$$\int_B (E_{iKjL} u_{i,K} u_{j,L} + A_{KL} \varphi_{,K} \varphi_{,L} + \xi \varphi^2 + 2D_{KiN} u_{i,K} \varphi_{,N} + \\ +2B_{Ki} u_{i,K} \varphi + 2B_K \varphi_{,K} \varphi) dV \geq d \int_B (u_{i,K} u_{i,K} + \varphi_{,K} \varphi_{,K} + \varphi^2) dV, \tag{4.2.2}$$

for all $u_i, \varphi \in C_0^\infty(B)$ and any $t \in [0, t_1]$; (iv) K_{MN} is positive definite in the sense that there exists a positive constant K_0 such that

$$\int_B \frac{1}{T} K_{MN} \theta_{,M} \theta_{,N} dV \geq K_0 \int_B \theta_{,M} \theta_{,M} dV, \tag{4.2.3}$$

for all $\theta \in C_0^\infty(B)$ and $t \in [0, t_1]$.

The field equations can be expressed in terms of u_i, φ and θ. Thus, substitution of (4.1.33), (4.1.46), (4.1.49), (4.1.60) into the equations (4.1.62)-(4.1.64) yields the equations

$$\begin{aligned} \ddot{u}_i &= \frac{1}{\rho_0} \left(E_{iKjL} u_{j,L} + D_{KiM} \varphi_{,M} + B_{Ki} \varphi - G_{Ki} \theta \right)_{,K} + F_i, \\ \ddot{\varphi} &= \frac{1}{\rho_0 \kappa} \Big[\left(D_{LiK} u_{i,L} + A_{KL} \varphi_{,L} + B_K \varphi - A_K \theta \right)_{,K} + B_0 \theta - \\ &\quad - B_{Li} u_{i,L} - B_K \varphi_{,K} - \xi \varphi \Big] + \frac{1}{\kappa} L, \\ \dot{\theta} &= \frac{1}{AT} \Big[\left(R_{MjN} u_{j,N} + L_{MK} \varphi_{,K} + E_M \varphi + K_{MN} \theta_{,N} + D_M \theta \right)_{,M} - \\ &\quad - \rho_0 \dot{\eta} \theta \Big] - \frac{1}{A} \Big(G_{Mi} \dot{u}_{i,M} + A_K \dot{\varphi}_{,K} + B_0 \dot{\varphi} + \\ &\quad + \dot{G}_{Mi} u_{i,M} + \dot{A}_K \varphi_{,K} + \dot{B}_0 \varphi + \dot{A} \theta \Big) + \frac{\rho_0 s}{AT}, \end{aligned} \tag{4.2.4}$$

on $B \times (0, t_1)$. We transform the boundary-initial-value problem into a temporally inhomogeneous abstract equation in a Hilbert space. Let $v_i = \dot{u}_i$, $\psi = \dot{\varphi}$ and $w = (u_1, u_2, u_3, v_1, v_2, v_3, \varphi, \psi, \theta) = (u_i, v_i, \varphi, \psi, \theta)$. We define

$$X = \{ w = (u_i, v_i, \varphi, \psi, \theta);\ u_i, \varphi \in W_0^{1,2}(B),\ v_i, \psi, \theta \in L_2(B) \},$$

where $W_0^{1,2}(B)$ is the well-known Sobolev space [2].

Let

$$\mathcal{M}_i w = v_i,\ \mathcal{N}_i w = \frac{1}{\rho_0}\,(E_{iKjL}u_{j,L} + D_{KiM}\varphi_{,M} + B_{Ki}\varphi - G_{Ki}\theta)_{,K},$$
$$\mathcal{P}w = \psi,\ \mathcal{R}w = \frac{1}{\rho_0\kappa}[(D_{LiK}u_{i,L} + A_{KL}\varphi_{,L} + B_K\varphi - A_K\theta)_{,K} +$$
$$+ B_0\theta - B_{Li}u_{i,L} - B_K\varphi_{,K} - \xi\varphi],$$
$$\mathcal{T}w = \frac{1}{AT}[(R_{MjN}u_{j,N} + L_{MK}\varphi_{,K} + E_M\varphi + K_{MN}\theta_{,N} +$$
$$+ D_M\theta)_{,M} - \rho_0\dot{\eta}\theta] - \frac{1}{A}(G_{Mi}v_{i,M} + A_K\psi_{,K} + B_0\psi +$$
$$+ \dot{G}_{Mi}u_{i,M} + \dot{A}_K\varphi_{,K} + \dot{B}_0\varphi + \dot{A}\theta). \tag{4.2.5}$$

We introduce the operator Q defined by

$$Qw = (\mathcal{M}_i w, \mathcal{N}_i w, \mathcal{P}w, \mathcal{R}w, \mathcal{T}w), \tag{4.2.6}$$

with the domain

$$D = \{w = (u_i, v_i, \varphi, \psi, \theta) \in X;\ Qw = X,\ \theta = 0 \text{ on } \partial B \times [0, t_1]\}.$$

We note that D is dense in X. The boundary-initial-value problem is reduced to the following abstract evolutionary equation

$$\frac{dw(t)}{dt} = Q(t)w(t) + f(t),\quad t \in [0, t_1],$$
$$w(0) = w_0, \tag{4.2.7}$$

where f and w_0 are nine-dimensional vector fields defined by

$$f = \left(0, 0, 0, F_1, F_2, F_3, 0, L, \frac{1}{AT}\rho_0 s\right),\quad w_0 = (u_i^0, v_i^0, \varphi^0, \psi^0, \theta^0).$$

Let $w = (u_i, v_i, \varphi, \psi, \theta)$ and $\overline{w} = (\overline{u}_i, \overline{v}_i, \overline{\varphi}, \overline{\psi}, \overline{\theta})$, $w, \overline{w} \in X$. We introduce in X the following inner product

$$\langle w, \overline{w}\rangle_t = \frac{1}{2}\int_B [E_{iKjL}u_{i,K}\overline{u}_{j,L} + A_{KL}\varphi_{,K}\overline{\varphi}_{,L} + \xi\varphi\overline{\varphi} + A\theta\overline{\theta} +$$
$$+ D_{KiN}(u_{i,K}\overline{\varphi}_{,N} + \overline{u}_{i,K}\varphi_{,N}) + B_{Ki}(u_{i,K}\overline{\varphi} + \overline{u}_{i,K}\varphi) + \tag{4.2.8}$$
$$+ B_K(\varphi_{,K}\overline{\varphi} + \overline{\varphi}_{,K}\varphi) + \rho_0 v_i\overline{v}_i + \rho_0\kappa\psi\overline{\psi}]dV,$$

for each $t \in [0, t_1]$. We can see that the norm $\|\cdot\|_t$ induced by this product is equivalent to the usual norm $\|\cdot\|$ in X.

Lemma 4.2.1. *For each $t \in [0, t_1]$ the operator $Q(t)$ is the generator of a quasi-contractive semigroup.*

Proof. By using the divergence theorem and the boundary conditions we obtain

$$\begin{aligned}\langle Qw, w\rangle_t = \frac{1}{2}\int_B [(K_{AM}T_{,A} - D_M T)\theta\theta_{,M} - TR_{AiN}u_{i,N}\theta_{,A} - \\ -TK_{AM}\theta_{,A}\theta_{,M} + (E_A T_{,A} - \dot{B}_0 T^2)\theta\varphi - TL_{AM}\theta_{,A}\varphi_{,M} + \\ +(R_{AiK}T_{,A} - \dot{G}_{Ki}T^2)u_{i,K}\theta - TE_A\theta_{,A}\varphi + (D_A T_{,A} - \dot{A}T^2 - \\ -\rho_0 T\dot{\eta})\theta^2 + (L_{AM}T_{,A} - \dot{A}_M T^2)\theta\varphi_{,M}]T^{-2}dV.\end{aligned}$$

We introduce the notations

$k_1^2 = 9R_{AiN}R_{AiN}T^{-2}$, $k_2^2 = (K_{AM}T_{,A} - D_M T)(K_{LM}T_{,L} - D_M T)T^{-4}$, $k_3^2 = E_K E_K T^{-2}$, $k_4^2 = 9L_{MN}L_{MN}T^{-2}$, $k_5^2 = (R_{AiK}T_{,A} - -\dot{G}_{Ki}T^2)(R_{NiK}T_{,N} - G_{Ki}T^2)T^{-4}$, $k_6^2 = (D_A T_{,A} - \dot{A}T^2 - \rho_0 T\dot{\eta})^2 T^{-4}$, $k_7^2 = (E_K T_{,K} - \dot{B}_0 T^2)^2 T^{-4}$, $k_8^2 = (L_{KM}T_{,K} - \dot{A}_M T^2)(L_{NM}T_{,N} - -\dot{A}_M T^2)T^{-4}$, $m_i = \max\limits_{[0,t_1]} \operatorname*{ess\,sup}\limits_{X\in B} k_i^2$, $(i = 1, 2, ..., 8)$.

Using the arithmetic-geometric mean inequality and Schwarz inequality we get

$$\begin{aligned}\langle Qw, w\rangle_t \le \frac{1}{4}\int_B \Bigg[\left(\frac{m_1}{\varepsilon_1} + \varepsilon_5\right)u_{i,K}u_{i,K} + \\ +(\varepsilon_1 + \varepsilon_2 + \varepsilon_3 + \varepsilon_4 - 2K_0)\theta_{,K}\theta_{,K} + \left(\frac{m_4}{\varepsilon_4} + \varepsilon_8\right)\varphi_{,N}\varphi_{,N} + \\ +\left(\frac{m_2}{\varepsilon_2} + \frac{m_5}{\varepsilon_5} + \frac{m_8}{\varepsilon_8} + 2m_6 + \varepsilon_7\right)\theta^2 + \left(\frac{m_3}{\varepsilon_3} + \frac{m_7}{\varepsilon_7}\right)\varphi^2\Bigg]dV,\end{aligned}$$

where ε_i are arbitrary positive constants. If we choose $\varepsilon_1, \varepsilon_2, \varepsilon_3$ and ε_4 such that $\varepsilon_1 + \varepsilon_2 + \varepsilon_3 + \varepsilon_4 - 2K_0 \le 0$ then we conclude that there exists a positive constant C_1 such that

$$\langle Qw, w\rangle_t \le C_1 \langle w, w\rangle_t. \tag{4.2.9}$$

Let us show that the operator Q satisfies the range condition Rang$(\lambda I - Q) = X$, where I is the identity operator and λ is a positive scalar. Let $\overline{w} = (\overline{u}_i, \overline{v}_i, \overline{\varphi}, \overline{\psi}, \overline{\theta}) \in X$. We must prove that the equation

$$\lambda w - Qw = \overline{w}, \tag{4.2.10}$$

has a solution $w = (u_i, v_i, \varphi, \psi, \theta) \in D$. By elliminating v_i and ψ, (4.2.10) yields the following system for u_i, φ and θ

$$\Lambda_i z = p_i, \ \Gamma z = p_4, \ \Phi z = p_5, \tag{4.2.11}$$

where

$$\begin{aligned}
&z = (u_1, u_2, u_3, \varphi, \theta) = (u_i, \varphi, \theta), \ p_i = \bar{v}_i + \lambda \bar{u}_i, \ p_4 = \overline{\psi} + \lambda \overline{\varphi},\\
&p_5 = \frac{1}{\lambda A}(A\overline{\theta} + G_{Nj}\bar{u}_{j,N} + A_K \overline{\varphi}_{,K} + B_0 \overline{\varphi}),\\
&\Lambda_i z = \lambda^2 u_i - \frac{1}{\rho_0}(E_{iKjL}u_{j,L} + D_{KiM}\varphi_{,M} + B_{Ki}\varphi - G_{Ki}\theta)_{,K},\\
&\Gamma z = \lambda^2 \varphi - \frac{1}{\rho_0 \kappa}[(D_{LiK}u_{i,L} + A_{KL}\varphi_{,L} + B_K\varphi - A_K\theta)_{,K} +\\
&\qquad\qquad + B_0\theta - B_{Li}u_{i,L} - B_K\varphi_{,K} - \xi\varphi],\\
&\Phi z = \theta - \frac{1}{\lambda A T}[(R_{MjN}u_{j,N} + L_{MK}\varphi_{,K} + E_M\varphi + K_{MN}\theta_{,N} +\\
&\qquad + D_M\theta)_{,M} - \rho_0\dot{\eta}\theta] + \frac{1}{A}(G_{Mi}u_{i,M} + A_K\varphi_{,K} + B\varphi) +\\
&\qquad + \frac{1}{A\lambda}(\dot{G}_{Mi}u_{i,M} + \dot{A}_K\varphi_{,K} + \dot{B}_0\varphi + \dot{A}\theta).
\end{aligned} \tag{4.2.12}$$

Let $z = (u_i, \varphi, \theta)$ and $\bar{z} = (\bar{u}_i, \overline{\varphi}, \overline{\theta})$. We consider the scalar product

$$[z, \bar{z}] = \int_{B_0} (\rho_0 u_i \bar{u}_i + \rho_0 \kappa \varphi \overline{\varphi} + A\theta\overline{\theta}) dV$$

and the bilinear form

$$F(z, \bar{z}) = [(\Lambda_i z, \Gamma z, \Phi z), (\bar{u}_i, \overline{\varphi}, \overline{\theta})]. \tag{4.2.13}$$

We find that

$$\begin{aligned}
F(z, z) = &\int_{B_0} \{\rho_0 \lambda^2 u_i u_i + E_{iKjN}u_{i,K}u_{j,N} + A_{KN}\varphi_{,K}\varphi_{,N} + 2B_K\varphi\varphi_{,K} +\\
&+ 2D_{KiN}u_{i,K}\varphi_{,N} + 2B_{Ki}u_{i,K}\varphi + (\rho_0\kappa\lambda^2 + \xi)\varphi^2 +\\
&+ \lambda^{-1}[(\dot{G}_{Ki}u_{i,K} + \dot{A}_K\varphi_{,K} + \dot{B}_0\varphi)\theta + (\rho_0\dot{\eta}T^{-1} + \dot{A} + \lambda A)\theta^2 +\\
&+ (R_{AiN}u_{i,N} + D_A\theta + K_{AM}\theta_{,M} + E_A\varphi + L_{AK}\varphi_{,K})(\theta T^{-1})_{,A}]\} dV.
\end{aligned}$$

It is not difficult to show that there exists a positive constant C_2 such that

$$F(z, z) \geq C_2 \|(\boldsymbol{u}, \varphi, \theta)\|^2_{\mathbf{W}_0^{1,2} \times W_0^{1,2} \times W_0^{1,2}}.$$

Thus, F determines a norm which is equivalent to the usual norm in $\boldsymbol{W}_0^{1,2} \times W_0^{1,2} \times W_0^{1,2}$. We can see that $(\boldsymbol{p}, p_4, p_5) \in \boldsymbol{W}^{-1,2} \times W^{-1,2} \times W^{-1,2}$. The Riesz theorem shows that there exists a unique solution $(\boldsymbol{u}, \varphi, \theta) \in \boldsymbol{W}_0^{1,2} \times W_0^{1,2} \times W_0^{1,2}$ to the system (4.2.11). From (4.2.10) we see that $\boldsymbol{v} \in \boldsymbol{W}_0^{1,2}$ and $\psi \in W_0^{1,2}$. The desired result is an immediate consequence of the Lummer-Phillips corollary to the Hille-Yosida theorem ([384], p. 343).

Lemma 4.2.2. *There exists a positive constant α such that*

$$\|w\|_t \leq \|w\|_s \exp(\alpha t_1), \tag{4.2.14}$$

for any $w = (u_i, v_i, \varphi, \psi, \theta) \in X$ and all $t, s \in [0, t_1]$.

Proof. We define for each $t \in [0, t_1]$, $y(t) = \|\cdot\|_t^2$ where $\|\cdot\|_t$ is the norm induced by the inner product (4.2.8). We can write [253]

$$\begin{aligned}\dot{y} \leq \frac{1}{2}\int_{B_0} &(\dot{E}_{iKjN} u_{i,K} u_{j,N} + \dot{A}_{KN} \varphi_{,K} \varphi_{,N} + \dot{\xi}\varphi^2 + \\ &+ \dot{A}\theta^2 + 2\dot{D}_{KiN} u_{i,K} \varphi_{,N} + 2\dot{B}_{Ki} u_{i,K} \varphi + 2\dot{B}_K \varphi_{,K} \varphi) dV.\end{aligned}$$

By using Schwarz inequality and the boundary conditions we can show that there exists a positive constant α such that

$$\dot{y} \leq 2\alpha y. \tag{4.2.15}$$

The desired result follows from (4.2.15). □

The preceding lemmas imply the following result

Lemma 4.2.3. *The family of operators $\{Q(t)\}$, $t \in [0, t_1]$, is stable in the sense of Kato with stability constants* $\exp(2\alpha t_1)$ *and* C_1.

The above lemmas and Kato theorem [215] lead to the following proposition

Theorem 4.2.1. *Assume that the hypotheses* (i)-(iv) *hold. Let F_i, $L\kappa^{-1} \in C^1([0, t_1], L^2(B)) \cap C^0([0, t_1], W_0^{1,2}(B) \cap W^{2,2}(B))$, $\rho_0 s(AT)^{-1} \in C^1([0, t_1], L^2(B)) \cap C^0([0, t_1], W^{2,2}(B))$. Then, for any $w_0 \in D$, there exists a unique solution to the boundary-initial-value problem such that $(u_i, v_i, \varphi, \psi, \theta) \in C^1([0, t_1]; X) \cap C^0([0, t_1]; D)$.*

Other uniqueness and continuous dependence results have been presented in [253], [321].

Chapter 5

THERMOELASTIC COSSERAT CONTINUA

5.1 The Linear Theory of Micropolar Thermoelasticity

In a remarkable monograph, E. and F. Cosserat [71] gave a systematic development of the mechanics of continuous media in which each point has the six degree of freedom of a rigid body. The orientation of a given particle of such a medium can be represented mathematically by the values of three mutually perpendicular unit vectors which Ericksen and Truesdell [98] called directors. In the 1960s the subject matter was reopened in various papers (see Eringen and Suhubi [100], Green and Rivlin [125], Mindlin [263]). These early theories are discussed in various works (see, e.g., Truesdell and Toupin [366], Eringen [105], Kunin [230], Capriz [38]).

In this section we establish the basic equations of the linear theory of a thermoelastic Cosserat medium. This theory is usually called "theory of micropolar thermoelasticity" (cf. Eringen [102], Nowacki [289]). An account of the historical developments of the theory as well as references to various contributions may be found in the books by Eringen and Kafadar [103], Nowacki [289], Dyszlewicz [96] and Eringen [105].

As remarked above, a Cosserat medium is a continuum in which each point has the degrees of freedom of a rigid body. Thus, the deformation of such a body is described by

$$\boldsymbol{u} = \boldsymbol{u}(\boldsymbol{X}, t), \;\; \boldsymbol{\varphi} = \boldsymbol{\varphi}(\boldsymbol{X}, t), \quad (\boldsymbol{X}, t) \in B \times I, \tag{5.1.1}$$

where $\boldsymbol{u}$ is the displacement vector field and $\boldsymbol{\varphi}$ is the microrotation vector field. The theory of a thermoelastic Cosserat continuum is based upon the following balance laws: the balance of mass, the balance of linear momentum, the balance of moment of momentum, the balance of energy and the second law of thermodynamics. The balance of mass, the balance of linear momentum and the second law of thermodynamics have the same form as in the classical theory of thermoelasticity. Let us consider an arbitrary region $\mathcal{P}$ of the continuum bounded by a surface $\partial\mathcal{P}$ at time t, and we suppose that P is the corresponding region at time t_0 bounded by the surface ∂P. We postulate the balance of moment of momentum in the form

$$\int_P \rho_0(\varepsilon_{ijk}x_j\ddot{u}_k + Y_{ij}\ddot{\varphi}_j)dV = \int_P \rho_0(\varepsilon_{ijk}x_jf_k + g_i)dV + \\ + \int_{\partial P}(\varepsilon_{ijk}x_jT_k + M_i)dA, \tag{5.1.2}$$

for every part P of B and every time. Here, T_i is the stress vector associated with the surface $\partial\mathcal{P}$ but measured per unit area of the surface ∂P, f_i is the body force per unit mass, g_i is the body couple per unit mass, M_i is the couple stress vector associated with the surface $\partial\mathcal{P}$ but measured per unit area of the surface ∂P, Y_{ij} are coefficients of inertia and ρ_0 is the density in the reference configuration. We note that the balance of linear momentum implies the Cauchy's formula

$$T_i = T_{Ki}N_K \tag{5.1.3}$$

and the following equations of motion

$$T_{Ki,K} + \rho_0 f_i = \rho_0\ddot{u}_i. \tag{5.1.4}$$

In view of (5.1.3) and (5.1.4) the relation (5.1.2) reduces to

$$\int_P \rho_0 Y_{ij}\ddot{\varphi}_j dV = \int_P (\rho_0 g_i + \varepsilon_{ijk}x_{j,L}T_{Lk})dV + \int_{\partial P} M_i dA. \tag{5.1.5}$$

We apply this equation to a region which in the reference state was a tetrahedron bounded by coordinate planes through the point X and by a plane whose unit normal is N_K. It follows that

$$M_i = M_{Ki}N_K, \tag{5.1.6}$$

where M_{Ki} is the couple stress vector associated with surfaces in the deformed body which were originally coordinate planes perpendicular to the X_K-axes through the point X, measured per unit area of these planes. M_{Ki} is called the couple stress tensor. In view of (5.1.6) and the divergence theorem, from (5.1.5) we obtain

$$M_{Ki,K} + \varepsilon_{ijr}x_{j,L}T_{Lr} + \rho_0 g_i = \rho_0 Y_{ij}\ddot{\varphi}_j. \tag{5.1.7}$$

We postulate the energy balance in the form

$$\int_P \rho_0(\dot{u}_i\ddot{u}_i + Y_{ij}\dot{\varphi}_i\ddot{\varphi}_j + \dot{e})dV = \int_P \rho_0(f_i\dot{u}_i + g_i\dot{\varphi}_i + S)dV + \\ + \int_{\partial P}(T_i\dot{u}_i + M_i\dot{\varphi}_i + Q)dA, \tag{5.1.8}$$

for every part P of B and every time. Here, e is the internal energy per unit mass, S is the heat supply per unit mass and unit time and Q is the heat flux across the surface $\partial\mathcal{P}$ measured per unit area of ∂P, per unit time. We assume that e is of class $C^{0,1}$, Q is of class $C^{1,0}$, and S is continuous on $B\times(t_0,t_1)$. In view of (5.1.3), (5.1.4), (5.1.6) and (5.1.7), the equation (5.1.8) becomes

$$\int_P \rho_0\dot{e}\,dV = \int_P [T_{Ki}(\dot{u}_{i,K} + \varepsilon_{ijs}x_{j,K}\dot{\varphi}_s) + M_{Ki}\dot{\varphi}_{i,K} + \rho_0 S]dV + \\ + \int_{\partial P} Q\,dA. \tag{5.1.9}$$

Next, we apply this equation to a region which in the reference state was a tetrahedron bounded by coordinate planes through the point X and by a plane whose unit normal is N_K, to obtain

$$Q = Q_K N_K. \tag{5.1.10}$$

Here Q_A is the heat flux across the surface that was originally the coordinate plane perpendicular to the X_A-axis through the point X, measured per unit area of this plane and per unit time. In view of (5.1.10), the equation (5.1.9) leads to the local equation of energy

$$\rho_0\dot{e} = T_{Ki}(\dot{u}_{i,K} + \varepsilon_{ijs}x_{j,K}\dot{\varphi}_s) + M_{Ki}\dot{\varphi}_{i,K} + \rho_0 S + Q_{L,L}. \tag{5.1.11}$$

We introduce the functions T_{KL}, M_{KL}, D_{KL} and Λ_{KL} by

$$T_{Ki} = x_{i,L}T_{KL}, \quad M_{Ki} = x_{i,L}M_{KL}, \\ D_{KL} = x_{i,L}(\dot{u}_{i,K} + \varepsilon_{ijs}x_{j,K}\dot{\varphi}_s), \quad \Lambda_{KL} = x_{i,L}\dot{\varphi}_{i,K}. \tag{5.1.12}$$

The equation (5.1.11) can be written in the form

$$\rho_0\dot{e} = T_{KL}D_{KL} + M_{KL}\Lambda_{KL} + \rho_0 S + Q_{L,L}. \tag{5.1.13}$$

We note that

$$x_{i,K} = u_{i,K} + \delta_{iK}, \tag{5.1.14}$$

where δ_{iK} is Kronecker's delta.

In the linear theory the displacement and microrotation fields have the forms $\boldsymbol{u} = \varepsilon \boldsymbol{u}'$ and $\varphi = \varepsilon\varphi'$ where ε is a constant small enough for squares and higher powers to be neglected, and $\boldsymbol{u}'$ and $\boldsymbol{\varphi}'$ are independent of ε. If we introduce the notations

$$u_K = \delta_{iK} u_i, \quad \varphi_K = \delta_{iK} \varphi_i, \tag{5.1.15}$$

then by (5.1.12), (5.1.14) and (5.1.15) we find that in the linear theory we have

$$D_{KL} = \dot{u}_{K,L} + \varepsilon_{KLM} \dot{\varphi}_M, \quad \Lambda_{KL} = \dot{\varphi}_{L,K}, \tag{5.1.16}$$

where $\varepsilon_{KLM} = \delta_{iK}\delta_{jL}\delta_{sM}\varepsilon_{ijs}$. We denote

$$e_{KL} = u_{L,K} + \varepsilon_{LKM}\varphi_M, \quad \kappa_{KL} = \varphi_{L,K}. \tag{5.1.17}$$

In view of (5.1.16) and (5.1.17), the equation (5.1.13) can be written in the form

$$\rho_o \dot{e} = T_{KL}\dot{e}_{KL} + M_{KL}\dot{\kappa}_{KL} + \rho_0 S + Q_{L,L}. \tag{5.1.18}$$

The geometrical meaning of the strain measures ε_{KL} and κ_{KL} has been discussed in various papers [105], [263]. The local form of the entropy production inequality is

$$\rho_0 T\dot{\eta} - \rho_0 S - Q_{L,L} + \frac{1}{T} Q_A T_{,A} \geq 0, \tag{5.1.19}$$

where η is the entropy per unit mass, and T is the absolute temperature, which is assumed to be positive. Introducing the Helmholtz free-energy $\psi = e - T\eta$, and combining (5.1.18) and (5.1.19) we arrive at the following local dissipation inequality

$$T_{KL}\dot{e}_{KL} + M_{KL}\dot{\kappa}_{KL} - \rho_0\eta\dot{T} - \rho_0\dot{\psi} + \frac{1}{T} Q_K T_{,K} \geq 0. \tag{5.1.20}$$

We introduce the notation

$$\theta = T - T_0,$$

where T_0 is the absolute temperature in the reference state. We assume that T_0 is a given constant. In the linear theory we assume that $\theta = \varepsilon\theta'$ where θ' is independent of ε.

To characterize a thermoelastic material we require constitutive equations for $\psi, T_{KL}, M_{KL}, \eta$ and Q_K. In the linear theory, a thermoelastic material is defined as one for which $\psi, T_{KL}, M_{KL}, \eta$ and Q_K are functions of $e_{AB}, \kappa_{MN}, \theta, \theta_{,N}$ and X_P, consistent with the assumption of the linear theory. If we use the constitutive assumptions and the standard arguments, then the inequality (5.1.20) implies that

$$\begin{aligned} &\psi = \Psi(e_{KL}, \kappa_{KL}, \theta, X_A), \\ &T_{KL} = \frac{\partial \sigma}{\partial e_{KL}}, \quad M_{KL} = \frac{\partial \sigma}{\partial \kappa_{KL}}, \quad \rho_0 \eta = -\frac{\partial \sigma}{\partial \theta} \end{aligned} \tag{5.1.21}$$

and

$$Q_K \theta_{,K} \geq 0. \tag{5.1.22}$$

Here we have used the notation $\sigma = \rho_0 \Psi$. Clearly, the inequality (5.1.22) requires the following property for the heat flux vector

$$Q_K = 0 \;\text{ if } \theta_{,M} = 0. \tag{5.1.23}$$

In the linear theory, and assuming that the body is free of initial stress and couple stress, we have

$$\begin{aligned} \sigma = &\frac{1}{2} A_{KLMN} e_{KL} e_{MN} + B_{KLMN} e_{KL} \kappa_{MN} + \\ &+ \frac{1}{2} C_{KLMN} \kappa_{KL} \kappa_{MN} - \frac{1}{2} a\theta^2 - \beta_{KL} e_{KL} \theta - \xi_{KL} \kappa_{KL} \theta, \end{aligned} \tag{5.1.24}$$

where the constitutive coefficients are prescribed functions of X_A. The coefficients in (5.1.24) have the following symmetries

$$A_{KLMN} = A_{MNKL}, \quad C_{KLMN} = C_{MNKL}. \tag{5.1.25}$$

If we take into account (5.1.21), (5.1.23)-(5.1.25) then we obtain the following constitutive equations for the linear theory of thermoelastic bodies

$$\begin{aligned} T_{KL} &= A_{KLMN} e_{MN} + B_{KLMN} \kappa_{MN} - \beta_{KL} \theta, \\ M_{KL} &= B_{MNKL} e_{MN} + C_{KLMN} \kappa_{MN} - \xi_{KL} \theta, \\ \rho_0 \eta &= \beta_{KL} e_{KL} + \xi_{KL} \kappa_{KL} + a\theta, \\ Q_M &= k_{MN} \theta_{,N}. \end{aligned} \tag{5.1.26}$$

In the linear theory, the energy equation reduces to

$$\rho_0 T_0 \dot{\eta} = Q_{A,A} + \rho_0 S. \tag{5.1.27}$$

Thus, the basic equations of the linear theory of micropolar thermoelasticity are: the equations of motion (5.1.4), (5.1.7), the energy equation (5.1.27), the constitutive equations (5.1.26) and the geometrical equations (5.1.17). These equations can be written using small indices. In the linear theory we emply the notations $X_i = \delta_{iK} X_K$ and denote $f_{,j} = \partial f / \partial X_j$. Thus the geometrical equations become

$$e_{ij} = u_{j,i} + \varepsilon_{jik} \varphi_k, \quad \kappa_{ij} = \varphi_{j,i}. \tag{5.1.28}$$

The constitutive equations take the form

$$\begin{aligned} t_{ij} &= A_{ijrs} e_{rs} + B_{ijrs} \kappa_{rs} - \beta_{ij} \theta, \\ m_{ij} &= B_{rsij} e_{rs} + C_{ijrs} \kappa_{rs} - \xi_{ij} \theta, \\ \rho_0 \eta &= \beta_{ij} e_{ij} + \xi_{ij} \kappa_{ij} + a\theta, \\ q_i &= k_{ij} \theta_{,j}, \end{aligned} \tag{5.1.29}$$

where we have used the notations

$$t_{ij} = \delta_{iK} \delta_{jL} T_{KL}, \quad m_{ij} = \delta_{iK} \delta_{jL} M_{KL}, \quad q_i = \delta_{iK} Q_K.$$

The equations of motion reduce to

$$\begin{aligned} & t_{ji,j} + \rho_0 f_i = \rho_0 \ddot{u}_i, \\ & m_{ji,j} + \varepsilon_{irs} t_{rs} + \rho_0 g_i = I_{ij} \ddot{\varphi}_j, \end{aligned} \tag{5.1.30}$$

where $I_{ij} = \rho_0 Y_{ij}$. The energy equation (5.1.18) can be written in the form

$$\rho_0 T_0 \dot{\eta} = q_{i,i} + \rho_0 S. \tag{5.1.31}$$

The inequality (5.1.22) becomes

$$k_{ij} \theta_{,i} \theta_{,j} \geq 0. \tag{5.1.32}$$

The constitutive coefficients satisfy the symmetry relations

$$A_{ijrs} = A_{rsij}, \quad C_{ijrs} = C_{rsij}. \tag{5.1.33}$$

We note that in the case of a centrosymmetric isotropic material the constitutive equations (5.1.29) take the form

$$\begin{aligned} t_{ij} &= \lambda e_{rr} \delta_{ij} + (\mu + \kappa) e_{ij} + \mu e_{ji} - b\theta \delta_{ij}, \\ m_{ij} &= \alpha \kappa_{rr} \delta_{ij} + \beta \kappa_{ji} + \gamma \kappa_{ij}, \\ \rho_0 \eta &= b e_{rr} + a\theta, \\ q_i &= k \theta_{,i}, \end{aligned} \tag{5.1.34}$$

where $\lambda, \mu, \kappa, \alpha, \beta, \gamma, b, a$ and k are constitutive coefficients. For isotropic bodies we shall assume that $I_{ij} = J\delta_{ij}$.

For convenience, in the linear theory we denote T_i, M_i and Q by t_i, m_i and q, respectively. It follows from (5.1.3), (5.1.6) and (5.1.10) that the components of the surface traction, the components of surface moment, and the heat flux at regular points of ∂B can be expressed in the form

$$t_i = t_{ji}n_j, \quad m_i = m_{ji}n_j, \quad q = q_jn_j,$$

where n_j are the components of the unit outward normal to ∂P.

Remark. It follows from (5.1.24) that the internal energy per unit of initial volume in the case of isothermal deformations is

$$W = \frac{1}{2}A_{ijrs}e_{ij}e_{rs} + B_{ijrs}e_{ij}\kappa_{rs} + \frac{1}{2}C_{ijrs}\kappa_{ij}\kappa_{rs}. \tag{5.1.35}$$

W is the potential energy density per unit of initial volume. For isotropic bodies we have

$$\begin{aligned} 2W &= \lambda e_{rr}e_{ss} + (\mu+\kappa)e_{ij}e_{ij} + \mu e_{ij}e_{ji}+ \\ &\quad + \alpha\kappa_{rr}\kappa_{ss} + \beta\kappa_{ij}\kappa_{ji} + \gamma\kappa_{ij}\kappa_{ij}. \end{aligned} \tag{5.1.36}$$

If we assume that W is a positive definite quadratic form, then we find that the constitutive coefficients of an isotropic body satisfy the inequalities [105]

$$\begin{aligned} &3\lambda + 2\mu + \kappa > 0, \quad 2\mu + \kappa > 0, \quad \kappa > 0, \\ &3\alpha + \beta + \gamma > 0, \quad \gamma + \beta > 0, \quad \gamma - \beta > 0. \end{aligned} \tag{5.1.37}$$

5.2 Thermoelastic Processes. Boundary-Initial-Value Problem

The fundamental system of field equations of the linear theory of micropolar thermoelasticity consists of the geometrical equations (5.1.28), the constitutive equations (5.1.29), the equations of motion (5.1.30) and the energy equation (5.1.31). We assume that $f_i, g_i, S, \rho_0, I_{ij}, T_0$ and the constitutive coefficients are prescribed functions with the following properties: (*i*) f_i, g_i and S are continuous on $\overline{B}\times[0, t_1)$; (*ii*) ρ_0 is continuous and strictly positive on $\overline{B}$; (*iii*) I_{ij} is continuous symmetric and positive definite tensor on $\overline{B}$; (*iv*) $A_{ijrs}, B_{ijrs}, C_{ijrs}, \beta_{ij}, \xi_{ij}$ and k_{ij} are smooth on $\overline{B}$ while a is continuous on $\overline{B}$; (*v*) A_{ijrs}, B_{ijrs} and k_{ij} satisfy the relations (5.1.32), (5.1.33); (*vi*) T_0 is a positive constant.

We say that $(\boldsymbol{u}, \boldsymbol{\varphi}, \theta)$ is a dynamically admissible state on $\overline{B}\times[0, t_1)$ provided: (α) $\boldsymbol{u}$ and $\boldsymbol{\varphi}$ are of class C^2 on $B\times(0, t_1)$, and θ is of class $C^{2,1}$ on $B\times(0, t_1)$; (β) $\boldsymbol{u}, \dot{\boldsymbol{u}}, \ddot{\boldsymbol{u}}, \boldsymbol{\varphi}, \dot{\boldsymbol{\varphi}}, \ddot{\boldsymbol{\varphi}}, \boldsymbol{e}, \dot{\boldsymbol{e}}, \boldsymbol{\kappa}, \dot{\boldsymbol{\kappa}}, \theta, \theta_{,i}$ and $\dot{\theta}$ are continuous on $\overline{B}\times[0, t_1)$. By an admissible system of stresses and couple stresses on $\overline{B}\times[0, t_1)$ we mean an ordered array of functions (t_{ij}, m_{ij}) with the properties: (a) t_{ij} and m_{ij} are of class $C^{1,0}$ on $B\times(0, t_1)$; (b) $t_{ij}, t_{ji,j}, m_{ij}$ and $m_{ji,j}$ are continuous on $\overline{B}\times[0, t_1)$. We say that $\boldsymbol{q}$ is an admissible heat flux vector on $\overline{B}\times[0, t_1)$ provided: (a$'$) q_k is of class $C^{1,0}$ on $B\times(0, t_1)$; (b$'$) q_i and $q_{k,k}$ are continuous on $\overline{B}\times[0, t_1)$.

By an admissible process on $\overline{B}\times[0, t_1)$ we mean an ordered array of functions $p = (u_i, \varphi_i, \theta, e_{ij}, \kappa_{ij}, t_{ij}, m_{ij}, \eta, q_i)$ with the following properties: (1) $(\boldsymbol{u}, \boldsymbol{\varphi}, \theta)$ is a dynamically admissible state on $\overline{B}\times[0, t_1)$; (2) e_{ij} and κ_{ij} are continuous tensor fields on $\overline{B}\times[0, t_1)$; (3) (t_{ij}, m_{ij}) is an admissible system of stresses and couple stresses; (4) $\boldsymbol{q}$ is an admissible heat flux vector on $\overline{B}\times[0, t_1)$. The set of all admissible processes on $\overline{B}\times[0, t_1)$ is a vector space provided addition and scalar multiplication are defined in the natural manner (see Section 1.5).

We say that $p = (u_i, \varphi_i, \theta, e_{ij}, \kappa_{ij}, t_{ij}, m_{ij}, \eta, q_i)$ is a thermoelastic process corresponding to the supply terms (f_i, g_i, S) if p is an admissible process that satisfies the fundamental system of field equations (5.1.28)-(5.1.31) on $B\times(0, t_1)$.

To the field equations we adjoin boundary conditions and initial conditions. Let S_m $(m = 1, 2, ..., 6)$ be subsets of ∂B such that $\overline{S}_1 \cup S_2 = \overline{S}_3 \cup S_4 = \overline{S}_5 \cup S_6 = \partial B$, $S_1 \cap S_2 = S_3 \cap S_4 = S_5 \cap S_6 = \emptyset$. In the case of the mixed boundary value problem the boundary conditions are

$$\begin{aligned}
&u_i = \widetilde{u}_i \quad \text{on } \overline{S}_1\times(0, t_1), \quad t_{ji}n_j = \widetilde{t}_i \quad \text{on } S_2\times(0, t_1),\\
&\varphi_i = \widetilde{\varphi}_i \quad \text{on } \overline{S}_3\times(0, t_1), \quad m_{ji}n_j = \widetilde{m}_i \quad \text{on } S_4\times(0, t_1),\\
&\theta = \widetilde{\theta} \quad \text{on } \overline{S}_5\times(0, t_1), \quad q_i n_i = \widetilde{q} \quad \text{on } S_6\times(0, t_1),
\end{aligned} \tag{5.2.1}$$

where $\widetilde{u}_i, \widetilde{\varphi}_i, \widetilde{\theta}, \widetilde{t}_i, \widetilde{m}_i$ and $\widetilde{q}$ are prescribed functions. We assume that $\widetilde{u}_i, \widetilde{\varphi}_i$ and $\widetilde{\theta}$ are continuous on $\overline{S}_1\times(0, t_1)$, $\overline{S}_3\times(0, t_1)$ and $\overline{S}_5\times(0, t_1)$, respectively, while $\widetilde{t}_i, \widetilde{m}_i$ and $\widetilde{q}$ are continuous in time and piecewise regular on $S_2\times(0, t_1)$, $S_4\times(0, t_1)$ and $S_6\times(0, t_1)$, respectively. In the linear theory the Lagrangian coordinates are usually indicated by x_k.

We consider the initial conditions in the form

$$\begin{aligned}
&u_i(\boldsymbol{x}, 0) = u_i^0(\boldsymbol{x}), \quad \dot{u}_i(\boldsymbol{x}, 0) = v_i^0(\boldsymbol{x}), \quad \varphi_i(\boldsymbol{x}, 0) = \varphi_i^0(\boldsymbol{x}),\\
&\dot{\varphi}_i(\boldsymbol{x}, 0) = \nu_i^0(\boldsymbol{x}), \quad \eta(\boldsymbol{x}, 0) = \eta^0(\boldsymbol{x}), \quad \boldsymbol{x} \in \overline{B},
\end{aligned} \tag{5.2.2}$$

where $u_i^0, v_i^0, \varphi_i^0, \nu_i^0$ and η^0 are continuous functions on $\overline{B}$.

The mixed problem of micropolar thermoelastodynamics consists in finding a thermoelastic process p corresponding to the supply terms $(\boldsymbol{f}, \boldsymbol{g}, S)$ that satisfies the boundary conditions (5.2.1) and the initial conditions (5.2.2). We call such a thermoelastic process a solution of the mixed problem.

We note that the initial conditions (5.2.2) imply that in the case $a \neq 0$, we obtain

$$\theta(\boldsymbol{x}, 0) = \theta^0(\boldsymbol{x}), \ \boldsymbol{x} \in \overline{B}, \tag{5.2.3}$$

where

$$\theta^0 = a^{-1}[\rho_0 \eta^0 - \beta_{ij}(u^0_{j,i} + \varepsilon_{jik}\varphi^0_k) - \xi_{ij}\varphi^0_{j,i}].$$

5.3 Reciprocity. Existence and Uniqueness Results

The first reciprocal theorem in the classical thermoelastodynamics is due to Ionescu-Cazimir [202]. This result has been extended by Nowacki [287] to the micropolar thermoelasticity. The proof of the theorem is based on the assumption of null initial data and systematic use of the Laplace transform. In [158], Ieşan established a reciprocal theorem without using the Laplace transform. The method is based on a characterization of the boundary-initial-value problem in which the initial conditions are incorporated into the field equations. Later, in [184] it was established a new reciprocal theorem in the classical thermoelasticity. The proof avoids both the use of the Laplace transform and the incorporation of the initial conditions into the equations of motion. Moreover, the theorem leads to a uniqueness result. The results of [184] have been extended to micropolar thermoelasticity by Scalia [327].

We first establish a reciprocity relation which involves two admissible processes at different times. In what follows we suppose that the underlying time interval is $I = [0, \infty)$. If g is a continuous function on $B \times I$, then we denote by $\widehat{g}$ the function defined by

$$\widehat{g}(\boldsymbol{x}, t) = \int_0^t g(\boldsymbol{x}, s)ds, \quad (\boldsymbol{x}, t) \in B \times I. \tag{5.3.1}$$

We introduce the notation

$$H = \rho_0 \widehat{S} + \rho_0 T_0 \eta^0. \tag{5.3.2}$$

An immediate consequence of the equation (5.1.31) and the initial conditions (5.2.2) is the following proposition: if $p = (u_i, \varphi_i, \theta, e_{ij}, \kappa_{ij}, t_{ij}, m_{ij}, \eta, q_i)$ is a solution of the mixed problem, then

$$\rho_0 T_0 \eta = \widehat{q}_{i,i} + H, \tag{5.3.3}$$

on $B\times I$. We consider two external data systems $L^{(\alpha)} = (f_i^{(\alpha)}, g_i^{(\alpha)}, S^{(\alpha)}, \widetilde{u}_i^{(\alpha)}, \widetilde{\varphi}_i^{(\alpha)}, \widetilde{\theta}^{(\alpha)}, \widetilde{t}_i^{(\alpha)}, \widetilde{m}_i^{(\alpha)}, \widetilde{q}^{(\alpha)}, u_i^{0(\alpha)}, v_i^{0(\alpha)}, \varphi_i^{0(\alpha)}, \eta^{0(\alpha)})$, $(\alpha = 1, 2)$. Let $p^{(\alpha)} = (u_i^{(\alpha)}, \varphi_i^{(\alpha)}, \theta^{(\alpha)}, e_{ij}^{(\alpha)}, \kappa_{ij}^{(\alpha)}, t_{ij}^{(\alpha)}, m_{ij}^{(\alpha)}, \eta^{(\alpha)}, q_i^{(\alpha)})$ be a solution corresponding to $L^{(\alpha)}$, $(\alpha = 1, 2)$. We introduce the notations

$$
\begin{aligned}
t_i^{(\alpha)} = t_{ji}^{(\alpha)} n_j, \quad m_i^{(\alpha)} = m_{ji}^{(\alpha)} n_j, \quad q^{(\alpha)} = q_i^{(\alpha)} n_i,\\
H^{(\alpha)} = \rho_0 \widehat{S}^{(\alpha)} + \rho_0 T_0 \eta^{0(\alpha)}, \quad (\alpha = 1, 2).
\end{aligned}
\tag{5.3.4}
$$

Theorem 5.3.1. *Assume that*

(i) A_{ijrs} *and* C_{ijrs} *satisfy the symmetry relations* (5.1.33);

(ii) *the conductivity tensor* k_{ij} *is symmetric.*

Let

$$
\begin{aligned}
D_{\alpha\beta}(r,s) = & \int_B [\rho_0 f_i^{(\alpha)}(\boldsymbol{x},r) u_i^{(\beta)}(\boldsymbol{x},s) + \rho_0 g_i^{(\alpha)}(\boldsymbol{x},r)\varphi_i^{(\beta)}(\boldsymbol{x},s) - \\
& - \frac{1}{T_0} H^{(\alpha)}(\boldsymbol{x},r)\theta^{(\beta)}(\boldsymbol{x},s)]dV + \int_{\partial B} [t_i^{(\alpha)}(\boldsymbol{x},r) u_i^{(\beta)}(\boldsymbol{x},s) + \\
& + m_i^{(\alpha)}(\boldsymbol{x},r)\varphi_i^{(\beta)}(\boldsymbol{x},s) - \frac{1}{T_0}\widehat{q}^{(\alpha)}(\boldsymbol{x},r)\theta^{(\beta)}(\boldsymbol{x},s)]dA - \\
& - \int_B [\rho_0 \ddot{u}_i^{(\alpha)}(\boldsymbol{x},r) u_i^{(\beta)}(\boldsymbol{x},s) + I_{ij}\ddot{\varphi}_j^{(\alpha)}(\boldsymbol{x},r)\varphi_i^{(\beta)}(\boldsymbol{x},s) - \\
& - \frac{1}{T_0}\widehat{q}_i^{(\alpha)}(\boldsymbol{x},r)\theta_{,i}^{(\beta)}(\boldsymbol{x},s)]dV,
\end{aligned}
\tag{5.3.5}
$$

for all $r, s \in I$. *Then*

$$
D_{\alpha\beta}(r,s) = D_{\beta\alpha}(s,r), \quad (a, \beta = 1, 2),
\tag{5.3.6}
$$

for all $r, s \in I$.

Proof. We introduce the notation

$$
K_{\alpha\beta}(r,s) = t_{ij}^{(\alpha)}(r) e_{ij}^{(\beta)}(s) + m_{ij}^{(\alpha)}(r)\kappa_{ij}^{(\beta)}(s) - \rho_0 \eta^{(\alpha)}(r)\theta^{(\beta)}(s),
\tag{5.3.7}
$$

where, for convenience, we have supressed the argument $\boldsymbol{x}$. From (5.1.29) and (5.3.7) we get

$$
\begin{aligned}
K_{\alpha\beta}(r,s) = & A_{ijmn} e_{mn}^{(\alpha)}(r) e_{ij}^{(\beta)}(s) + C_{ijmn}\kappa_{mn}^{(\alpha)}(r)\kappa_{ij}^{(\beta)}(s) - \\
& - a\theta^{(\alpha)}(r)\theta^{(\beta)}(s) + B_{ijmn}\left(\kappa_{mn}^{(\alpha)}(r) e_{ij}^{(\beta)}(s) + e_{ij}^{(\alpha)}(r)\kappa_{mn}^{(\beta)}(s)\right) - \\
& - \beta_{ij}\left(\theta^{(\alpha)}(r) e_{ij}^{(\beta)}(s) + e_{ij}^{(\alpha)}(r)\theta^{(\beta)}(s)\right) - \\
& - \xi_{ij}\left(\theta^{(\alpha)}(r)\kappa_{ij}^{(\beta)}(s) + \kappa_{ij}^{(\alpha)}(r)\theta^{(\beta)}(s)\right).
\end{aligned}
\tag{5.3.8}
$$

On the basis of (5.1.33) we conclude that

$$K_{\alpha\beta}(r,s) = K_{\beta\alpha}(s,r), \quad (\alpha,\beta = 1,2), \tag{5.3.9}$$

for all $r, s \in I$.

On the other hand, in view of (5.1.28), (5.1.30), (5.3.3) and (5.3.7), we find that

$$K_{\alpha\beta}(r,s) = \left[t_{ij}^{(\alpha)}(r)u_j^{(\beta)}(s) + m_{ij}^{(\alpha)}(r)\varphi_j^{(\beta)}(s) - \frac{1}{T_0}\widehat{q}_i^{(\alpha)}(r)\theta^{(\beta)}(s)\right]_{,i} -$$
$$-u_i^{(\beta)}(s)\left[\rho_0\ddot{u}_i^{(\alpha)}(r) - \rho_0 f_i^{(\alpha)}(r)\right] - \varphi_i^{(\beta)}(s)\left[I_{ij}\ddot{\varphi}_j^{(\alpha)}(r) - \rho_0 g_i^{(\alpha)}(r)\right] -$$
$$-\frac{1}{T_0}H^{(\alpha)}(r)\theta^{(\beta)}(s) + \frac{1}{T_0}\widehat{q}_i^{(\alpha)}(r)\theta_{,i}^{(\beta)}(s).$$

If we integrate the above relation on B and use the divergence theorem and (5.3.9), then we obtain (5.3.6). □

We introduce the notations

$$\begin{aligned} f_i^{(\alpha)} &= \rho_0\left(j * f_i^{(\alpha)} + tv_i^{0(\alpha)} + u_i^{0(\alpha)}\right), \\ G_i^{(\alpha)} &= \rho_0 j * g_i^{(\alpha)} + I_{ij}\left(t\nu_j^{0(\alpha)} + \varphi_j^{0(\alpha)}\right), \ j(t) = t, \ t \in I. \end{aligned} \tag{5.3.10}$$

Theorem 5.3.1 implies the following reciprocal theorem.

Theorem 5.3.2. *Assume that k_{ij} and I_{ij} are symmetric tensors and the relations (5.1.33) hold. Let $p^{(\alpha)} = (u_i^{(\alpha)}, \varphi_i^{(\alpha)}, \theta^{(\alpha)}, e_{ij}^{(\alpha)}, \kappa_{ij}^{(\alpha)}, t_{ij}^{(\alpha)}, m_{ij}^{(\alpha)}, \eta^{(\alpha)}, q_i^{(\alpha)})$ be a solution corresponding to the external data system $L^{(\alpha)}$, $(\alpha = 1,2)$. Then*

$$\begin{aligned} &\int_B \left[F_i^{(1)} * u_i^{(2)} + G_i^{(1)} * \varphi_i^{(2)} - \frac{1}{T_0} j * H^{(1)} * \theta^{(2)}\right] dV + \\ &+\int_{\partial B} j * \left[t_i^{(1)} * u_i^{(2)} + m_i^{(1)} * \varphi_i^{(2)} - \frac{1}{T_0}\widehat{q}^{(1)} * \theta^{(2)}\right] dA = \\ &= \int_B \left[F_i^{(2)} * u_i^{(1)} + G_i^{(2)} * \varphi_i^{(1)} - \frac{1}{T_0} j * H^{(2)} * \theta^{(1)}\right] dV + \\ &+\int_{\partial B} j * \left[t_i^{(2)} * u_i^{(1)} + m_i^{(2)} * \varphi_i^{(1)} - \frac{1}{T_0}\widehat{q}^{(2)} * \theta^{(1)}\right] dA. \end{aligned} \tag{5.3.11}$$

Proof. We take in (5.3.6), $r = \tau$ and $s = t - \tau$ and integrate from 0 to t. With the aid of (5.3.5), the constitutive equations and symmetry of conductivity tensor, we arrive at

$$
\begin{aligned}
&\int_{\partial B}\left[t_i^{(1)} * u_i^{(2)} + m_i^{(1)} * \varphi_i^{(2)} - \frac{1}{T_0}\widehat{q}^{(1)} * \theta^{(2)}\right] dA+\\
&+\int_B\left[\rho_0 f_i^{(1)} * u_i^{(2)} + \rho_0 g_i^{(1)} * \varphi_i^{(2)} - \frac{1}{T_0} H^{(1)} * \theta^{(2)}\right] dV-\\
&-\int_B\left[\rho_0 \ddot{u}_i^{(1)} * u_i^{(2)} + I_{ij}\ddot{\varphi}_j^{(1)} * \varphi_i^{(2)}\right] dV =\\
&=\int_{\partial B}\left[t_i^{(2)} * u_i^{(1)} + m_i^{(2)} * \varphi_i^{(1)} - \frac{1}{T_0}\widehat{q}^{(2)} * \theta^{(1)}\right] dA+\\
&+\int_B\left[\rho_0 f_i^{(2)} * u_i^{(1)} + \rho_0 g_i^{(2)} * \varphi_i^{(1)} - \frac{1}{T_0} H^{(2)} * \theta^{(1)}\right] dV-\\
&-\int_B\left[\rho_0 \ddot{u}_i^{(2)} * u_i^{(1)} + I_{ij}\ddot{\varphi}_j^{(2)} * \varphi_i^{(1)}\right] dV.
\end{aligned}
\tag{5.3.12}
$$

If we take convolution of the relation (5.3.12) with $j(t) = t$, $t \in I$, and use the relations

$$
j * \ddot{u}_i^{(\alpha)} = u_i^{(\alpha)} - tv_i^{0(\alpha)} - u_i^{0(\alpha)}, \quad j * \ddot{\varphi}_i^{(\alpha)} = \varphi_i^{(\alpha)} - t\nu_i^{0(\alpha)} - \varphi_i^{0(\alpha)},
$$

then we obtain (5.3.11). □

The reciprocity relation (5.3.11) has been established in [160] by using a different method. Various applications of the reciprocity relation have been presented in [289], p. 218.

Now we use the reciprocity relation (5.3.6) to derive the following result.

Lemma 5.3.1. *Assume that* $p = (u_i, \varphi_i, \theta, e_{ij}, \kappa_{ij}, t_{ij}, m_{ij}, \eta, q_i)$ *is a solution corresponding to the external data system* $(f_i, g_i, S, \widetilde{u}_i, \widetilde{\varphi}_i, \widetilde{\theta}, \widetilde{t}_i, \widetilde{m}_i, \widetilde{q}, u_i^0, v_i^0, \varphi_i^0, \eta^0)$. *Let the constitutive coefficients be as in Theorem* 5.3.1 *and let*

$$
\begin{aligned}
D(r,s) = &\int_B\left[\rho_0 f_i(r)u_i(s) + \rho_0 g_i(r)\varphi_i(s) - \frac{1}{T_0} H(r)\theta(s)\right] dV+\\
&+\int_{\partial B}\left[t_i(r)u_i(s) + m_i(r)\varphi_i(s) - \frac{1}{T_0}\widehat{q}(r)\theta(s)\right] dA,
\end{aligned}
$$

for all $r, s \in I$. *Then*

$$
\begin{aligned}
&\frac{d}{dt}\left\{\int_B(\rho_0 u_i u_i + I_{ij}\varphi_i\varphi_j)\, dV + \frac{1}{T_0}\int_0^t\int_B k_{ij}\widehat{\theta}_{,i}\widehat{\theta}_{,j} dt\, dV\right\} =\\
&=\int_0^t [D(t-s,t+s) - D(t+s,t-s)]ds + \int_B\{\rho_0[\dot{u}_i(2t)u_i(0)+\\
&+\dot{u}_i(0)u_i(2t)] + I_{ij}[\dot{\varphi}_j(2t)\varphi_i(0) + \dot{\varphi}_j(0)\varphi_i(2t)]\}dV.
\end{aligned}
\tag{5.3.13}
$$

Proof. From (5.3.6) we obtain

$$\int_0^t D_{11}(t+s,t-s)ds = \int_0^t D_{11}(t-s,t+s)ds. \tag{5.3.14}$$

We apply relation (5.3.14) to the process $p^{(1)} = p$. It follows from (5.3.5) that

$$\begin{aligned}\int_0^t D_{11}(t+s,t-s)ds &= \int_0^t D(t+s,t-s)ds- \\ &-\int_0^t\int_B [\rho_0\ddot{u}_i(t+s)u_i(t-s)+I_{ij}\ddot{\varphi}_j(t+s)\varphi_i(t-s)+ \\ &+\frac{1}{T_0}k_{ij}\widehat{\theta}_{,j}(t+s)\theta_{,i}(t-s)]ds\,dV.\end{aligned} \tag{5.3.15}$$

Similarly,

$$\begin{aligned}\int_0^t D_{11}(t-s,t+s)ds &= \int_0^t D(t-s,t+s)ds- \\ &-\int_0^t\int_B [\rho_0\ddot{u}_i(t-s)u_i(t+s)+I_{ij}\ddot{\varphi}_j(t-s)\varphi_i(t+s)+ \\ &+\frac{1}{T_0}k_{ij}\widehat{\theta}_{,j}(t-s)\theta_{,i}(t+s)]ds\,dV.\end{aligned} \tag{5.3.16}$$

With the help of the identities

$$\begin{aligned}&\int_0^t \ddot{f}(t+s)g(t-s)ds = \dot{f}(2t)g(0)-\dot{f}(t)g(t)+\int_0^t \dot{f}(t+s)\dot{g}(t-s), \\ &\int_0^t \ddot{g}(t-s)f(t+s)ds = \dot{g}(t)f(t)-\dot{g}(0)f(2t)+\int_0^t \dot{g}(t-s)\dot{f}(t+s)ds, \\ &\int_0^t f(t+s)\dot{g}(t-s)ds = f(t)g(t)-g(0)f(2t)+\int_0^t \dot{f}(t+s)g(t-s)ds,\end{aligned} \tag{5.3.17}$$

the relations (5.3.15) and (5.3.16) can be expressed in the form

$$\begin{aligned}\int_0^t D_{11}(t+s,t-s)ds &= \int_0^t D(t+s,t-s)ds- \\ &-\int_B\left\{\rho_0\left[\dot{u}_i(2t)u_i(0)-\dot{u}_i(t)u_i(t)+\int_0^t \dot{u}_i(t-s)\dot{u}_i(t+s)ds\right]+\right.\end{aligned}$$

$$
+I_{ij}\left[\dot{\varphi}_j(2t)\varphi_i(0)-\dot{\varphi}_j(t)\varphi_i(t)+\int_0^t\dot{\varphi}_j(t-s)\dot{\varphi}_i(t+s)ds\right]-
$$
$$
-\frac{1}{T_0}k_{ij}\left[\widehat{\theta}_{,j}(t)\widehat{\theta}_{,i}(t)+\int_0^t\theta_{,j}(t+s)\widehat{\theta}_{,i}(t-s)ds\right]\Bigg\}dV,
$$

$$
\int_0^t D_{11}(t-s,t+s)ds=\int_0^t D(t-s,t+s)ds-
$$
$$
-\int_B\Bigg\{\rho_0\left[\dot{u}_i(t)u_i(t)-\dot{u}_i(0)u_i(2t)+\int_0^t\dot{u}_i(t-s)\dot{u}_i(t+s)ds\right]+
$$
$$
+I_{ij}\left[\dot{\varphi}_j(t)\varphi_i(t)-\dot{\varphi}_j(0)\varphi_i(2t)+\int_0^t\dot{\varphi}_j(t-s)\dot{\varphi}_i(t+s)ds\right]-
$$
$$
-\frac{1}{T_0}\int_0^t k_{ij}\widehat{\theta}_{,j}(t-s)\theta_{,i}(t+s)\Bigg\}dV.
\tag{5.3.18}
$$

It follows from (5.3.14) and (5.3.18) that (5.3.13) holds. □

Lemma 5.3.1 forms the basis of the following uniqueness result.

Theorem 5.3.3. *Assume that*

(i) ρ_0 *is strictly positive;*
(ii) I_{ij} *is symmetric and positive definite;*
(iii) k_{ij} *is symmetric and positive definite;*
(iv) $a \neq 0$ *on* B*;*
(v) *the symmetry relations* (5.1.33) *hold.*

Then the mixed problem of thermoelastodynamics has at most one solution.

Proof. Suppose that there are two solutions. Then their difference $(u_i,\varphi_i,\theta,e_{ij},\kappa_{ij},t_{ij},m_{ij},\eta,q_i)$ corresponds to null data. Thus, from (5.3.13) we get

$$
\frac{d}{dt}\left\{\int_B(\rho_0 u_i u_i+I_{ij}\varphi_i\varphi_j)dV+\frac{1}{T_0}\int_0^t\int_B k_{ij}\widehat{\theta}_{,j}\widehat{\theta}_{,i}dt\,dV\right\}=0.
$$

Since u_i and φ_i vanish initially, the above relation implies

$$
\int_B(\rho_0 u_i u_i+I_{ij}\varphi_i\varphi_j)dV+\frac{1}{T_0}\int_0^t\int_B k_{ij}\widehat{\theta}_{,j}\widehat{\theta}_{,i}dt\,dV=0. \tag{5.3.19}
$$

In view of hypotheses (i)-(iii), from (5.3.19) we obtain

$$
u_i=0,\quad \varphi_i=0,\quad \widehat{\theta}_{,i}=0\quad \text{on } B\times[0,\infty). \tag{5.3.20}
$$

By (5.3.20) we obtain $\theta_{,i} = 0$ on $B \times I$ so that $q_i = 0$ on $B \times I$ and $\theta(x,t) = G(t)$, $x \in B$, $t \in I$. It follows from the constitutive equations and the initial conditions that $\rho_0 \eta = aG$ on $B \times I$. The energy equation implies that $a_0 T_0 \dot{G} = 0$ on $B \times I$. Since T_0 and a are different from zero we obtain $G =$ constant on $B \times I$. Since $\rho_0 \eta = a\theta$, from the initial conditions and the hypothesis (iv) we conclude that $G(0) = 0$. Thus, $\theta = 0$ on $B \times I$ and the proof is complete. □

Other uniqueness results have been presented by Nowacki [289] and Eringen [102], [105]. Existence and spatial decay of solutions have been investigated in various papers (see, e.g. [69], [309], [185]).

In what follows we use the results established by Dafermos in [78] to prove the existence of a generalized solution. We assume that B is a properly regular region [110] of three-dimensional Euclidean space and consider the following boundary conditions

$$u_i = 0, \quad \varphi_i = 0, \quad \theta = 0 \quad \text{on } \partial B \times (0, t_1), \tag{5.3.21}$$

where $(0, t_1)$ is a given finite interval of time. To the field equations we add the initial conditions (5.2.2) and the boundary conditions (5.3.21). Let $z = (\boldsymbol{u}, \boldsymbol{\varphi}, \theta)$ be a dynamically admissible state corresponding to a solution of the problem and let $y = (\boldsymbol{u}^*, \boldsymbol{\varphi}^*, \theta^*)$ be a dynamically admissible state. It follows from (5.1.30) and (5.1.31) that

$$\begin{aligned} \rho_0 \left(f_i \dot{u}_i^* + g_i \dot{\varphi}_i^* + \frac{1}{T_0} S\theta^* \right) &= \rho_0 \ddot{u}_i \dot{u}_i^* + I_{ij} \ddot{\varphi}_j \dot{\varphi}_i^* + a\dot{\theta}\theta^* - \\ - \left(t_{ji} \dot{u}_i^* + m_{ji} \dot{\varphi}_i^* + \frac{1}{T_0} q_j \theta^* \right)_{,j} &+ \frac{1}{T_0} k_{ij} \theta_{,j} \theta^*_{,i} + \Lambda(z,y) + N(z,y), \end{aligned} \tag{5.3.22}$$

where

$$\begin{aligned} \Lambda(z,y) &= A_{ijmn} e_{ij} \dot{e}^*_{mn} + C_{ijmn} \kappa_{ij} \dot{\kappa}^*_{mn} + \\ &\qquad + B_{ijmn} (e_{ij} \dot{\kappa}^*_{mn} + \dot{e}^*_{mn} \kappa_{ij}), \\ N(z,y) &= \beta_{ij} (\dot{e}_{ij} \theta^* - \dot{e}^*_{ij} \theta) + \xi_{ij} (\dot{\kappa}_{ij} \theta^* - \dot{\kappa}^*_{ij} \theta), \\ e^*_{ij} &= e_{ij}(y) = u^*_{j,i} + \varepsilon_{jik} \varphi^*_k, \ \kappa^*_{ij} = \kappa_{ij}(y) = \varphi^*_{j,i}. \end{aligned} \tag{5.3.23}$$

We introduce the notation

$$2\mathcal{E}(y) = A_{ijmn} e^*_{ij} e^*_{mn} + C_{ijmn} \kappa^*_{ij} \kappa^*_{mn} + 2B_{ijmn} e^*_{ij} \kappa^*_{mn}. \tag{5.3.24}$$

It follows from (5.3.23) and (5.3.24) that

$$\Lambda(y,y) = \dot{\mathcal{E}}(y), \quad N(y,y) = 0. \tag{5.3.25}$$

Let $C^m(\overline{B})$ be the class of scalar fields possessing derivatives up to the m-th order in B which are continuous on $\overline{B}$. We denote by $\boldsymbol{C}^m(\overline{B})$ the class of vector fields with components in $C^m(\overline{B})$. We shall use the notations $\widehat{C}^1(B) = \{g \in C^1(\overline{B}) : g = 0 \text{ on } \partial B\}$, $\widehat{\boldsymbol{C}}^1(B) = \{\boldsymbol{u} \in \boldsymbol{C}^1(\overline{B}) : u_i = 0 \text{ on } \partial B\}$. Let H be a Banach space. By $C^m([0,t_1];H)$ we denote the set of functions from $[0,t_1]$ to H which possess on $[0,t_1]$ time derivatives in H up to the m-th order, continuous on $[0,t_1]$. In an analogous way we introduce the symbols $L_1([0,t_1];H)$, $L_2([0,t_1];H)$. We assume that: (i) the symmetry relations (5.1.33) hold; (ii) ρ_0 and a are strictly positive; (iii) $\mathcal{E}$ is a positive definite quadratic form; (iv) k_{ij} is symmetric and positive definite; (v) I_{ij} is symmetric and positive definite.

Let $\Omega = B\times(0,t_1)$. We consider the sets

$$\widetilde{C}^0(B) = \{(\boldsymbol{f},\boldsymbol{g},S) : (\boldsymbol{f},\boldsymbol{g},S)) \in \boldsymbol{C}^0(B)\times\boldsymbol{C}^0(B)\times C^0(B)\}\,,$$

$$\widetilde{C}^1(B) = \left\{(\boldsymbol{u},\boldsymbol{\varphi},\theta) : (\boldsymbol{u},\boldsymbol{\varphi},\theta)) \in \widehat{\boldsymbol{C}}^1(B)\times\widehat{\boldsymbol{C}}^1(B)\times\widehat{C}^1(B)\right\},$$

$$\mathcal{F}(\Omega) = C^\infty([0,t_1];\widetilde{C}^1(B)),$$

$$\overset{\circ}{\mathcal{F}}(\Omega) = \{(\boldsymbol{u},\boldsymbol{\varphi},\theta) : (\boldsymbol{u},\boldsymbol{\varphi},\theta) \in \mathcal{F}(\Omega),\ \boldsymbol{u} = \boldsymbol{0},\ \boldsymbol{\varphi} = \boldsymbol{0} \text{ on } B\times 0\}\,.$$

Let $z = (\boldsymbol{u},\boldsymbol{\varphi},\theta)$ be a dynamically admissible state corresponding to a solution of the considered problem. Let $y = (\overline{\boldsymbol{u}},\overline{\boldsymbol{\varphi}},\overline{\theta}) \in\overset{\circ}{\mathcal{F}}(\Omega)$. If we take in (5.3.22), $\dot{\boldsymbol{u}}^* = (t-t_1)\dot{\overline{\boldsymbol{u}}}$, $\dot{\boldsymbol{\varphi}}^* = (t-t_1)\dot{\overline{\boldsymbol{\varphi}}}$, $\theta^* = (t-t_1)\overline{\theta}$ and integrate over Ω, then we obtain

$$A(z,y) = P(\sigma,y) + J(\chi,y), \quad \forall\, y \in\overset{\circ}{\mathcal{F}}(\Omega), \tag{5.3.26}$$

where

$$\begin{aligned} A(z,y) = &\int_0^{t_1}\int_B \{\rho_0 \dot{u}_i \dot{\overline{u}}_i + I_{jk}\dot{\varphi}_k\dot{\overline{\varphi}}_j + a\theta\overline{\theta} + \\ &+\frac{1}{T_0}\int_0^t k_{ij}\theta_{,j}\overline{\theta}_{,i}ds + \beta_{ij}(u_{j,i}+\varepsilon_{jik}\varphi_k)\overline{\theta} + \xi_{ij}\varphi_{j,i}\overline{\theta} + \\ &+(t-t_1)[\rho_0\dot{u}_i\ddot{\overline{u}}_i + I_{jk}\dot{\varphi}_k\ddot{\overline{\varphi}}_j + a\theta\dot{\overline{\theta}} - \Lambda(z,y) + \\ &+\beta_{ij}(u_{j,i}+\varepsilon_{jik}\varphi_k)\dot{\overline{\theta}} + \xi_{ij}\varphi_{j,i}\dot{\overline{\theta}} + \beta_{ij}(\dot{\overline{u}}_{j,i}+\varepsilon_{jik}\dot{\overline{\varphi}}_k)\theta + \\ &+\xi_{ij}\dot{\overline{\varphi}}_{j,i}\theta]\}dV\,dt, \end{aligned} \tag{5.3.27}$$

$$P(\sigma,y) = -\int_0^{t_1}\int_B (t-t_1)\rho_0\left(f_i\dot{\overline{u}}_i + g_i\dot{\overline{u}}_i + \frac{1}{T_0}S\overline{\theta}\right)dV\,dt,$$

$$J(\chi,y) = t_1\int_B [\rho_0 v_i^0\dot{\overline{u}}_i + I_{jk}\nu_k^0\dot{\overline{\varphi}}_j + a\theta^0\overline{\theta} +$$

$$+\beta_{ij}(u^0_{j,i}+\varepsilon_{jik}\varphi^0_k)\overline{\theta}+\xi_{ij}\varphi^0_{j,i}\overline{\theta}\Big]\Big|_{t=0}dV$$

and

$$\sigma=(f_i,g_i,S),\quad \chi=(u^0_i,v^0_i,\varphi^0_i,\nu^0_i,\theta^0), \tag{5.3.28}$$

where θ^0 is given in (5.2.3).

We note that $A(z,y)$ makes sense for $z\in\mathcal{F}(\Omega)$ and $y\in\overset{\circ}{\mathcal{F}}(\Omega)$. Moreover, $P(\sigma,y)$ and $J(\chi,y)$ can be defined for $\sigma\in C^\infty([0,t_1];\widetilde{C}^0(B))$, $y\in\overset{\circ}{\mathcal{F}}(\Omega)$ and χ with the properties that $(\boldsymbol{u}^0,\boldsymbol{\varphi}^0,\theta^0)\in\widetilde{C}^1(B)$, $v^0_i,\nu^0_i\in\widehat{C}^1(B)$.

It follows from (5.3.25) and (5.3.27) that

$$\begin{aligned}A(y,y)=&\int_0^{t_1}\int_B\Big\{\frac{1}{2}\left(\rho_0\dot{\overline{u}}_i\dot{\overline{u}}_i+I_{ij}\dot{\overline{\varphi}}_j\dot{\overline{\varphi}}_i+a\overline{\theta}^2\right)+\\&+\mathcal{E}(y)+\frac{1}{T_0}\int_0^t k_{ij}\overline{\theta}_{,j}\overline{\theta}_{,i}ds\Big\}dV\,dt+\\&+\frac{1}{2}t_1\int_B\left(\rho_0\dot{\overline{u}}_i\dot{\overline{u}}_i+I_{ij}\dot{\overline{\varphi}}_j\dot{\overline{\varphi}}_i+a\overline{\theta}^2\right)\Big|_{t=0}dV,\end{aligned}\tag{5.3.29}$$

for any $y\in\overset{\circ}{\mathcal{F}}(\Omega)$. We denote by $\boldsymbol{W}_m(B)$ the completion of $\boldsymbol{C}^m(\overline{B})$ by means of the norm induced by the inner product

$$(\boldsymbol{v},\boldsymbol{u})_{\mathbf{W}(B)}=\int_B\sum_{k=0}^m v_{i,p_1p_2\ldots p_k}u_{i,p_1p_2\ldots p_k}dV.$$

Let $V(\Omega)$ be the Hilbert space obtained as the completion of $\mathcal{F}(\Omega)$ by means of the norm $|\cdot|$ induced by the inner product

$$\begin{aligned}<(\boldsymbol{u},\boldsymbol{\varphi},\theta),(\overline{\boldsymbol{u}},\overline{\boldsymbol{\varphi}},\overline{\theta})>=&\int_0^{t_1}\int_B\{u_i\overline{u}_i+\varphi_i\overline{\varphi}_i+\theta\overline{\theta}+u_{i,j}\overline{u}_{i,j}+\\&+\varphi_{i,j}\overline{\varphi}_{i,j}+\dot{u}_i\dot{\overline{u}}_i+\dot{\varphi}_i\dot{\overline{\varphi}}_i+\int_0^t\theta_{,i}\overline{\theta}_{,i}ds\}dV\,dt.\end{aligned}\tag{5.3.30}$$

By $\overset{\circ}{V}(\Omega)$ we denote the closed linear manifold of $V(\Omega)$ obtained as the completion of $\overset{\circ}{\mathcal{F}}(\Omega)$ by means of $|\cdot|$. We shall denote by $\overset{\circ}{U}(\Omega)$ the Hilbert space obtained as the completion of $\overset{\circ}{\mathcal{F}}(\Omega)$ by means of the norm generated by the inner product

$$[y_1,y_2]=\langle y_1,y_2\rangle+\langle\dot{y}_1,\dot{y}_2\rangle,$$

where $y_1 = (\boldsymbol{u}, \boldsymbol{\varphi}, \theta)$ and $y_2 = (\overline{\boldsymbol{u}}, \overline{\boldsymbol{\varphi}}, \overline{\theta})$. Let $G(B)$ be the completion of $\widetilde{C}^0(B)$ by means of the norm $\|\cdot\|_{\mathbf{W}_0(B)\times\mathbf{W}_0(B)\times W_0(B)}$. On the basis of Sobolev's embedding theorem, A and P can be extended by continuity onto $V(\Omega)\times \overset{\circ}{U}(\Omega)$ and $L_1([0,t_1]; G(B))\times \overset{\circ}{U}(\Omega)$, respectively. By $H_0(B)$ we denote the completion of the set $\{(\boldsymbol{u}, \boldsymbol{v}, \boldsymbol{\varphi}, \boldsymbol{w}, \theta) : (\boldsymbol{u}, \boldsymbol{\varphi}, \theta) \in \widetilde{C}^1(B),\ \boldsymbol{v}, \boldsymbol{w} \in \widehat{\boldsymbol{C}}^1(B)\}$ by means of the norm

$$|(\boldsymbol{u}, \boldsymbol{v}, \boldsymbol{\varphi}, \boldsymbol{w}, \theta)|_0 = \left\{\frac{1}{2}\int_B [\rho_0 \boldsymbol{v}^2 + I_{ij}w_j w_i + a\theta^2 + 2\mathcal{E}(y)]dV\right\}^{1/2},$$

where $y = (\boldsymbol{u}, \boldsymbol{\varphi}, \theta)$. We note that $J(\mathcal{X}, y)$ makes sense for $\mathcal{X} \in H_0(B)$ and $y \in \overset{\circ}{U}(\Omega)$. In view of (5.3.29) we find that

$$\begin{aligned} A(y,y) \geq \frac{1}{2}\int_0^{t_1}\int_B [\rho_0 \dot{\overline{\boldsymbol{u}}}^2 + I_{ij}\dot{\overline{\varphi}}_j\dot{\overline{\varphi}}_i + a\overline{\theta}^2 + 2\mathcal{E}(y)+ \\ + \frac{2}{T_0}\int_0^t k_{ij}\overline{\theta}_{,j}\overline{\theta}_{,i}ds]dV\,dt, \end{aligned} \tag{5.3.31}$$

for any $y \in \overset{\circ}{\mathcal{F}}(\Omega)$. This inequality can be extended by continuity to $\overset{\circ}{U}(\Omega)$. We note that

$$\int_B \mathcal{E}(y)dV$$

is coercive on $\|(\boldsymbol{u}, \boldsymbol{\varphi})\|_{\mathbf{W}_1(B)\times\mathbf{W}_1(B)}$. This result can be derived by using the positive definiteness of $\mathcal{E}(y)$ and the results of Hlavácek and Nečas [153] on inequalities of Korn's type. It follows from (5.3.31) that there exists a positive constant c_1 depending only on B, T_0, ρ_0, I_{rs} and the constitutive coefficients such that

$$|y|^2 \leq c_1 A(y,y), \quad \forall\, y \in \overset{\circ}{U}(\Omega). \tag{5.3.32}$$

$z = (\boldsymbol{u}, \boldsymbol{\varphi}, \theta) \in V(\Omega)$ is called a finite energy solution of the boundary-initial-value problem with the boundary conditions (5.3.21), the initial data $\mathcal{X} = (u_i^0, v_i^0, \varphi_i^0, \nu_i^0, \theta^0) \in H_0(B)$ and the source $\sigma = (\boldsymbol{f}, \boldsymbol{g}, S) \in L_1([0,t_1]; G(B))$, if z satisfies the conditions

$$\begin{aligned} & A(z,y) = P(\sigma, y) + J(\mathcal{X}, y), \quad \forall\, y \in \overset{\circ}{U}(\Omega), \\ & \boldsymbol{u}(t) \overset{\mathbf{W}_0(B)}{\longrightarrow} \boldsymbol{u}^0, \quad \boldsymbol{\varphi} \overset{\mathbf{W}_0(B)}{\longrightarrow} \boldsymbol{\varphi}^0, \quad t \longrightarrow 0. \end{aligned} \tag{5.3.33}$$

It can be shown that if z is a finite energy solution which is smooth enough, then it is also a solution in the classical sense.

Lemma 5.3.2. *Let $z = (\boldsymbol{u}, \boldsymbol{\varphi}, \theta) \in V(\Omega)$ be a finite energy solution in Ω, with the initial data $\chi = (u_i^0, v_i^0, \varphi_i^0, \nu_i^0, \theta^0)$ and source $\sigma = (f_i, g_i, S)$. Then $\widetilde{z} = (\widetilde{\boldsymbol{u}}, \widetilde{\boldsymbol{\varphi}}, \widetilde{\theta})$ defined by*

$$\widetilde{z} = \int_0^t z\, d\tau, \tag{5.3.34}$$

is also a finite energy solution in Ω, with initial data $\widetilde{\chi} = (0, u_i^0, 0, \varphi_i^0, 0)$ and the source

$$\widetilde{\sigma} = \int_0^t \sigma\, d\tau + \sigma^*, \tag{5.3.35}$$

where

$$\begin{aligned} &\sigma^* = (f_i^*, g_i^*, S^*), \quad f_i^* = v_i^0, \quad g_i^* = I_{ij}\nu_j^0, \\ &\rho_0 S^* = T_0[\beta_{ij}(u_{j,i}^0 + \varepsilon_{jik}\varphi_k^0) + \xi_{ij}\varphi_{j,i}^0 + a\theta^0]. \end{aligned} \tag{5.3.36}$$

Proof. We have to show that

$$A(\widetilde{z}, y) = P(\widetilde{\sigma}, y) + J(\widetilde{\chi}, y), \tag{5.3.37}$$

for any $y = (\overline{\boldsymbol{u}}, \overline{\boldsymbol{\varphi}}, \overline{\theta}) \in \overset{\circ}{U}(\Omega)$. It follows from (5.3.27) that

$$\begin{aligned} A(\widetilde{z}, y) = &\int_0^{t_1}\int_B \{\rho_0 \dot{\widetilde{u}}_i \frac{d}{dt}[(t - t_1)\dot{\overline{u}}_i] + I_{jk}\dot{\widetilde{\varphi}}_k \frac{d}{dt}[(t - t_1)\dot{\overline{\varphi}}_j] + \\ &+ a\widetilde{\theta}\frac{d}{dt}[(t - t_1)\overline{\theta}] + \frac{1}{T_0}\int_0^t k_{ij}\widetilde{\theta}_{,j}\overline{\theta}_{,i} ds - (t - t_1)\Lambda(\widetilde{z}, y) + \\ &+ [\beta_{ij}(\widetilde{u}_{j,i} + \varepsilon_{jik}\widetilde{\varphi}_k) + \xi_{ij}\widetilde{\varphi}_{j,i}]\frac{d}{dt}[(t - t_1)\overline{\theta}] + \\ &+ (t - t_1)[\beta_{ij}(\dot{\overline{u}}_{j,i} + \varepsilon_{jik}\dot{\overline{\varphi}}_k)\widetilde{\theta} + \xi_{ij}\dot{\overline{\varphi}}_{j,i}\widetilde{\theta}]\} dV\, dt. \end{aligned}$$

We define $\widetilde{y} = (\boldsymbol{u}^*, \boldsymbol{\varphi}^*, \theta^*)$ by

$$\begin{aligned} u_i^* &= -\int_0^t \frac{1}{s - t_1}\int_{t_1}^s (\tau - t_1)\dot{\overline{u}}_i(\tau) d\tau\, ds, \\ \varphi_i^* &= -\int_0^t \frac{1}{s - t_1}\int_{t_1}^s (\tau - t_1)\dot{\overline{\varphi}}_i(\tau) d\tau\, ds, \\ \theta^* &= -\frac{1}{t - t_1}\int_{t_1}^t (\tau - t_1)\overline{\theta}(\tau) d\tau. \end{aligned}$$

We note that

$$\frac{d}{dt}[(t-t_1)\dot{u}_i^*] = -(t-t_1)\dot{\overline{u}}_i,\ \frac{d}{dt}[(t-t_1)\dot{\varphi}_i^*] = -(t-t_1)\dot{\overline{\varphi}}_i,$$
$$\frac{d}{dt}[(t-t_1)\theta^*] = -(t-t_1)\overline{\theta},\ u_i^*(\boldsymbol{x},0)=0,\ \varphi_i^*(\boldsymbol{x},0)=0,\ \widetilde{y}\in\overset{\circ}{U}(\Omega).$$

An integration by parts with respect to t shows that

$$\begin{aligned}
&\int_0^{t_1}(t-t_1)\Lambda(\widetilde{z},y)dt = \int_0^{t_1}(t-t_1)\Lambda(z,\widetilde{y})dt,\\
&A(\widetilde{z},y) = A(z,\widetilde{y}) + t_1\int_B(\rho_0 u_i^0\dot{u}_i^* + I_{jk}\varphi_k^0\dot{\varphi}_j^*)\Big|_{t=0}dV,\\
&P(\widetilde{\sigma},y) = P(\sigma,\widetilde{y}) + J(\chi,\widetilde{y}),\\
&J(\widetilde{\chi},y) = t_1\int_B(\rho_0 u_i^0\dot{\overline{u}}_i + I_{ij}\varphi_j^0\dot{\overline{\varphi}}_i)dV.
\end{aligned} \tag{5.3.38}$$

With the help of (5.3.38) the relation (5.3.37) reduces to

$$A(z,\widetilde{y}) = P(\sigma,\widetilde{y}) + J(\chi,\widetilde{y}).$$

We note that this equation is verified since z is a finite energy solution of the problem and $\widetilde{y}\in\overset{\circ}{U}(\Omega)$. □

Theorem 5.3.4. *Assume that the hypotheses* (i)-(v) *hold and* $(\boldsymbol{f},\boldsymbol{g},S)\in L_1([0,t_1];G(B))$, $(\boldsymbol{u}^0,\boldsymbol{v}^0,\boldsymbol{\varphi}^0,\boldsymbol{\nu}^0,\theta^0)\in H_0(B)$. *Then the boundary-initial-value problem has at most one finite energy solution.*

Proof. We assume that there exist two solutions $z_1, z_2\in\overset{\circ}{V}(\Omega)$. Clearly, z_1-z_2 is a solution corresponding to null data. Moreover, $z_1-z_2\in\overset{\circ}{V}(\Omega)$. If we define

$$z(\boldsymbol{x},t) = \int_0^t[z_1(\boldsymbol{x},\tau) - z_2(\boldsymbol{x},\tau)]d\tau,$$

then, in view of Lemma 5.3.2, z is a finite energy solution corresponding to null data,

$$A(z,y) = P(0,y) + J(0,y),\quad \forall y\in\overset{\circ}{U}(\Omega). \tag{5.3.39}$$

Since $z\in\overset{\circ}{U}(\Omega)$ we can take $y=z$ and from (5.3.32) and (5.3.39) we obtain $z=0$. □

Theorem 5.3.5. *Assume that the hypotheses* (i)-(v) *hold and* $\sigma = (\boldsymbol{f}, \boldsymbol{g}, S) \in L_2([0, t_1]; G(B))$. *Then there exists a finite energy solution in* Ω *with zero initial data and source* σ. *Moreover, we have*

$$|z| \leq c_0 \|\sigma\|_{\mathcal{L}}, \quad c_0 > 0, \tag{5.3.40}$$

where c_0 *is a constant and* $\mathcal{L} = \boldsymbol{L}_2(\Omega) \times \boldsymbol{L}_2(\Omega) \times L_2(\Omega)$.

Proof. Let us consider $y \in \overset{\cup}{U}(\Omega)$, y fixed. Then by Riesz-Fréchet representation theorem, there exists a linear map $F : \overset{\circ}{U}(\Omega) \longrightarrow \overset{\circ}{V}(\Omega)$ such that

$$A(z, y) = < z, Fy >, \quad \forall\, z \in \overset{\circ}{V}(\Omega). \tag{5.3.41}$$

Let us show that the map F is one to one. If there exists $y_0 \in \overset{\circ}{U}(\Omega)$ such that $Fy_0 = 0$, then from (5.3.41) we obtain $A(z, y_0)$ for any $z \in \overset{\circ}{V}(\Omega)$. In particular we have $A(y_0, y_0) = 0$ and (5.3.32) implies $y_0 = 0$. We conclude that F is one to one. Let $\mathcal{R}(F)$ be the range of F in $\overset{\circ}{V}(\Omega)$. We denote by F^{-1} the inverse map of F from $\mathcal{R}(F)$ to $\overset{\circ}{U}(\Omega)$. From (5.3.32) and Schwarz's inequality we obtain

$$|y|^2 \leq c_1 A(y, y) = c_1 \langle y, Fy \rangle \leq c_1 |y|\, |Fy|,$$

so that

$$|y| \leq c_1 |Fy|, \quad \forall\, y \in \overset{\circ}{U}(\Omega). \tag{5.3.42}$$

Let us show that $\mathcal{R}(F)$ is dense in $\overset{\circ}{V}(\Omega)$. Suppose this is not true. Thus, there exists $z_0 \in \overset{\circ}{V}(\Omega) \backslash \mathcal{R}(F)$, $z_0 \neq 0$, such that $< z_0, Fy >= 0$ for any $y \in \overset{\circ}{U}(\Omega)$. From (5.3.41) we get $A(z_0, y) = 0$ for any $y \in \overset{\circ}{U}(\Omega)$ so that z_0 is a finite energy solution of the problem corresponding to null data. By Theorem 5.3.4 we conclude that $z_0 = 0$. It follows that $\mathcal{R}(F)$ is dense in $\overset{\circ}{V}(\Omega)$. Then, the domain of F^{-1} may be extended by continuity from $\mathcal{R}(F)$ onto $\overset{\circ}{V}(\Omega)$ so that $F^{-1} : \overset{\circ}{V}(\Omega) \longrightarrow \overset{\circ}{V}(\Omega)$ becomes a bounded operator with the norm

$$\|F^{-1}\| \leq c_1. \tag{5.3.43}$$

Let us consider the bilinear form $P(\sigma, y)$. With the help of (5.3.43) and Schwarz's inequality we conclude that there exists a constant c_2 such that

$$\left|P(\sigma, F^{-1}p)\right| \leq c_2 \left|F^{-1}(p)\right| \|\sigma\|_{\mathcal{L}} \leq c_1 c_2 |p|\, \|\sigma\|_{\mathcal{L}},$$

for any $p \in \overset{\circ}{V}(\Omega)$. Hence $P(\sigma, F^{-1}p)$ is bounded on $\overset{\circ}{V}(\Omega)$. By the Riesz-Fréchet theorem, there exists $\widehat{z}(\sigma) \in \overset{\circ}{V}(\Omega)$ such that

$$P(\sigma, F^{-1}p) = < \widehat{z}(\sigma), p >, \quad \forall p \in \overset{\circ}{V}(\Omega).$$

In particular, for any $y \in \overset{\circ}{U}(\Omega)$, $Fy \in \overset{\circ}{V}(\Omega)$ we have

$$P(\sigma, y) = < \widehat{z}(\sigma), Fy > . \tag{5.3.44}$$

It follows from (5.3.41) and (5.3.44) that

$$A(\widehat{z}(\sigma), y) = P(\sigma, y), \quad \forall y \in \overset{\circ}{U}(\Omega).$$

We conclude that $\widehat{z}(\sigma)$ is the desired solution. From (5.3.41)-(5.3.44) we get

$$|\widehat{z}|^2 = < \widehat{z}, \widehat{z} > = A(\widehat{z}, F^{-1}\widehat{z}) = P(\sigma, F^{-1}\widehat{z}) \leq c_1 c_2 |\widehat{z}| \, \|\sigma\|_{\mathcal{L}},$$

so that the estimate (5.3.40) holds. □

Theorem 5.3.5 can be extended to the case of nonzero initial data and mixed boundary conditions (see [78]).

5.4 Variational Theorem

With a view toward establishing a variational characterization of the solution we first give an alternative formulation of the boundary-initial-value problem of thermoelastodynamics. We introduce the functions $\boldsymbol{F}, \boldsymbol{G}$ and H on $\overline{B} \times [0, t_1)$ defined by

$$\begin{gathered} F_i = \rho_0(j * f_i + t v_i^0 + u_i^0), \; G_i = \rho_0 j * g_i + I_{ij}(t \nu_j^0 + \varphi_j^0), \\ H = \rho_0(1 * S + T_0 \eta^0), \end{gathered} \tag{5.4.1}$$

where $j(t) \equiv t$, $1(t) \equiv 1$, $t \in [0, t_1)$, and $*$ denotes the operation of convolution (see Section 2.2).

Theorem 5.4.1. *Let* $p = (u_i, \varphi_i, \theta, e_{ij}, \kappa_{ij}, t_{ij}, m_{ij}, \eta, q_i)$ *be an admissible process on* $\overline{B} \times [0, t_1)$. *Then* p *satisfies the equations of motion* (5.1.30), *the equation of energy* (5.1.31) *and the initial conditions* (5.2.2) *if and only if*

$$\begin{aligned} & j * t_{ki,k} + F_i = \rho_0 u_i, \\ & j * (m_{ki,k} + \varepsilon_{irs} t_{rs}) + G_i = I_{ik} \varphi_k, \\ & \rho_0 T_0 \eta = 1 * q_{i,i} + H, \end{aligned} \tag{5.4.2}$$

on $B \times [0, t_1)$.

Proof. Assume that (5.1.30),(5.1.31) and the initial conditions (5.2.2) hold. Taking the convolution of the equation (5.1.30) with j, and the convolution of the equation (5.1.31) with 1, we conclude with the aid of (5.4.1) that the equations (5.4.2) hold. The converse is verified by reversing the foregoing argument. □

In the classical elasticity this result was established by Ignaczak [191] and Gurtin [138]. Theorem 5.4.1 was derived in [158]. This result furnishes an alternative characterization of the solution to the mixed problem in which the initial conditions are incorporated into the field equations. Thus, the admissible process p is a solution of the mixed problem of thermoelastodynamics if and only if p satisfies the equations (5.1.28), (5.1.29), (5.4.2) and the boundary conditions (5.2.1).

Let $\mathcal{A}$ be a linear space, K a subset of $\mathcal{A}$ and Ω a functional on K. Further, let

$$y, y' \in \mathcal{A}, \quad y + \lambda y' \in K, \tag{5.4.3}$$

for every scalar λ, and formally define $\delta_{y'}\Omega\{y\}$ by

$$\delta_{y'}\Omega\{y\} = \frac{d}{d\lambda}\Omega\{y + \lambda y'\}\bigg|_{\lambda=0}.$$

We say that the variation of Ω is zero at y over K, and write $\delta\Omega\{y\} = 0$, if and only if $\delta_{y'}\Omega\{y\}$ exists and equals zero for all y' consistent with (5.4.3).

If the conductivity tensor k_{rs} is invertible then we define λ_{ij} by

$$k_{im}\lambda_{mj} = \delta_{ij}, \tag{5.4.4}$$

where δ_{ij} is the Kronecker delta.

In the context of the classical theory of thermoelasticity, variational theorems have been presented in various works (see, e.g., Carlson [40], p. 338, Lebon [239], p. 370). Ieşan [158] established the following variational theorem in micropolar thermoelasticity.

Theorem 5.4.2. *Assume that*

(i) *the conductivity tensor is symmetric and invertible;*

(ii) $a \neq 0$ *on* B;

(iii) I_{ij} *is a symmetric tensor;*

(iv) *the symmetry relations* (5.1.33) *hold.*

Let M be the set of all admissible processes, and for each $t \in [0,t_1)$, *define the functional* $\Lambda_t\{\cdot\}$ *on M by*

$$\Lambda_t\{p\} = \int_B \Big\{ j * \Big[\frac{1}{2} A_{pqrs} e_{pq} * e_{rs} + B_{pqrs} e_{pq} * \kappa_{rs} +$$
$$+\frac{1}{2} C_{pqrs}\kappa_{pq} * \kappa_{rs} + \frac{1}{2a}(\rho_0\eta - \beta_{ij}e_{ij} - \xi_{ij}\kappa_{ij}) * (\rho_0\eta - \beta_{mn}e_{mn} -$$
$$-\xi_{mn}\kappa_{mn}) - j * (t_{rs} * e_{rs} + m_{rs} * \kappa_{rs} + \rho_0\eta * \theta) + \frac{1}{2}(\rho_0 u_i * u_i +$$
$$+I_{ik}\varphi_i * \varphi_k + \frac{1}{T_0} j * 1 * \lambda_{rs} q_s * q_r) - (j * t_{ki,k} + F_i) * u_i -$$
$$-[j * (m_{ki,k} + \varepsilon_{irs}t_{rs}) + G_i] * \varphi_i + \frac{1}{T_0} j * (H * \theta - 1 * q_i * \theta_{,i}) \Big\} dV +$$
$$+ \int_{S_1} j * t_i * \widetilde{u}_i \, dA + \int_{S_2} j * (t_i - \widetilde{t}_i) * u_i dA + \int_{S_3} j * m_i * \widetilde{\varphi}_i dA +$$
$$+ \int_{S_4} j * (m_i - \widetilde{m}_i) * \varphi_i dA + \frac{1}{T_0}\int_{S_5} j * 1 * (\theta - \widetilde{\theta}) * q \, dA +$$
$$+\frac{1}{T_0}\int_{S_6} j * 1 * \widetilde{q} * \theta \, dA, \tag{5.4.5}$$

for every $p = (u_i, \varphi_i, \theta, e_{ij}, \kappa_{ij}, t_{ij}, m_{ij}, \eta, q_i) \in M$. *Then*

$$\delta\Lambda_t\{p\} = 0, \quad t \in [0,t_1),$$

at an admissible process p if and only if p is a solution of the mixed problem.

Proof. Let $p' = (u_i', \varphi_i', \theta', e_{ij}', \kappa_{ij}', t_{ij}', m_{ij}', \eta', q_i') \in M$ and $p \in M$. Then $p + \lambda p' \in M$ for every scalar λ. A straightforward calculation yields

$$\delta_{p'}\Lambda_t\{p\} = \int_B j * \{[A_{rsmn}e_{mn} + B_{rsmn}\kappa_{mn} - \frac{1}{a}\beta_{rs}(\rho_0\eta -$$
$$-\beta_{mn}e_{mn} - \xi_{mn}\kappa_{mn}) - t_{rs}] * e_{rs}' + [B_{mnrs}e_{mn} + C_{rsmn}\kappa_{mn} -$$
$$-\frac{1}{a}\xi_{rs}(\rho_0\eta - \beta_{mn}e_{mn} - \xi_{mn}\kappa_{mn}) - m_{rs}] * \kappa_{rs}' +$$
$$+\frac{1}{a}(\rho_0\eta - \beta_{mn}e_{mn} - \xi_{mn}\kappa_{mn} - a\theta) * \rho_0\eta' + 1 * (\lambda_{im}q_m - \theta_{,i}) * \frac{q_i'}{T_0}\} dV -$$
$$-\int_B \{(j * t_{ki,k} + F_i - \rho_0 u_i) * u_i' + [j * (m_{ki,k} + \varepsilon_{irs}t_{rs}) +$$

$$
\begin{aligned}
&+G_i - I_{im}\varphi_m] * \varphi_i' + j * [\rho_0\eta - \frac{1}{T_0}(H + 1 * q_{i,i})] * \theta'\}dV - \\
&-\int_B j * [(e_{mn} - u_{n,m} + \varepsilon_{mnk}\varphi_k) * t_{mn}' + (\kappa_{rs} - \varphi_{s,r}) * m_{rs}']dV + \\
&+\int_{S_1} j * (\widetilde{u}_i - u_i) * t_i' dA + \int_{S_2} j * (t_i - \widetilde{t}_i) * u_i' dA + \\
&+\int_{S_3} j * (\widetilde{\varphi}_i - \varphi_i) * m_i' dA + \int_{S_4} j * (m_i - \widetilde{m}_i) * \varphi_i' dA + \\
&+\frac{1}{T_0}\int_{S_5} j * 1 * (\theta - \widetilde{\theta}) * q' dA + \frac{1}{T_0}\int_{S_6} j * 1 * (\widetilde{q} - q) * \theta' dA,
\end{aligned}
\tag{5.4.6}
$$

for $t \in [0, t_1)$. If p is a solution of the problem, then on the basis of (5.4.6) and Theorem 5.4.1 we obtain

$$
\delta_{p'}\Lambda_t\{p\} = 0, \quad t \in [0, t_1), \tag{5.4.7}
$$

for every $p' \in M$, and therefore

$$
\delta\Lambda_t\{p\} = 0, \quad t \in [0, t_1). \tag{5.4.8}
$$

Conversely, suppose that (5.4.8) and hence (5.4.7) holds. On making suitable choices of p' and appealling Lemmas 64.1-64.3 established by Gurtin [139], p. 224, we see that p is a solution of the mixed problem. □

Clearly, the functional whose variation vanishes at a solution can be simplified by putting restrictions on its domain. Other variational theorems in micropolar thermoelasticity have been presented by Nowacki [289], p. 247 and Eringen [105], p. 132.

5.5 Homogeneous and Isotropic Solids. Plane Waves

In the case of centrosymmetric homogeneous and isotropic bodies, from (5.1.28), (5.1.30), (5.1.31) and (5.1.34) we obtain the following field equations in terms of the displacement vector, microrotation vector and temperature variation

$$
\begin{aligned}
&(\mu + \kappa)\Delta u_i + (\lambda + \mu)u_{j,ji} + \kappa\varepsilon_{irs}\varphi_{s,r} - b\theta_{,i} + \rho_0 f_i = \rho_0\ddot{u}_i, \\
&\gamma\Delta\varphi_i + (\alpha + \beta)\varphi_{j,ji} + \kappa\varepsilon_{imn}u_{n,m} - 2\kappa\varphi_i + \rho_0 g_i = J\ddot{\varphi}_i, \\
&k\Delta\theta - bT_0\dot{u}_{j,j} - aT_0\dot{\theta} = -\rho_0 S,
\end{aligned}
\tag{5.5.1}
$$

on $B\times I$. We introduce the operators

$$
\begin{aligned}
&D_1 = (\lambda+2\mu+\kappa)\Delta - \rho_0\frac{\partial^2}{\partial t^2}, \qquad D_2 = (\mu+\kappa)\Delta - \rho_0\frac{\partial^2}{\partial t^2},\\
&D_3 = (\alpha+\beta+\gamma)\Delta - J\frac{\partial^2}{\partial t^2} - 2\kappa, \quad D_4 = \gamma\Delta - J\frac{\partial^2}{\partial t^2} - 2\kappa, \qquad (5.5.2)\\
&D_5 = k\Delta - c\frac{\partial}{\partial t},
\end{aligned}
$$

where $c = aT_0$. The equations (5.5.1) can be expressed in the form

$$
\begin{aligned}
&D_2\boldsymbol{u} + (\lambda+\mu)\operatorname{grad}\operatorname{div}\boldsymbol{u} + \kappa\operatorname{curl}\boldsymbol{\varphi} - b\operatorname{grad}\theta + \rho_0\boldsymbol{f} = \boldsymbol{0},\\
&D_4\boldsymbol{\varphi} + (\alpha+\beta)\operatorname{grad}\operatorname{div}\boldsymbol{\varphi} + \kappa\operatorname{curl}\boldsymbol{u} + \rho_0\boldsymbol{g} = \boldsymbol{0}, \qquad (5.5.3)\\
&D_5\theta - bT_0\operatorname{div}\dot{\boldsymbol{u}} = -\rho_0 S.
\end{aligned}
$$

To these equations we must adjoin initial conditions and boundary conditions.

We now investigate the propagation of the harmonic plane waves in a homogeneous and isotropic space with zero body loads. In what follows we assume that the inequalities (5.1.37) hold. Moreover, we suppose that $k>0$ and $J>0$. We consider plane waves of the form

$$
\begin{aligned}
\boldsymbol{u} &= \operatorname{Re}\{C_1\boldsymbol{A}\exp[i(\zeta\boldsymbol{x}\cdot\boldsymbol{p}-\omega t]\},\\
\boldsymbol{\varphi} &= \operatorname{Re}\{C_2\boldsymbol{B}\exp[i(\zeta\boldsymbol{x}\cdot\boldsymbol{p}-\omega t]\}, \qquad (5.5.4)\\
\theta &= \operatorname{Re}\{C_3\exp[i(\zeta\boldsymbol{x}\cdot\boldsymbol{p}-\omega t)]\},
\end{aligned}
$$

where $\boldsymbol{A},\boldsymbol{B}$ and $\boldsymbol{p}$ are unit vectors and C_i,ζ and ω are constants. The vectors $\boldsymbol{A}$ and $\boldsymbol{B}$ define the directions of the displacement vector and the microrotation vector, and $\boldsymbol{p}$ defines the direction of propagation. The functions (5.5.4) satisfy the equations (5.5.3) with $\boldsymbol{f}=\boldsymbol{g}=\boldsymbol{0}$, $S=0$ only if

$$
\begin{aligned}
&[\rho_0\omega^2-(\mu+\kappa)\zeta^2]C_1\boldsymbol{A}-(\lambda+\mu)\zeta^2C_1(\boldsymbol{A}\cdot\boldsymbol{p})\boldsymbol{p}+\\
&\qquad +i\kappa\zeta C_2\boldsymbol{p}\times\boldsymbol{B}-ib\zeta C_3\boldsymbol{p}=\boldsymbol{0},\\
&(J\omega^2-2\kappa-\gamma\zeta^2)C_2\boldsymbol{B}-(\alpha+\beta)\zeta^2C_2(\boldsymbol{B}\cdot\boldsymbol{p})\boldsymbol{p}+i\kappa\zeta C_1\boldsymbol{p}\times\boldsymbol{A}=\boldsymbol{0}, \qquad (5.5.5)\\
&-bT_0\omega\zeta C_1\boldsymbol{A}\cdot\boldsymbol{p}+(i\omega c-k\zeta^2)C_3=0.
\end{aligned}
$$

i) Longitudinal waves. If $\boldsymbol{p}=\boldsymbol{A}$ and $\boldsymbol{B}=\boldsymbol{0}$, then we have longitudinal displacement waves. In this case the system (5.5.5) becomes

$$
\begin{aligned}
&[\rho_0\omega^2-(\lambda+2\mu+\kappa)\zeta^2]C_1-ib\zeta C_3=0, \qquad (5.5.6)\\
&-bT_0\omega\zeta C_1+(i\omega c-k\zeta^2)C_3=0.
\end{aligned}
$$

We introduce the notations

$$c_1 = \left(\frac{\lambda + 2\mu + \kappa}{\rho_0}\right)^{1/2}, \ \omega^* = \frac{c_1^2 c}{k}, \ \chi = \frac{\omega}{\omega^*},$$
$$\xi = \frac{c_1}{\omega^*}\zeta, \ \varepsilon = \frac{b^2 T_0}{c c_1^2 \rho_0}. \tag{5.5.7}$$

The system (5.5.6) has non-trivial solutions if and only if

$$\xi^4 - [\chi^2 + (1+\varepsilon)i\chi]\xi^2 + i\chi^3 = 0. \tag{5.5.8}$$

It follows from (5.5.7) and (5.5.8) that ζ is complex. The longitudinal displacement waves are damped and dispersive.

If $\boldsymbol{p} = \boldsymbol{B}$ and $\boldsymbol{A} = \boldsymbol{0}$ we have longitudinal microrotational waves. In this case the equations (5.5.5) reduce to $C_1 = C_3 = 0$ and

$$J\omega^2 - 2\kappa - (\alpha + \beta + \gamma)\zeta^2 = 0. \tag{5.5.9}$$

We introduce the notations

$$\omega_0 = \left(\frac{2\kappa}{J}\right)^{1/2}, \quad c_3 = \left(\frac{\alpha + \beta + \gamma}{J}\right)^{1/2}. \tag{5.5.10}$$

From (5.5.9) we obtain that the phase velocity is given by $c_3(1 - \omega_0^2/\omega^2)^{-1/2}$. The phase velocity is real only for $\omega > \omega_0$.

ii) Transverse waves. In the case of transverse waves we have $\boldsymbol{A}\cdot\boldsymbol{p} = 0$, $\boldsymbol{B}\cdot\boldsymbol{p} = 0$, $\boldsymbol{A}\cdot\boldsymbol{B} = 0$ and from (5.5.5) we obtain $C_3 = 0$ and the equations

$$[\rho_0\omega^2 - (\mu+\kappa)\zeta^2]C_1\boldsymbol{A} + i\kappa\zeta C_2(\boldsymbol{p}\times\boldsymbol{B}) = \boldsymbol{0},$$
$$(J\omega^2 - 2\kappa - \gamma\zeta^2)C_2\boldsymbol{B} + i\kappa\zeta C_1(\boldsymbol{p}\times\boldsymbol{A}) = \boldsymbol{0}. \tag{5.5.11}$$

In view of the relation $\boldsymbol{p}\times(\boldsymbol{p}\times\boldsymbol{A}) = -\boldsymbol{A}$, we obtain

$$[\rho_0\omega^2 - (\mu+\kappa)\zeta^2](J\omega^2 - 2\kappa - \gamma\zeta^2) - \kappa^2\zeta^2 = 0. \tag{5.5.12}$$

If we introduce the notations

$$c_2 = \left(\frac{\mu+\kappa}{\rho_0}\right)^{1/2}, \quad c_4 = \left(\frac{\gamma}{J}\right)^{1/2}, \quad \sigma_2 = \frac{\omega}{c_2}, \quad \sigma_4 = \frac{\omega}{c_4},$$
$$p_0 = \frac{\kappa}{J c_4^2}, \quad s = \frac{\kappa}{\rho_0 c_2^2}, \tag{5.5.13}$$

then the equation (5.5.12) can be written in the form

$$\zeta^4 - [\sigma_2^2 + \sigma_4^2 + p_0(s-2)]\zeta^2 + \sigma_2^2(\sigma_4^2 - 2p_0) = 0. \tag{5.5.14}$$

From (5.5.14) we find that

$$\zeta_{1,2}^2 = \frac{1}{2}[\sigma_2^2+\sigma_4^2+p_0(s-2) \pm \{[\sigma_4^2-\sigma_2^4+p_0(s-2)]^2+4p_0 s\sigma_2^2\}^{1/2}] =$$
$$= \frac{1}{2}[\sigma_2^2+\sigma_4^2+p_0(s-2) \pm \{[\sigma_2^2+\sigma_4^2+p_0(s-2)]^2-4\sigma_2^2(\sigma_4^2-2p_0)\}^{1/2}],$$

so that ζ_1^2 and ζ_2^2 are real. Moreover, for $\sigma_4^2 > 2p_0$, i.e. for $\omega^2 > \omega_0^2$, we have $\zeta_1^2 > 0$, $\zeta_2^2 > 0$ which correspond to two real phase velocities. For $\omega^2 < \omega_0^2$ we have $\zeta_1^2 > 0$, $\zeta_2^2 < 0$, so that there is one phase velocity only, the second being imaginary. We note that the phase velocities which appear in (5.5.12) depend on the frequency.

We conclude that the only waves coupled with the temperature field are the longitudinal displacement waves.

5.6 A Representation of Galerkin Type. Fundamental Solutions

In the first part of this section we present a solution of Galerkin type of the field equations (5.5.1). Then we use this solution to derive the fundamental solutions in the theory of steady vibrations.

We introduce the operators

$$\begin{aligned} &\Gamma = D_2D_4+\kappa^2\Delta, \quad \Lambda = D_1D_5 - b^2T_0\Delta\frac{\partial}{\partial t}, \quad A = (\lambda+\mu)D_4 - \kappa^2,\\ &B = (\alpha+\beta)D_2 - \kappa^2, \quad P = AD_5 - b^2T_0\frac{\partial}{\partial t}D_4, \end{aligned} \tag{5.6.1}$$

where D_i are defined in (5.5.2).

Theorem 5.6.1. *Let*

$$\begin{aligned} &\boldsymbol{u} = (P\,\mathrm{grad}\,\mathrm{div} - \Lambda D_4)\,\boldsymbol{G} + \kappa D_3\,\mathrm{curl}\,\boldsymbol{H} - b\,\mathrm{grad}\,\Phi,\\ &\boldsymbol{\varphi} = \kappa\Lambda\,\mathrm{curl}\,\boldsymbol{G} + (B\,\mathrm{grad}\,\mathrm{div} - D_2D_3)\boldsymbol{H},\\ &\theta = -bT_0\frac{\partial}{\partial t}\Gamma\,\mathrm{div}\,\boldsymbol{G} - D_1\Phi, \end{aligned} \tag{5.6.2}$$

where the fields G_i of class C^8, H_i of class C^6 and Φ of class C^4 satisfy the equations

$$\Gamma\Lambda G_i = \rho_0 f_i, \quad D_3\Gamma H_i = \rho_0 g_i, \quad \Lambda\Phi = \rho_0 S, \tag{5.6.3}$$

on $B\times I$. Then u_i, φ_i and θ satisfy the equations (5.5.3).

Proof. Clearly,

$$D_5\Gamma+P\Delta=\Lambda D_4,\ A\Delta-D_1D_4=-\Gamma,\ D_1-D_2=(\lambda+\mu)\Delta,$$
$$D_4B+(\alpha+\beta)(B\Delta-D_2D_3)+\kappa^2D_3=0, \tag{5.6.4}$$
$$D_2P-(\lambda+\mu)D_5\Gamma+\kappa^2\Lambda+b^2T_0\dot{\Gamma}=0.$$

If we substitute $\boldsymbol{u},\boldsymbol{\varphi}$ and θ given by (5.6.2) into the left-hand side of (5.5.3), then in view of (5.6.4) and (5.6.3) we find that

$$\begin{aligned}
&D_2\boldsymbol{u}+(\lambda+\mu)\operatorname{grad}\operatorname{div}\boldsymbol{u}+\kappa\operatorname{curl}\boldsymbol{\varphi}-b\operatorname{grad}\theta=\\
&\quad=-\Lambda D_2D_4\boldsymbol{G}+\{D_2P-(\lambda+\mu)D_5\Gamma+\\
&\quad+\kappa^2\Lambda+b^2T_0\dot{\Gamma}\}\operatorname{grad}\operatorname{div}\boldsymbol{G}=-\Lambda D_2D_4\boldsymbol{G}=-\rho_0\boldsymbol{f},
\end{aligned}$$

$$\begin{aligned}
&D_4\boldsymbol{\varphi}+(\alpha+\beta)\operatorname{grad}\operatorname{div}\boldsymbol{\varphi}+\kappa\operatorname{curl}\boldsymbol{u}=-D_3(D_2D_4+\kappa^2\Delta)\boldsymbol{H}+\\
&+\{D_4B+(\alpha+\beta)(B\Delta-D_2D_3)+\kappa^2D_3\}\operatorname{grad}\operatorname{div}\boldsymbol{H}=-\rho_0\boldsymbol{g}.
\end{aligned}$$

Similarly,

$$D_5\theta-bT_0\operatorname{div}\dot{\boldsymbol{u}}=-bT_0(\Gamma D_5+P\Delta-\Lambda D_4)\operatorname{div}\dot{\boldsymbol{G}}-\Lambda\Phi=-\rho S.\ \square$$

The representations of Galerkin type in micropolar thermoelasticity have been investigated in various papers (see, for example, Stefaniak [349], Dyszlewicz [96], Nowacki [289], Ieşan [159]).

We now use the solution (5.6.2) to derive fundamental solutions of the system (5.5.3) in the case of steady vibrations. We assume that

$$\begin{aligned}
&\boldsymbol{f}=\operatorname{Re}[\boldsymbol{f}^*(\boldsymbol{x})\exp(-i\omega t)],\ \boldsymbol{g}=\operatorname{Re}[\boldsymbol{g}^*(\boldsymbol{x})\exp(-i\omega t)],\\
&S=\operatorname{Re}[S^*(\boldsymbol{x})\exp(-i\omega t)],
\end{aligned} \tag{5.6.5}$$

where ω is the frequency of vibration and $i=(-1)^{1/2}$. If we seek for solutions to (5.5.3) of the form

$$\begin{aligned}
&\boldsymbol{u}=\operatorname{Re}[\boldsymbol{u}^*(\boldsymbol{x};\omega)\exp(-i\omega t)],\ \boldsymbol{\varphi}=\operatorname{Re}[\boldsymbol{\varphi}^*(\boldsymbol{x};\omega)\exp(-i\omega t)],\\
&\theta=\operatorname{Re}[\theta^*(\boldsymbol{x};\omega)\exp(-i\omega t)],
\end{aligned} \tag{5.6.6}$$

then the field equations reduce to a differential system for the amplitudes u_i^*,φ_i^* and θ^*. We introduce the notations

$$\begin{aligned}
&G_j=\operatorname{Re}[G_j^*(\boldsymbol{x};\omega)\exp(-i\omega t)],\\
&H_j=\operatorname{Re}[H_j^*(\boldsymbol{x};\omega)\exp(-i\omega t)],\\
&\Phi=\operatorname{Re}[\Phi^*(\boldsymbol{x};\omega)\exp(-i\omega t)].
\end{aligned} \tag{5.6.7}$$

It follows from (5.6.2) that $\boldsymbol{u}^*, \boldsymbol{\varphi}^*$ and θ^* can be expressed in terms of functions $\boldsymbol{G}^*, \boldsymbol{H}^*$ and S^* by the relations

$$\begin{aligned} u_j^* &= P^* G_{s,sj}^* - \Lambda^* D_4^* G_j^* + \kappa D_3^* \varepsilon_{jrs} H_{s,r}^* - b\Phi_{,j}^*, \\ \varphi_j^* &= \kappa \Lambda^* \varepsilon_{jrs} G_{s,r}^* + B^* H_{s,sj}^* - D_2^* D_3^* H_j^*, \\ \theta^* &= i\omega b T_0 \Gamma^* G_{m,m}^* - D_1^* \Phi^*, \end{aligned} \tag{5.6.8}$$

where

$$\begin{aligned} &P^* = A^* D_5^* + i\omega b^2 T_0 D_4^*, \ \Gamma^* = D_2^* D_4^* + \kappa^2 \Delta, \\ &\Lambda^* = D_1^* D_5^* + i\omega b^2 T_0 \Delta, \ A^* = (\lambda + \mu) D_4^* - \kappa^2, \\ &B^* = (\alpha + \beta) D_2^* - \kappa^2, \ D_1^* = (\lambda + 2\mu + \kappa)\Delta + \rho_0 \omega^2, \\ &D_2^* = (\mu + \kappa)\Delta + \rho_0 \omega^2, \ D_5^* = k\Delta + i\omega c, \\ &D_3^* = (\alpha + \beta + \gamma)\Delta + J\omega^2 - 2\kappa, \ D_4^* = \gamma\Delta + J\omega^2 - 2\kappa. \end{aligned} \tag{5.6.9}$$

In view of (5.6.3), (5.6.5) and (5.6.7), we find that the functions G_j^*, H_j^* and θ^* satisfy the equations

$$\Gamma^* \Lambda^* G_j^* = \rho_0 f_j^*, \ D_3^* \Gamma^* H_j^* = \rho_0 g_j^*, \ \Lambda^* \Phi^* = \rho_0 S^*. \tag{5.6.10}$$

In what follows we shall assume that (5.1.37) holds and we shall use the notations (5.5.7), (5.5.10), (5.5.13). We note that

$$\begin{aligned} D_3^* &= (\alpha + \beta + \gamma)(\Delta + \tau^2), \\ \Gamma^* &= \gamma(\mu + \kappa)(\Delta + k_1^2)(\Delta + k_2^2), \\ \Lambda^* &= k(\lambda + 2\mu + \kappa)(\Delta + k_3^2)(\Delta + k_4^2), \end{aligned} \tag{5.6.11}$$

where

$$\tau = \sigma_3 \left(1 - \frac{\omega_0^2}{\omega^2}\right)^{1/2}, \quad \sigma_3 = \frac{\omega}{c_3}, \tag{5.6.12}$$

k_1^2, k_2^2 are the roots of the equation

$$y^4 - [\sigma_2^2 + \sigma_4^2 + p_0(s-2)]y^2 + \sigma_2^2(\sigma_4^2 - 2p_0) = 0,$$

and k_3^2, k_4^2 are the roots of the equation

$$z^4 - \left[\sigma_1^2 + \frac{i\omega c}{k}(1+\varepsilon)\right] z^2 + \frac{i\omega c}{k}\sigma_1^2 = 0.$$

The roots k_1^2 and k_2^2 are real (see Section 5.5). The roots k_3^2 and k_4^2 are complex. In what follows we denote by k_s, $(s = 1, 2, 3, 4)$, the roots with positive real parts and assume that τ, k_1, k_2, k_3 and k_4 are distinct.

The equations (5.6.10) can be written in the form

$$\begin{aligned}(\Delta + k_1^2)(\Delta + k_2^2)(\Delta + k_3^2)(\Delta + k_4^2)G_j^* &= \rho_1 f_j^*,\\ (\Delta + k_1^2)(\Delta + k_2^2)(\Delta + \tau^2)H_j^* &= \rho_2 g_j^*,\\ (\Delta + k_3^2)(\Delta + k_4^2)\Phi^* &= \rho_3 S^*,\end{aligned} \tag{5.6.13}$$

where

$$\begin{aligned}\rho_1 &= \rho_0[k\gamma(\mu+\kappa)(\lambda+2\mu+\kappa)]^{-1},\\ \rho_2 &= \rho_0[\gamma(\mu+k)(\alpha+\beta+\gamma)]^{-1},\\ \rho_3 &= \rho_0[k(\lambda+2\mu+\kappa)]^{-1}.\end{aligned} \tag{5.6.14}$$

If $f_i^* = f^*\delta_{ij}$, (j fixed), $g_i^* = 0$, $S^* = 0$, then we take $G_i^* = \psi\delta_{ij}$, $H_i^* = 0$, $S^* = 0$. It follows from (5.6.13) that ψ satisfies the equation

$$(\Delta + k_1^2)(\Delta + k_2^2)(\Delta + k_3^2)(\Delta + k_4^2)\psi = \rho_1 f^*. \tag{5.6.15}$$

If the functions ψ_s, $(s = 1, 2, 3, 4)$, satisfy the equations

$$(\Delta + k_j^2)\psi_j = \rho_1 f^*, \text{ (no sum; } j = 1, 2, 3, 4), \tag{5.6.16}$$

then the function ψ can be taken in the form

$$\psi = \sum_{s=1}^{4} a_s \psi_s,$$

where

$$a_s^{-1} = \prod_{j=1\ (j\neq s)}^{4} (k_s^2 - k_j^2), \quad (s = 1, 2, 3, 4). \tag{5.6.17}$$

Let us assume that $f^* = \delta(x - y)$ where δ is the Dirac delta and y is a fixed point. Then, from (5.6.16) we obtain

$$\psi_s = -\frac{\rho_1}{4\pi r}\, e^{ik_s r}, \quad (s = 1, 2, 3, 4),$$

where $r = |\boldsymbol{x} - \boldsymbol{y}|$. Thus, in this case we get

$$G_i^* = E\delta_{ij}, \quad H_i^* = 0, \quad \Phi^* = 0, \tag{5.6.18}$$

where

$$E = -\frac{1}{4\pi r}\,\rho_1 \sum_{s=1}^{4} a_s e^{ik_s r}.$$

From (5.6.8) and (5.6.18) we obtain

$$\begin{aligned} u_k^{*(j)} &= P^* E_{,jk} - \delta_{jk}\Lambda^* D_4^* E, \\ \varphi_k^{*(j)} &= \kappa\Lambda^* \varepsilon_{ksj} E_{,s}, \\ \theta^{*(j)} &= i\omega T_0 b\Gamma^* E_{,j}, \ (j = 1, 2, 3). \end{aligned} \tag{5.6.19}$$

Let us assume now that $f_i^* = 0$, $g_i^* = \delta_{ij}\delta(x - y)$, $S^* = 0$. In this case we take $G_i^* = 0$, $H_i^* = e\delta_{ij}$, $\Phi^* = 0$. From (5.6.13) we obtain

$$(\Delta + k_1^2)(\Delta + k_2^2)(\Delta + \tau^2)e = \rho_2 \delta.$$

By using the same procedure, we get

$$e = -\frac{\rho_2}{4\pi r} \sum_{s=1}^{3} b_s e^{i\lambda_s r}, \tag{5.6.20}$$

where $\lambda_1 = k_1$, $\lambda_2 = k_2$, $\lambda_3 = \tau$ and

$$b_s^{-1} = \prod_{i=1\,(i\neq s)}^{3} (\lambda_s^2 - \lambda_i^2).$$

In this case, from (5.6.8) we obtain the amplitudes

$$\begin{gathered} u_k^{*(3+j)} = \kappa D_3^* \varepsilon_{krj} e_{,r}, \ \varphi_k^{*(3+j)} = B^* e_{,jk} - D_2^* D_3^* e \delta_{jk}, \\ \theta^{*(3+j)} = 0. \end{gathered} \tag{5.6.21}$$

Finally, let us assume that $f_i^* = 0$, $g_i^* = 0$, $S^* = \delta(x - y)$. Now we take $G_i^* = 0$, $H_i^* = 0$, $\Phi^* = h$, where h satisfies the equation

$$(\Delta + k_3^2)(\Delta + k_4^2)h = \rho_3 \delta.$$

In this case we obtain

$$h = \frac{\rho_3}{4\pi r(k_3^2 - k_4^2)}[\exp(ik_3 r) - \exp(ik_4 r)]. \tag{5.6.22}$$

From (5.6.8) we find the solution

$$u_k^{*(7)} = -bh_{,k}, \ \varphi_k^{*(7)} = 0, \ \theta^{*(7)} = -D_1^* h. \tag{5.6.23}$$

The functions $u_k^{*(s)}$, $\varphi_k^{*(s)}, \theta^{*(s)}$, $(s = 1, 2, ..., 7)$, represent the fundamental solutions of the system of field equations in the case of steady vibrations. These solutions have been established by Nowacki [288]. Shanker and Dhalival [333] and Crăciun [76] have used these solutions and the method of potentials to reduce boundaryvalue problems to singular integral equations for which Fredholm's theorems hold.

5.7 Transient Waves

In this section we study the propagation of discontinuities in the thermoelastic space under the action of concentrated body loads. This problem has been considered by various authors (see, e.g., Shanker and Dhalival [334], [335], Chiriţă [52]). Here we present some of the results established in [52].

We consider a homogeneous and isotropic material which occupies the three-dimensional space. In this section we shall designate by $f^{(1)}$ the Laplace transform with respect to time of the function f,

$$f^{(1)}(\boldsymbol{x},p)=\int_0^\infty f(\boldsymbol{x},t)\exp(-pt)dt.$$

We introduce the Fourier-Laplace transform of the function f by

$$f^{(2)}(\boldsymbol{\xi},p)=\frac{1}{(2\pi)^{3/2}}\int_{E^3} f^{(1)}(\boldsymbol{x},p)e^{ix_j\xi_j}dV.$$

If we use the notations (5.5.7), (5.5.10), (5.5.13), the equations (5.5.1) can be written in the form

$$\begin{aligned}
&c_2^2\Delta u_i+(c_1^2-c_2^2)u_{j,ji}+sc_2^2\varepsilon_{ijk}\varphi_{k,j}-\frac{b}{\rho_0}\,\theta_{,i}+f_i=\ddot{u}_i,\\
&c_4^2\Delta\varphi_i+(c_3^2-c_4^2)\varphi_{j,ji}+\sigma c_4^2\varepsilon_{ijk}u_{k,j}-2\sigma c_4^2\varphi_i+\frac{\rho_0}{J}\,g_i=\ddot{\varphi}_i,\\
&\Delta\theta-\nu\dot{u}_{j,j}-\frac{c}{k}\,\dot{\theta}=-\frac{\rho_0}{k}\,S,
\end{aligned}\tag{5.7.1}$$

where

$$\sigma=\frac{\kappa}{Jc_4^2},\quad \nu=\frac{bT_0}{k}.\tag{5.7.2}$$

Taking the Fourier-Laplace transform of equations (5.7.1) we obtain the following algebraic equations

$$\begin{aligned}
&c_2^2(\xi^2+\beta_2^2)u_k^{(2)}+(c_1^2-c_2^2)\xi_k\xi_j u_j^{(2)}+isc_2^2\varepsilon_{krs}\xi_r\varphi_s^{(2)}-\frac{ib}{\rho_0}\,\xi_k\theta^{(2)}=f_k^{(2)},\\
&c_4^2(\xi^2+2\sigma+\beta_4^2)\varphi_k^{(2)}+(c_3^2-c_4^2)\xi_k\xi_j\varphi_j^{(2)}+\sigma c_4^2\xi_r\varepsilon_{krs}u_s^{(2)}=\frac{\rho_0}{J}\,g_k^{(2)},\\
&(\xi^2+q)\theta^{(2)}-i\nu p\xi_j u_j^{(2)}=\frac{\rho_0}{k}\,S^{(2)},
\end{aligned}\tag{5.7.3}$$

where

$$\xi^2=\xi_1^2+\xi_2^2+\xi_3^2,\ \beta_2=\frac{p}{c_2},\ \beta_4=\frac{p}{c_4},\ q=\frac{c}{k}\,p.\tag{5.7.4}$$

We denote

$$\begin{aligned}
&\beta_1 = \frac{p}{c_1}, \quad \beta_3 = \frac{1}{c_3}\,(p^2 + 2\sigma c_4^2)^{1/2}, \\
&P_1 = (\xi^2 + \beta_1^2)(\xi^2 + q) + q\varepsilon\xi^2, \\
&P_2 = (\xi^2 + \beta_2^2)(\xi^2 + 2\sigma + \beta_4^2) - s\sigma\xi^2,
\end{aligned} \tag{5.7.5}$$

where ε is defined in (5.5.7). It follows from (5.7.3) that

$$\begin{aligned}
\xi_j u_j^{(2)} &= \frac{1}{c_1^2 P_1}(\xi^2 + q)\xi_k f_k^{(2)} + \frac{im}{P_1}\,\xi^2 S^{(2)}, \\
\xi_j \varphi_j^{(2)} &= \frac{\rho_0}{Jc_3^2(\xi^2 + \beta_3^2)}\,\xi_k g_k^{(2)}, \\
\theta^{(2)} &= \frac{i\nu p}{c_1^2 P_1}\,\xi_k f_k^{(2)} + \frac{\rho_0}{kP_1}\,(\xi^2 + \beta_1^2) S^{(2)},
\end{aligned} \tag{5.7.6}$$

where

$$m = \frac{b}{kc_1^2}.$$

We also obtain

$$\begin{aligned}
i\varepsilon_{jrs}\xi_r u_s^{(2)} &= \frac{1}{\xi^2 + \beta_2^2}[s(\xi_j\xi_r\varphi_r^{(2)} - \xi^2\varphi_j^{(2)}) + \frac{i}{c_2^2}\,\varepsilon_{jrs}\xi_r f_s^{(2)}], \\
i\varepsilon_{jrs}\xi_r \varphi_s^{(2)} &= \frac{1}{\xi^2 + 2\sigma + \beta_4^2}[\sigma(\xi_j\xi_s u_s^{(2)} - \xi^2 u_j^{(2)}) + \frac{i\rho_0}{\gamma}\,\varepsilon_{jrs}\xi_r g_s^{(2)}].
\end{aligned} \tag{5.7.7}$$

By using (5.7.6) and (5.7.7), from (5.7.3) we find that

$$\begin{aligned}
u_k^{(2)} &= \frac{im}{P_1}\,\xi_k S^{(2)} - \frac{is\rho_0}{\gamma P_2}\,\varepsilon_{krj}\xi_r g_j^{(2)} + \frac{\xi^2 + q}{c_1^2\xi^2 P_1}\,\xi_k\xi_j f_j^{(2)} + \\
&+ \frac{\xi^2 + 2\sigma + \beta_4^2}{c_2^2\xi^2 P_2}\,(\xi^2 f_k^{(2)} - \xi_k\xi_j f_j^{(2)}), \\
\varphi_k^{(2)} &= -\frac{i\sigma}{c_2^2 P_2}\,\varepsilon_{krs}\xi_r f_s^{(2)} + \frac{c_4^2\rho_0\xi_k\xi_j}{c_3^2\xi^2(\xi^2 + \beta_3^2)}\,g_j^{(2)} + \\
&+ \frac{\rho_0(\xi^2 + \beta_2^2)}{\gamma\xi^2 P_2}\,(\xi^2 g_k^{(2)} - \xi_k\xi_j g_j^{(2)}).
\end{aligned} \tag{5.7.8}$$

From (5.1.28), (5.1.33), (5.7.6) and (5.7.8) we get

$$
\begin{aligned}
t_{jk}^{(2)} = & \left[(2\mu+\kappa)m(\xi_j\xi_k - \xi^2\delta_{jk}) - \frac{b}{k}\,\rho_0\beta_1^2\delta_{jk}\right]\frac{1}{P_1}\,S^{(2)} - \\
& -\frac{s}{\gamma}\,\rho_0\,[(\mu+\kappa)\varepsilon_{krs}\xi_j + \mu\varepsilon_{jrs}\xi_k]\,\xi_r g_s^{(2)} - \\
& -\frac{\kappa\rho_0}{\gamma}\,\varepsilon_{mjk}\left\{\left[\frac{c_4^2}{c_3^2\xi^2(\xi^2+\beta_3^2)} - \frac{\xi^2+\beta_2^2}{P_2\xi^2}\right]\xi_m\xi_s g_s^{(2)} + \frac{\xi^2+\beta_2^2}{P_2}\,g_m^{(2)}\right\} - \\
& -(2\mu+\kappa)i\left(\frac{\xi^2+q}{c_1^2\xi^2P_1} - \frac{\xi^2+2\sigma+\beta_4^2}{c_2^2\xi^2P_2}\right)\xi_k\xi_j\xi_s f_s^{(2)} - \\
& -(2\mu+\kappa)\,\frac{\xi^2+2\sigma+\beta_4^2}{2c_2^2P_2}\,i(\xi_j f_k^{(2)} + \xi_k f_j^{(2)}) - \\
& -\frac{i\kappa}{2c_2^2P_2}\,(\xi^2+\beta_4^2)(\xi_j f_k^{(2)} - \xi_k f_j^{(2)}) - \\
& -i\lambda\delta_{jk}\,\frac{\xi^2+q}{c_1^2P_1}\,\xi_s f_s^{(2)} - \frac{i\varepsilon}{P_1}\,\rho_0 q\delta_{jk}\xi_s f_s^{(2)}, \\
m_{jk}^{(2)} = & -\frac{\sigma}{c_2^2P_2}(\gamma\varepsilon_{krs}\xi_j + \beta\varepsilon_{jrs}\xi_k)\xi_r f_s^{(2)} - \\
& -\frac{1}{\gamma}\,(\beta+\gamma)i\left[\frac{c_4^2}{c_3^2\xi^2(\xi^2+\beta_3^2)} - \frac{\xi^2+\beta_2^2}{\xi^2P_2}\right]\xi_j\xi_k\xi_s g_s^{(2)} - \\
& -\frac{i\rho_0}{\gamma P_2}\,(\xi^2+\beta_2^2)[\gamma\xi_j g_k^{(2)} + \beta\xi_k g_j^{(2)}] - \frac{i\alpha c_4^2}{\gamma c_3^2}\,\rho_0\delta_{jk}\,\frac{\xi_s g_s^{(2)}}{\xi^2+\beta_3^2}, \\
q_k^{(2)} = & \frac{\nu kp}{c_1^2P_1}\xi_k\xi_s f_s^{(2)} - \frac{i\rho_0}{P_1}\,\xi_k(\xi^2+\beta_1^2)S^{(2)}.
\end{aligned}
\tag{5.7.9}
$$

a) **The effect of an impulsive body force.** We assume that the body force is applied impulsively at origin, in the absence of body couple and heat supply. We take

$$2\pi^2\rho_0 f_i = \delta_{i3}F\delta(x_1)\delta(x_2)\delta(x_3)\delta(t), \quad g_i = 0, \quad S = 0,$$

where F is a given constant. Clearly,

$$(2^5\pi^7)^{1/2}\rho_0 f_i^{(2)} = F\delta_{i3}, \quad g_i^{(2)} = 0, \quad S^{(2)} = 0. \tag{5.7.10}$$

We introduce the notations

$$
\begin{aligned}
2\lambda_{1,2}^2 &= q(1+\varepsilon) + \beta_1^2 \pm [(q+\varepsilon q+\beta_1^2)^2 - 4q\beta_1^2]^{1/2}, \\
2\mu_{1,2}^2 &= \beta_2^2 + \beta_4^2 + \sigma(2-s) \pm [(\beta_2^2+\beta_4^2+2\sigma - s\sigma)^2 - \\
&\qquad -4\beta_2^2(2\sigma+\beta_4^2)]^{1/2}.
\end{aligned}
\tag{5.7.11}
$$

If we substitute (5.7.10) into (5.7.6) and (5.7.8), then, by inverting the Fourier transforms, we get

$$
\begin{aligned}
du_k^{(1)} =& -\frac{\partial^2}{\partial x_k \partial x_3}\left\{\frac{1}{c_1^2(\lambda_1^2-\lambda_2^2)\lambda_1^2\lambda_2^2 r}[\lambda_2^2(q-\lambda_1^2)e^{-\lambda_1 r}-\lambda_1^2(q-\lambda_2^2)e^{-\lambda_2 r}]-\right.\\
&\left.-\frac{1}{c_2^2(\mu_1^2-\mu_2^2)\mu_1^2\mu_2^2 r}[\mu_2^2(2\sigma+\beta_4^2-\mu_1^2)e^{-\mu_1 r}-\mu_2^2(2\sigma+\beta_4^2-\mu_2^2)e^{-\mu_2 r}]\right\}+\\
&+\frac{\delta_{k3}}{c_2^2(\mu_1^2-\mu_2^2)r}[(2\sigma+\beta_4^2-\mu_2^2)e^{-\mu_2 r}-(2\sigma+\beta_4^2-\mu_1^2)e^{-\mu_1 r}],\\
d\varphi_\alpha^{(1)} =& \frac{\sigma}{c_2^2(\mu_1^2-\mu_2^2)}\varepsilon_{3\alpha\beta}\frac{\partial}{\partial x_\beta}\left[\frac{1}{r}\left(e^{-\mu_2 r}-e^{-\mu_1 r}\right)\right], \quad \varphi_3^{(1)}=0,\\
d\theta^{(1)} =& -\frac{\nu p}{c_1^2(\lambda_1^2-\lambda_2^2)}\frac{\partial}{\partial x_3}\left[\frac{1}{r}(e^{-\lambda_2 r}-e^{-\lambda_1 r})\right],
\end{aligned}
\tag{5.7.12}
$$

where λ_α and μ_α are roots with positive real parts, and

$$d=(2\pi)^3\rho_0/F, \quad r=(x_1^2+x_2^2+x_3^2)^{1/2}.$$

As in classical thermoelasticity [145], [23], we restrict our attention to the case of short time approximations. For large values of p we obtain

$$
\begin{aligned}
\lambda_1^2 &= \frac{p^2}{c_1^2}+\varepsilon\tau p+\varepsilon\tau^2 c_1^2-c_1^4\tau^3\varepsilon(\varepsilon-1)\frac{1}{p}+O(p^{-2}),\\
\lambda_2^2 &= \tau p-c_1^2\tau^2\varepsilon+c_1^4\tau^3\varepsilon(\varepsilon-1)\frac{1}{p}+O(p^{-2}),\\
\mu_1^2 &= \frac{p^2}{c_2^2}-\frac{s\sigma}{c_4^2-c_2^2}+O(p^{-2}),\\
\mu_2^2 &= \frac{p^2}{c_4^2}+2\sigma+\frac{2s\sigma}{c_4^2-c_2^2}c_2^2+O(p^{-2}),\\
\lambda_1 &= \frac{p}{c_1}+\frac{1}{2}\varepsilon\tau c_1+\frac{1}{8p}c_1^3\tau^2\varepsilon(4-\varepsilon)+O(p^{-2}),\\
\lambda_2 &= (\tau p)^{1/2}-\frac{1}{2}\left(\frac{\tau^3}{p}\right)^{1/2}\varepsilon c_1^2+\\
&\quad+\frac{1}{8}\left(\frac{\tau^5}{p^3}\right)^{1/2}c_1^4\varepsilon(3\varepsilon-4)+O(p^{-\frac{5}{2}}),\\
\mu_1 &= \frac{p}{c_2}-\frac{s\sigma c_2 c_4^2}{2(c_4^2-c_2^2)p}+O(p^{-3}),\\
\mu_2 &= \frac{p}{c_4}+\left[\sigma c_4+\frac{s\sigma c_4 c_2^2}{2(c_4^2-c_2^2)}\right]\frac{1}{p}+O(p^{-3}),
\end{aligned}
\tag{5.7.13}
$$

where $\tau = c/k$.

If we expand the functions $u_k^{(1)}, \varphi_\alpha^{(1)}$ and $\theta^{(1)}$, given by (5.7.12), in the ascending powers of p^{-1}, then we find

$$
\begin{aligned}
du_k^{(1)} = & \left[\frac{1}{c_1^2 r^3}\, x_i x_3 + \frac{1}{p c_1 r^4}\,(3x_i x_3 - r^2 \delta_{i3}) + O(p^{-2})\right] e^{-\lambda_1 r} + \\
& + O(p^{-2}) e^{-\lambda_2 r} - \Big[\frac{1}{c_2^2 r^3}\,(x_i x_3 - r^2 \delta_{i3}) + \frac{1}{p c_2 r^4}\,(3x_i x_3 - \\
& - r^2 \delta_{i3}) + O(p^{-2})\Big] e^{-\mu_1 r} + O(p^{-2}) e^{-\mu_2 r}, \\
d\varphi_\alpha^{(1)} = & \left[\frac{\sigma c_4^2}{c_2(c_4^2 - c_2^2) p r^2}\, \varepsilon_{3\alpha\beta} x_\beta + O(p^{-2})\right] e^{-\mu_1 r} + \\
& + \left[\frac{\sigma c_4}{(c_2^2 - c_4^2) p r^2}\, \varepsilon_{3\alpha\beta} x_\beta + O(p^{-2})\right] e^{-\mu_2 r}, \\
d\theta^{(1)} = & \left\{-\frac{\nu}{c_1 r^2}\, x_3 + \frac{\nu x_3}{2 r^3 p}\,[c_1 \tau(\varepsilon - 2) r - 2] + O(p^{-2})\right\} e^{-\lambda_1 r} + \\
& + \left[\frac{\nu x_3}{r^2} \left(\frac{\tau}{p}\right)^{1/2} + O(p^{-3/2})\right] e^{-\lambda_2 r}.
\end{aligned}
\tag{5.7.14}
$$

The above expansions are suitable for the Laplace inversion. The analysis of discontinuities can be made by the methods of classical thermoelasticity. By using the results established by Boley and Hetnarski [25] and Boley [26], we find that the discontinuities of u_i, φ_j and θ can only occur at $t = r/c_1$, $t = r/c_2$ and $t = r/c_4$. At the wave front $r = c_1 t$ the microrotations φ_i are continuous functions but the functions u_i and θ have discontinuities. The jumps of these functions are

$$
\frac{1}{c_1 d r^4}\,(3x_i x_3 - r^2 \delta_{i3}) \exp\left(-\frac{1}{2}\,\varepsilon c_1 \tau r\right)
$$

and

$$
\frac{\nu x_3}{2 d r^3}\,[c_1 \tau(\varepsilon - 2) r - 2] \exp\left(-\frac{1}{2}\,\varepsilon c_1 \tau r\right),
$$

respectively. These jumps decrease rapidly with the distance from the origin of the disturbance. At $t = r/c_2$, the jumps of the functions u_i and φ_α are

$$
-\frac{1}{d c_2 r^4}\,(3x_i x_3 - r^2 \delta_{i3})
$$

and

$$
\frac{\sigma}{c_2(c_4^2 - c_2^2) r^2}\, c_4^2 \varepsilon_{3\alpha\beta} x_\beta,
$$

respectively. We note that φ_3 and θ are continuous functions. At $t = r/c_4$, u_i, φ_3 and θ have no discontinuity. The microrotations φ_α have the jumps

$$\frac{\sigma}{d(c_2^2 - c_4^2)r^2}\, c_4 \varepsilon_{3\alpha\beta} x_\beta.$$

The discontinuities of the functions t_{ij}, m_{ij} and q_i are presented in [52].

b) **The effect of an impulsive body couple.** We assume that

$$f_i = 0, \quad 2\pi^2 \rho_0 g_i = M\delta(x_1)\delta(x_2)\delta(x_3)\delta(t)\delta_{i3}, \quad S = 0,$$

where M is a constant. In this case we get

$$f_i^{(2)} = 0, \ (2^5\pi^7)^{1/2} \rho_0 g^{(2)} = M\delta_{i3}, \ S^{(2)} = 0. \tag{5.7.15}$$

If we substitute (5.7.15) into (5.7.6) and (5.7.8), and perform the Fourier-inversion, then we obtain

$$\begin{aligned}
\ell u_\alpha^{(1)} &= \frac{s}{c_4^2(\mu_1^2 - \mu_2^2)}\, \varepsilon_{3\alpha\beta} \frac{\partial}{\partial x_\beta}\Big[\frac{1}{r}\,(e^{-\mu_2 r} - e^{-\mu_1 r})\Big], \ u_3^{(1)} = 0,\\
\ell \varphi_i^{(1)} &= \frac{\partial^2}{\partial x_i \partial x_3}\Big\{\frac{1}{c_3^2 \beta_3^2 r}\, e^{-\beta_3 r} + \frac{1}{c_4^2 \mu_1^2 \mu_2^2 (\mu_1^2 - \mu_2^2) r}\big[(\beta_2^2 - \mu_1^2)\mu_2^2 e^{-\mu_1 r} -\\
&\quad -(\beta_2^2 - \mu_2^2)\mu_1^2 e^{-\mu_2 r}\big]\Big\} + \frac{1}{c_4^2(\mu_1^2 - \mu_2^2) r}\big[(\beta_2^2 - \mu_2^2) e^{-\mu_2 r} -\\
&\quad -(\beta_2^2 - \mu_1^2) e^{-\mu_1 r}\big], \ \theta^{(1)} = 0,
\end{aligned} \tag{5.7.16}$$

where $\ell = (2\pi)^3 \rho_0 / M$. We note that for large values of p we have

$$\begin{aligned}
\beta_3 &= \frac{p}{c_3} + \frac{\kappa}{J c_3 p} + O(p^{-3}),\\
\ell u_\alpha^{(1)} &= \Big[\frac{s}{(c_4^2 - c_2^2) r^2 p}\, c_2 \varepsilon_{3\alpha\beta} x_\beta + O(p^{-2})\Big] e^{-\mu_1 r} +\\
&\quad + \Big[\frac{\sigma}{c_4(c_2^2 - c_4^2) r^2 p}\, c_2^2 \varepsilon_{3a\beta} x_\beta + O(p^{-2})\Big] e^{-\mu_2 r},\\
\ell \varphi_i^{(1)} &= \Big[\frac{1}{c_3^2 r^3}\, x_i x_3 + \frac{1}{c_3 r^4 p}\,(3x_i x_3 - r^2 \delta_{i3}) + O(p^{-2})\Big] e^{-\beta_3 r} +\\
&\quad + O(p^{-2}) e^{-\mu_1 r} - \Big[\frac{1}{c_4^2 r^3}\,(x_i x_3 - r\delta_{i3}) + \frac{1}{c_4 r^4 p}\,(3x_i x_3 -\\
&\quad - r^2 \delta_{i3}) + O(p^{-2})\Big] e^{-\mu_2 r}.
\end{aligned} \tag{5.7.17}$$

It follows from (5.7.17) that discontinuities can only occur at $t = r/c_2$, $t = r/c_3$ and $t = r/c_4$. At $t = r/c_2$ the microrotation vector is a continuous function, and the functions u_α have the jumps

$$\frac{s}{\ell(c_4^2 - c_2^2)r^2}\,\varepsilon_{3\alpha\beta}x_\beta.$$

At $t = r/c_3$, u_α have no discontinuity. The jumps of the functions φ_i are given by

$$\frac{1}{c_3\ell r^4}\,(3x_i x_3 - r^2\delta_{i3}).$$

The jumps of the functions u_α and φ_i across the wave front $r = c_4 t$ are

$$\frac{s}{\ell c_4(c_2^2 - c_4^2)r^2}\,c_2^2\varepsilon_{3\alpha\beta}x_\beta \quad \text{and} \quad \frac{1}{\ell c_4 r^4}\,(r^2\delta_{i3} - 3x_i x_3),$$

respectively.

c) **Impulsive heat source.** We now consider the loads

$$f_i = 0, \quad g_i = 0, \quad 2\pi^2 S = Q\delta(x_1)\delta(x_2)\delta(x_3)\delta(t),$$

where Q is a prescribed constant. It follows from (5.7.6) and (5.7.8) that for large values of p we obtain

$$\begin{aligned}
\chi u_i^{(1)} &= \left[\frac{m\tau}{r^2}\,c_1^2\left(\frac{\tau}{p^3}\right)^{1/2} x_i + O(p^{-2})\right]e^{-\lambda_2 r} - \\
&\quad - \left[\frac{m\tau}{pr^2}\,c_1 x_i + O(p^{-2})\right]e^{-\lambda_1 r},
\end{aligned}$$

$$\begin{aligned}
\varphi_i^{(1)} &= 0, \quad \chi\theta^{(1)} = \left[\frac{\tau}{r} - \frac{\varepsilon\tau^2}{pr}\,c_1^2 + O(p^{-2})\right]e^{-\lambda_2 r} + \\
&\quad + \left[\frac{\omega\tau^2}{pr}\,c_1^2 + O(p^{-2})\right]e^{-\lambda_1 r},
\end{aligned}$$

where $\chi = (2\pi)^3/Q$. In this case we can have discontinuities only at $t = r/c_1$. The jumps of the functions u_i and θ are

$$-\frac{m}{r^2}\,c_1 x_i \exp\left(-\frac{1}{2}\,\varepsilon c_1\tau r\right) \quad \text{and} \quad \frac{\varepsilon\tau^2}{r}\,c_1^2\exp\left(-\frac{1}{2}\,\varepsilon c_1\tau r\right),$$

respectively.

The case of suddenly applied body loads has been investigated in [52], [335].

5.8 Plane Strain Problem

Throughout this section the region B refers to the interior of a right cylinder with the generic cross-section Σ and the lateral boundary Π. We denote by L the boundary of Σ. The rectangular Cartesian coordinate frame is supposed to be chosen in such a way that the x_3-axis is parallel to the generators of B. We assume that the domain B is occupied by a homogeneous and isotropic material. The thermoelastic process $p = (u_i, \varphi_i, \theta, e_{ij}, \kappa_{ij}, t_{ij}, m_{ij}, \eta, q_i)$ is a process of plane strain parallel to the x_1, x_2-plane, provided

$$\begin{aligned} &u_a = u_\alpha(x_1, x_2, t), \; u_3 = 0, \; \varphi_a = 0, \; \varphi_3 = \varphi(x_1, x_2, t), \\ &\theta = \theta(x_1, x_2, t), \; (x_1, x_2) \in \Sigma, \; t \in I. \end{aligned} \tag{5.8.1}$$

From (5.1.28) and (5.8.1) we obtain

$$e_{\alpha\beta} = u_{\beta,\alpha} + \varepsilon_{\beta\alpha 3}\varphi, \; \kappa_{\alpha\beta} = \varphi_{,\alpha} \tag{5.8.2}$$

and $e_{i3} = 0, \; e_{3\alpha} = 0, \; \kappa_{\alpha\beta} = 0, \; \kappa_{3i} = 0$. In view of (5.1.34) we get

$$\begin{aligned} &t_{\alpha\beta} = \lambda e_{\rho\rho}\delta_{\alpha\beta} + (\mu + \kappa)e_{\alpha\beta} + \mu e_{\beta\alpha} - b\theta\delta_{\alpha\beta}, \\ &m_{\alpha\beta} = \gamma\kappa_{\alpha 3}, \\ &\rho_0\eta = be_{\rho\rho} + a\theta, \; q_\alpha = k\theta_{,\alpha} \end{aligned} \tag{5.8.3}$$

and

$$t_{\alpha 3} = t_{3\alpha} = 0, \; t_{33} = \lambda e_{\rho\rho} - b\theta, \; m_{\alpha\beta} = 0, \; m_{33} = 0, \; m_{3\alpha} = \beta\kappa_{\alpha 3}.$$

We note that $t_{ij}, m_{ij}, \rho_0\eta$ and q_i are independent of x_3. We assume that $\boldsymbol{f}, \boldsymbol{g}$ and S are independent of x_3 and that $f_3 = 0$ and $g_\alpha = 0$. The equations of motion (5.1.30) reduce to

$$\begin{aligned} &t_{\beta\alpha,\beta} + \rho_0 f_\alpha = \rho_0\ddot{u}_a, \\ &m_{\alpha 3,\alpha} + t_{12} - t_{21} + \rho_0 g_3 = J\ddot{\varphi}, \end{aligned} \tag{5.8.4}$$

on $\Sigma \times I$. The energy equation (5.1.31) becomes

$$\rho_0 T_0 \dot{\eta} = q_{\alpha,\alpha} + \rho_0 S, \tag{5.8.5}$$

on $\Sigma \times I$. We consider Neumann boundary conditions on Π. Given $\widetilde{\boldsymbol{t}}, \widetilde{\boldsymbol{m}}$ and $\widetilde{q}$ on Π, with $\widetilde{\boldsymbol{t}}, \widetilde{\boldsymbol{m}}$ and $\widetilde{q}$ independent of x_3 and $\widetilde{t}_3 = 0, \widetilde{m}_\alpha = 0$, the boundary conditions on the lateral surface become

$$t_{\beta\alpha}n_\beta = \widetilde{t}_\alpha, \; m_{\alpha 3}n_\alpha = \widetilde{m}_3, \; q_\alpha n_\alpha = \widetilde{q} \text{ on } L \times I. \tag{5.8.6}$$

The initial conditions (5.2.2) take the form

$$\begin{aligned}&u_\alpha(\boldsymbol{x},0)=u^0_\alpha(\boldsymbol{x}),\quad \dot{u}_\alpha(\boldsymbol{x},0)=v^0_\alpha(\boldsymbol{x}),\quad \varphi(\boldsymbol{x},0)=\varphi^0(\boldsymbol{x}),\\&\dot{\varphi}(\boldsymbol{x},0)=\nu^0(\boldsymbol{x}),\qquad \eta(\boldsymbol{x},0)=\eta^0(\boldsymbol{x}),\qquad \boldsymbol{x}\in\overline{\Sigma},\end{aligned}\tag{5.8.7}$$

where $u^0_\alpha, v^0_\alpha, \varphi^0, \nu^0$ and η^0 are prescribed functions.

From (5.8.2)-(5.8.5) we obtain the field equations expressed in terms of the displacement vector, microrotation vector and temperature field

$$\begin{aligned}&(\mu+\kappa)\Delta u_\alpha+(\lambda+\mu)u_{\beta,\beta\alpha}+\kappa\varepsilon_{\alpha\beta3}\varphi_{,\beta}-b\theta_{,\alpha}+\\&\qquad\qquad+\rho_0 f_\alpha=\rho_0\ddot{u}_\alpha,\\&\gamma\Delta\varphi-2\kappa\varphi+\kappa\varepsilon_{3\alpha\beta}u_{\beta,\alpha}+\rho_0 g_3=J\ddot{\varphi},\\&k\Delta\theta-c\dot{\theta}-bT_0\dot{u}_{\rho,\rho}=-\rho_0 S,\end{aligned}\tag{5.8.8}$$

on $\Sigma\times I$. The plane problem of micropolar thermoelasticity has been investigated in various papers (see, e.g., Nowacki [289], Ieşan [159], [162], Shanker and Dhalival [334], Dyszlewicz [96], and the literature cited therein).We introduce the notations

$$\begin{aligned}&A_1=(\lambda+2\mu+\kappa)\Delta-\rho_0\frac{\partial^2}{\partial t^2},\ A_2=(\mu+\kappa)\Delta-\rho_0\frac{\partial^2}{\partial t^2},\\&A_3=\gamma\Delta-J\frac{\partial^2}{\partial t^2}-2\kappa,\ A_4=k\Delta-c\frac{\partial}{\partial t},\\&Q=A_1A_4-b^2T_0\Delta\frac{\partial}{\partial t},\ M=A_2A_3+\kappa^2\Delta,\end{aligned}\tag{5.8.9}$$

where Δ is the two-dimensional Laplacian.

The following solution of the field equations (5.8.8) has been established in [159].

Theorem 5.8.1. *Let*

$$\begin{aligned}u_\alpha&=QA_3\Gamma_\alpha-\big[(\lambda+\mu)A_3A_4-b^2T_0A_3\frac{\partial}{\partial t}-\\&\quad-\kappa^2A_4\big]\Gamma_{\rho,\rho\alpha}-\kappa\varepsilon_{3\alpha\beta}U_{,\beta}+bV_{,\alpha},\\\varphi&=-\kappa\varepsilon_{3\alpha\beta}Q\Gamma_{\beta,\alpha}+A_2U,\\\theta&=bT_0M\dot{\Gamma}_{\rho,\rho}+A_1V,\end{aligned}\tag{5.8.10}$$

where the functions Γ_α of class C^8, U of class C^4 and V of class C^4 satisfy the equations

$$Q M\Gamma_\alpha=-\rho_0 f_\alpha,\ MU=-\rho_0 g_3,\ QV=-\rho_0 S,\tag{5.8.11}$$

on $\Sigma \times I$. Then u_α, φ and θ satisfy the equations (5.8.8).

Proof. In view of (5.8.10) we find that

$$\begin{aligned} u_{\rho,\rho} &= A_4 M \Gamma_{\rho,\rho} + b\Delta V, \\ \varepsilon_{3\alpha\beta}\varphi_{,\beta} &= \kappa Q(\Delta\Gamma_\alpha - \Gamma_{\rho,\rho\alpha}) + \varepsilon_{3\alpha\beta} A_2 U_{,\beta}, \\ \varepsilon_{3\alpha\beta} u_{\beta,\alpha} &= \varepsilon_{3\alpha\beta} Q A_3 \Gamma_{\beta,\alpha} + \kappa \Delta U. \end{aligned} \tag{5.8.12}$$

If we substitute u_α, φ and θ given by (5.8.10) into (5.8.8) and use (5.8.12) we obtain

$$\begin{aligned} &(\mu+\alpha)\Delta u_\alpha - \rho_0 \ddot{u}_\alpha + (\lambda+\mu) u_{\beta,\beta\alpha} + \kappa\varepsilon_{3\alpha\beta}\varphi_{,\beta} - b\theta_{,\alpha} = \\ &\quad = A_2 A_3 Q \Gamma_\alpha - \Big[(\lambda+\mu)A_3 A_4 - b^2 T_0 A_3 \frac{\partial}{\partial t} - \\ &\quad -\kappa^2 A_4\Big] A_2 \Gamma_{\rho,\rho\alpha} - \kappa\varepsilon_{3\alpha\beta} A_2 U_{,\beta} + b A_2 V_{,\alpha} + \\ &\quad +(\lambda+\mu)(A_4 M \Gamma_{\rho,\rho\alpha} + b\Delta V_{,\alpha}) + \\ &\quad +\kappa^2 Q(\Delta\Gamma_\alpha - \Gamma_{\rho,\rho\alpha}) + \kappa\varepsilon_{3\alpha\beta} A_2 U_{,\beta} - b^2 T_0 M \dot{\Gamma}_{\rho,\rho\alpha} - \\ &\quad -b_1 A_1 V_{,\alpha} = Q M \Gamma_\alpha. \end{aligned} \tag{5.8.13}$$

Similarly,

$$\begin{aligned} A_3\varphi + \kappa\varepsilon_{3\alpha\beta} u_{\beta,\alpha} &= MU, \\ A_4\theta - bT_0 \dot{u}_{\rho,\rho} &= QV. \end{aligned} \tag{5.8.14}$$

It follows from (5.8.11), (5.8.13) and (5.8.14) that u_α, φ and θ satisfy the equations (5.8.8). □

We now apply Theorem 5.8.1 to investigate the fundamental solutions of the equations (5.8.8). Consider the case of steady vibrations,

$$\begin{aligned} &f_\alpha = \mathrm{Re}[f_\alpha^*(\boldsymbol{x})\exp(-i\omega t)], \ g_3 = \mathrm{Re}[g_3^*(\boldsymbol{x})\exp(-i\omega t)], \\ &S = \mathrm{Re}[S^*(\boldsymbol{x})\exp(-i\omega t)], \end{aligned}$$

where ω is the given frequency. We seek solutions of the form

$$\begin{aligned} &u_\alpha = \mathrm{Re}[u_\alpha^*(\boldsymbol{x};\omega)\exp(-i\omega t)], \ \varphi = \mathrm{Re}[\varphi^*(\boldsymbol{x};\omega)\exp(-i\omega t)], \\ &\theta = \mathrm{Re}[\theta^*(\boldsymbol{x};\omega)\exp(-i\omega t)]. \end{aligned}$$

If we introduce the notations

$$\begin{aligned} &\Gamma_\alpha = \mathrm{Re}[\Gamma_\alpha^*(\boldsymbol{x};\omega)\exp(-i\omega t)], \ U = \mathrm{Re}[U^*(\boldsymbol{x};\omega)\exp(-i\omega t)], \\ &V = \mathrm{Re}[V^*(\boldsymbol{x};\omega)\exp(-i\omega t)], \ A_1^* = (\lambda+2\mu+\kappa)\Delta + \rho_0\omega^2, \\ &A_2^* = (\mu+\kappa)\Delta + \rho_0\omega^2, \ A_3^* = \gamma\Delta + J\omega^2 - 2\kappa, \ A_4^* = \gamma\Delta + i\omega c, \\ &Q^* = A_1^* A_2^* + i\omega b^2 T_0 \Delta, \ M^* = A_2^* A_3^* + \kappa^2\Delta, \end{aligned}$$

from (5.8.10) we obtain

$$\begin{aligned}
u_a^* &= A_3^* Q^* \Gamma_\alpha^* - [(\lambda+\mu)A_3^* A_4^* + i\omega b^2 T_0 A_3^* - \\
&\quad -\kappa^2 A_4^*]\Gamma^*_{\rho,\rho\alpha} - \kappa\varepsilon_{3\alpha\beta} U^*_{,\beta} + bV^*_{,\alpha}, \\
\varphi^* &= -\kappa\varepsilon_{3\alpha\beta} Q^* \Gamma^*_{\beta,\alpha} + A_2^* U^*, \\
\theta^* &= -i\omega b T_0 M^* \Gamma^*_{\rho,\rho} + A_1^* V^*.
\end{aligned} \tag{5.8.15}$$

It follows from (5.8.11) that the functions Γ_α^*, U^* and V^* satisfy the equations

$$Q^* M^* \Gamma_\alpha^* = -\rho_0 f_\alpha^*,\ M^* U^* = -\rho_0 g_3^*,\ Q^* V^* = -\rho_0 S^*. \tag{5.8.16}$$

We denote by κ_1^2 and κ_2^2 the roots of the equation

$$y^4 + \left[\sigma_1^2 + \frac{i}{k}(1+\varepsilon)\omega c\right] y^2 + \frac{i\omega c}{k}\sigma_1^2 = 0$$

and by κ_3^2 and κ_4^2 the roots of the equation

$$x^4 + \left[\sigma_2^2 + \sigma_4^2 + p_0(s-2)\right]x^2 + \left(\sigma_4^2 - 2p_0\right)\sigma_2^2 = 0.$$

Here we have used the notations (5.5.7), (5.5.10), (5.5.13). Clearly, we have

$$\begin{aligned}
Q^* &= (\lambda + 2\mu + \kappa)k(\Delta - \kappa_1^2)(\Delta - \kappa_2^2), \\
M^* &= \gamma(\mu+\kappa)(\Delta - \kappa_3^2)(\Delta - \kappa_4^2).
\end{aligned} \tag{5.8.17}$$

Thus, the equations (5.8.16) become

$$\begin{aligned}
(\Delta - \kappa_1^2)(\Delta - \kappa_2^2)(\Delta - \kappa_3^2)(\Delta - \kappa_4^2)\Gamma_\alpha^* &= -d_1 f_\alpha^*, \\
(\Delta - \kappa_3^2)(\Delta - \kappa_4^2)U^* &= -d_2 g_3^*, \\
(\Delta - \kappa_1^2)(\Delta - \kappa_2^2)V^* &= -d_3 S^*,
\end{aligned} \tag{5.8.18}$$

where

$$\begin{aligned}
&d_1 = \rho_0[k\gamma(\mu+\kappa)(\lambda+2\mu+\kappa)]^{-1},\ d_2 = \rho_0[\gamma(\mu+\kappa)]^{-1}, \\
&d_3 = \rho_0[k(\lambda+2\mu+\kappa)]^{-1}.
\end{aligned}$$

We first examine the effect of a body force acting in an infinite medium along the axis x_β. We assume that

$$f_\alpha^* = P\delta_{\alpha\beta},\ g_3^* = 0,\ S^* = 0.$$

From (5.8.18) we see that $\Gamma^*_\alpha = G\delta_{\alpha\beta}$, $U^* = 0$, $V^* = 0$, where G satisfies the equation

$$(\Delta - \kappa_1^2)(\Delta - \kappa_2^2)(\Delta - \kappa_3^2)(\Delta - \kappa_4^2)G = -d_1 P.$$

The function G can be written in the form

$$G = \sum_{k=1}^{4} \eta_k G_k,$$

where G_j satisfy the equations

$$(\Delta - \kappa_s^2)G_s = -d_1 P, \text{ (no sum; } s = 1,2,3,4) \tag{5.8.19}$$

and

$$\eta_r^{-1} = \prod_{i=1\,(i\neq r)}^{4} (\kappa_i^2 - \kappa_r^2), \; (r = 1,2,3,4). \tag{5.8.20}$$

We now suppose that $P = \delta(x - y)$, where δ is the Dirac delta and y is a fixed point. From (5.8.19) we obtain

$$\begin{aligned} G_j &= \frac{1}{2\pi} d_1 K_0(\kappa_j r), \; (j = 1,2,3,4), \\ r &= [(x_\alpha - y_\alpha)(x_\alpha - y_\alpha)]^{1/2}, \end{aligned} \tag{5.8.21}$$

where K_0 denotes the modified Bessel function of the third kind, and zeroth order. Here, we have denoted by κ_s the roots with positive real parts. Thus, we find that

$$\Gamma^*_\alpha = \psi\delta_{\alpha\beta}, \quad U^* = V^* = 0, \tag{5.8.22}$$

where

$$\psi = \frac{1}{2\pi} d_1 \sum_{j=1}^{4} \eta_j K_0(\kappa_j r). \tag{5.8.23}$$

It follows from (5.8.15) and (5.8.22) that the corresponding amplitudes are given by

$$\begin{aligned} u_\alpha^{*(\beta)} &= \delta_{\alpha\beta} A_3^* Q^* \psi - [(\lambda + \mu) A_3^* A_4^* + i\omega b^2 T_0 A_3^* - \kappa^2 A_4^*]\psi_{,\alpha\beta}, \\ \varphi^{*(\beta)} &= -\kappa\varepsilon_{3\alpha\beta} Q^* \psi_{,\alpha}, \; \theta^{*(\beta)} = -i\omega b T_0 M^* \psi_{,\beta}. \end{aligned}$$

Let us assume that

$$f^*_\alpha = 0, \; g^*_3 = \delta(x - y), \; S^* = 0.$$

In this case we get

$$\Gamma_\alpha^* = 0, \quad V^* = 0,$$
$$U^* = \frac{1}{2\pi(\kappa_3^2 - \kappa_4^2)} d_2[K_0(\kappa_3 r) - K_0(\kappa_4 r)]. \tag{5.8.24}$$

From (5.8.15) and (5.8.24) we find the following amplitudes

$$u_\alpha^{*(3)} = -\kappa\varepsilon_{3\alpha\beta}U^*_{,\beta}, \ \varphi^{*(3)} = A_2^* U^*, \ \theta^{*(3)} = 0. \tag{5.8.25}$$

If $f_i^* = 0$, $g_3^* = 0$ and $S^* = \delta(x - y)$, then the functions Γ_α^*, U^* and V^* are given by

$$\Gamma_\alpha^* = 0, \ U^* = 0, \ V^* = \frac{1}{2\pi(\kappa_1^2 - \kappa_2^2)} d_3[K_0(\kappa_1 r) - K_0(\kappa_2 r)]. \tag{5.8.26}$$

It follows from (5.8.15) and (5.8.26) that the amplitudes have the form

$$u_\alpha^{*(4)} = bV^*_{,\alpha}, \ \varphi^{*(4)} = 0, \ \theta^{*(4)} = A_1^* V^*. \tag{5.8.27}$$

In the case of steady vibrations, the functions $u_\alpha^{*(k)}$, $\varphi^{*(k)}$, $\theta^{*(k)}$, $(k = 1, 2, 3, 4)$, represent the fundamental solutions of the field equations in the plane strain problem. These solutions have been used in [159] to reduce boundary value problems to singular integral equations for which Fredholm's theorems are valid.

In [334], Shanker and Dhalival have studied the dynamic thermoelastic plane strain problem by using Laplace-Fourier transforms. The short time solutions in the case of impulsive as well as sudden, concentrated body loads have been derived. Other applications of the thermoelastic plane strain were presented by Nowacki [289].

5.9 Bending of Micropolar Thermoelastic Plates

In [105], Eringen presented a micropolar plate theory and established uniqueness results in the context of isothermal case. In this section we use the results of Eringen [105], Nowacki [285] and Inan and Eringen [201] to derive a theory of bending of micropolar thermoelastic plates. We assume that the region B refers to the interior of a right cylinder of length $2h$ with the open cross-section Σ and the smooth lateral boundary Π. Let Γ be the boundary of Σ. The rectangular coordinate frame is supposed to be chosen in such a way that the plane x_1Ox_2 is middle plane. Thus, we have

$$B = \{\boldsymbol{x} : (x_1, x_2) \in \Sigma, \ -h < x_3 < h\},$$
$$\Pi = \{\boldsymbol{x} : (x_1, x_2) \in \Gamma, \ -h < x_3 < h\}.$$

We assume that the domain B is occupied by a homogeneous and isotropic material.

Let $\mathcal{T}$ be a given interval of time. The dynamically admissible state $\mathcal{P} = (u_i, \varphi_i, \theta)$ is a *state of bending* on $B \times \mathcal{T}$ provided [105]

$$\begin{aligned}
&u_\alpha(x_1,x_2,x_3,t) = -\, u_\alpha(x_1,x_2,-x_3,t), \quad u_3(x_1,x_2,x_3,t) = \\
&= u_3(x_1,x_2,-x_3,t), \quad \varphi_\alpha(x_1,x_2,x_3,t) = \varphi_3(x_1,x_2,-x_3,t), \\
&\varphi_3(x_1,x_2,x_3,t) = -\varphi_3(x_1,x_2,-x_3,t), \\
&\theta(x_1,x_2,x_3,t) = -\theta(x_1,x_2,-x_3,t), \; (x_1,x_2,x_3) \in B, \; t \in \mathcal{T}.
\end{aligned} \tag{5.9.1}$$

In view of (5.1.28), (5.1.34) and (5.9.1) we find that

$$\begin{aligned}
&t_{\alpha\beta}(x_\rho,x_3,t) = -t_{\alpha\beta}(x_\rho,-x_3,t), t_{33}(x_\rho,x_3,t) = -t_{33}(x_\rho,-x_3,t), \\
&t_{\alpha 3}(x_\rho,x_3,t) = t_{\alpha 3}(x_\rho,-x_3,t), t_{3\alpha}(x_\rho,x_3,t) = t_{3\alpha}(x_\rho,-x_3,t), \\
&m_{\alpha\beta}(x_\rho,x_3,t) = m_{\alpha\beta}(x_\rho,-x_3,t), m_{33}(x_\rho,x_3,t) = m_{33}(x_\rho,-x_3,t), \\
&m_{\alpha 3}(x_\rho,x_3,t) = -m_{\alpha 3}(x_\rho,-x_3,t), m_{3\alpha}(x_\rho,x_3,t) = -m_{3\alpha}(x_\rho,-x_3,t), \\
&\eta(x_\rho,x_3,t) = -\eta(x_\rho,-x_3,t), q_\alpha(x_\rho,x_3,t) = -q_\alpha(x_\rho,-x_3,t), \\
&q_3(x_\rho,x_3,t) = q_3(x_\rho,-x_3,t), (x_1,x_2,x_3) \in B, \; t \in \mathcal{T}.
\end{aligned} \tag{5.9.2}$$

We say that the system of body loads $(\boldsymbol{f}, \boldsymbol{g}, S)$ is compatible with a state of bending if

$$\begin{aligned}
&f_\alpha(x_\rho,-x_3,t) = -f_\alpha(x_\rho,x_3,t), && f_3(x_\rho,-x_3,t) = f_3(x_\rho,x_3,t), \\
&g_\alpha(x_\rho,-x_3,t) = g_\alpha(x_\rho,x_3,t), && g_3(x_\rho,-x_3,t) = -g_3(x_\rho,x_3,t), \\
&S(x_\rho,-x_3,t) = -S(x_\rho,x_3,t), && (x_1,x_2,x_3) \in B, \; t \in \mathcal{T}.
\end{aligned} \tag{5.9.3}$$

In what follows we assume that the body loads satisfy the restrictions (5.9.3). We derive a theory of thermoelastic thin plates of uniform thickness assuming that the fields $\boldsymbol{u}$, $\boldsymbol{\varphi}$ and θ do not vary violently with respect to x_3. We denote

$$\begin{aligned}
&\tau_{ij} = \frac{1}{2h}\int_{-h}^{h} t_{ij}dx_3, \; \mu_{ij} = \frac{1}{2h}\int_{-h}^{h} m_{ij}dx_3, \; H_i = \frac{1}{2h}\int_{-h}^{h} q_i dx_3, \\
&\eta^* = \frac{1}{2h}\int_{-h}^{h} \eta\, dx_3, \; f_i^* = \frac{1}{2h}\int_{-h}^{h} f_i\, dx_3, \\
&g_i^* = \frac{1}{2h}\int_{-h}^{h} g_i\, dx_3, \; S^* = \frac{1}{2h}\int_{-h}^{h} S\, dx_3.
\end{aligned} \tag{5.9.4}$$

From (5.9.2) and (5.9.3) we get

$$\begin{aligned}&\tau_{\alpha\beta}=0,\ \tau_{33}=0,\ \mu_{\alpha 3}=0,\ H_\alpha=0,\ \eta^*=0,\\&f_\alpha^*=0,\ g_3^*=0,\ S^*=0.\end{aligned} \tag{5.9.5}$$

We assume that the functions t_i, m_i and q are prescribed on the surfaces $x_3=\pm h$. We integrate the equations (5.1.30) with respect to x_3 between the limits $-h$ and h. On the basis of (5.9.1) and (5.9.5) we obtain the equations

$$\begin{aligned}&\tau_{\alpha 3,\alpha}+F=\rho_0\ddot{w},\\&\mu_{\beta\alpha,\beta}+\varepsilon_{3\rho\alpha}(\tau_{3\rho}-\tau_{\rho 3})+G_\alpha=J\ddot{\psi}_\alpha,\ \text{on }\Sigma\times\mathcal{T},\end{aligned} \tag{5.9.6}$$

where

$$\begin{aligned}&w=\frac{1}{2h}\int_{-h}^{h}u_3\,dx_3,\ \psi_\alpha=\frac{1}{2h}\int_{-h}^{h}\varphi_\alpha\,dx_3,\\&F=f_3^*+\frac{1}{h}t_{33}(x_1,x_2,h,t),\ G_\alpha=g_\alpha^*+\frac{1}{h}m_{3\alpha}(x_1,x_2,h,t),\\&(x_1,x_2,t)\in\Sigma\times\mathcal{T}.\end{aligned} \tag{5.9.7}$$

If we multiply by x_3 the equations $(5.1.30)_1$ and $(5.1.30)_3$ and integrate from $x_3=-h$ to $x_3=h$, then we obtain

$$M_{\beta\alpha,\beta}-2h\tau_{3\alpha}+L_\alpha=\rho_0 I\ddot{v}_\alpha, \tag{5.9.8}$$

where

$$\begin{aligned}&M_{\alpha\beta}=\int_{-h}^{h}x_3t_{\alpha\beta}dx_3,\ Iv_\alpha=\int_{-h}^{h}x_3u_\alpha dx_3,\ I=\frac{2}{3}h^3,\\&L_\alpha=2ht_{3\alpha}(x_1,x_2,h,t)+\int_{-h}^{h}x_3f_\alpha dx_3,\ (x_1,x_2,t)\in\Sigma\times\mathcal{T}.\end{aligned} \tag{5.9.9}$$

Now multiply by x_3 the equation (5.1.31) and integrate from $x_3=-h$ to $x_3=h$. In view of (5.9.2)-(5.9.4) we obtain

$$\rho_0T_0\dot{\sigma}=Q_{\alpha,\alpha}-2hR+s, \tag{5.9.10}$$

where

$$\begin{aligned}&\sigma=\int_{-h}^{h}x_3\eta dx_3,\ Q_\alpha=\int_{-h}^{h}x_3q_\alpha dx_3,\ R=H_3,\\&s=2hq_3(x_1,x_2,h,t)+\int_{-h}^{h}\rho_0x_3S\,dx_3,\ (x_1,x_2,t)\in\Sigma\times\mathcal{T}.\end{aligned} \tag{5.9.11}$$

The functions F, G_α, L_α and s are prescribed. Following [105] we restrict our attention to the state of bending characterized by

$$u_\alpha = x_3 v_\alpha(x_1,x_2,t),\ u_3 = w(x_1,x_2,t),\ \varphi_\alpha = \psi_\alpha(x_1,x_2,t),\\ \varphi_3 = 0,\ \theta = x_3 T(x_1,x_2,t),\ \ (x_1,x_2,x_3)\in B,\ t\in \mathcal{T}. \tag{5.9.12}$$

In view of (5.9.12) we have

$$e_{\alpha\beta} = x_3\varepsilon_{\alpha\beta},\quad e_{\alpha 3} = \varepsilon_{\alpha 3},\qquad e_{3\alpha} = \varepsilon_{3\alpha},\quad e_{33} = 0,\\ \kappa_{\alpha\beta} = \eta_{\alpha\beta},\qquad \kappa_{\alpha 3} = \kappa_{3\alpha} = 0,\quad \kappa_{33} = 0, \tag{5.9.13}$$

where

$$\varepsilon_{\alpha\beta} = v_{\beta,\alpha},\ \varepsilon_{\alpha 3} = w_{,\alpha} + \varepsilon_{3\alpha\beta}\psi_\beta,\\ \varepsilon_{3\alpha} = v_\alpha - \varepsilon_{3\alpha\beta}\psi_\beta,\ \eta_{\alpha\beta} = \psi_{\beta,\alpha}. \tag{5.9.14}$$

From (5.1.34) and (5.9.13) it follows that

$$\begin{aligned} &t_{\alpha\beta} = x_3[\lambda\varepsilon_{\rho\rho}\delta_{\alpha\beta} + (\mu+\kappa)\varepsilon_{\alpha\beta} + \mu\varepsilon_{\beta\alpha} - bT\delta_{\alpha\beta}],\\ &t_{\alpha 3} = (\mu+\kappa)\varepsilon_{\alpha 3} + \mu\varepsilon_{3\alpha},\ t_{3\alpha} = (\mu+\kappa)\varepsilon_{3\alpha} + \mu\varepsilon_{\alpha 3},\\ &t_{33} = x_3(\lambda\varepsilon_{\rho\rho} - bT),\\ &m_{\kappa\nu} = \alpha\eta_{\rho\rho}\delta_{\kappa\nu} + \beta\eta_{\nu\kappa} + \gamma\eta_{\kappa\nu},\\ &m_{\alpha 3} = m_{3\alpha} = 0,\ m_{33} = \alpha\eta_{\rho\rho},\ q_\alpha = kx_3T_{,\alpha},\\ &q_3 = kT,\ \rho_0\eta = x_3(b\varepsilon_{\rho\rho} + aT). \end{aligned} \tag{5.9.15}$$

It follows from (5.9.4), (5.9.9), (5.9.11) and (5.9.15) that

$$\begin{aligned} &\tau_{\alpha 3} = (\mu+\kappa)\varepsilon_{\alpha 3} + \mu\varepsilon_{3\alpha},\ \tau_{3\alpha} = (\mu+\kappa)\varepsilon_{3\alpha} + \mu\varepsilon_{\alpha 3},\\ &\mu_{\kappa\nu} = \alpha\eta_{\rho\rho}\delta_{\kappa\nu} + \beta\eta_{\nu\kappa} + \gamma\eta_{\kappa\nu},\\ &M_{\alpha\beta} = I[\lambda\varepsilon_{\rho\rho}\delta_{\alpha\beta} + (\mu+\kappa)\varepsilon_{\alpha\beta} + \mu\varepsilon_{\beta\alpha} - bT\delta_{\alpha\beta}],\\ &Q_\alpha = IkT_{,\alpha},\ R = kT. \end{aligned} \tag{5.9.16}$$

The equations of the theory consist of the equations of motion (5.9.6), (5.9.8), the energy equation (5.9.10), the constitutive equations (5.9.16) and the geometrical equations (5.9.14). These equations can be expressed in terms of the functions v_α, w, ψ_α and T. Thus, we obtain

$$\begin{aligned} &I[(\mu+\kappa)\Delta v_\alpha + (\lambda+\mu)v_{\rho,\rho\alpha} - bT_{,\alpha}]-\\ &\qquad -2h[(\mu+\kappa)v_\alpha + \mu w_{,\alpha}] + 2h\kappa\varepsilon_{3\alpha\beta}\psi_\beta + L_\alpha = \rho_0 I\ddot{v}_\alpha,\\ &(\mu+\kappa)\Delta w + \mu v_{\rho,\rho} + \kappa\varepsilon_{3\alpha\beta}\psi_{\beta,\alpha} + F = \rho_0\ddot{w},\\ &\gamma\Delta\psi_\alpha + (\alpha+\beta)\psi_{\rho,\rho\alpha} + \kappa\varepsilon_{3\rho\alpha}(v_\rho - w_{,\rho}) - 2\kappa\psi_\alpha + G_\alpha = J\ddot{\psi}_\alpha,\\ &Ik\Delta T - IT_0(b\dot{v}_{\rho,\rho} + a\dot{T}) - 2hkT = -s \ \text{ on } \Sigma\times\mathcal{T}. \end{aligned} \tag{5.9.17}$$

To the field equations we adjoin initial conditions and boundary conditions. The initial conditions are

$$\begin{aligned}
&v_\alpha(x_1,x_2,0)=v_\alpha^0(x_1,x_2),\ w(x_1,x_2,0)=w^0(x_1,x_2),\\
&\psi_\alpha(x_1,x_2,0)=\psi_\alpha^0(x_1,x_2),\ T(x_1,x_2,0)=T^0(x_1,x_2),\\
&\dot v_\alpha(x_1,x_2,0)=\nu_\alpha(x_1,x_2),\ \dot w(x_1,x_2,0)=\omega(x_1,x_2),\\
&\dot\psi_\alpha(x_1,x_2,0)=\chi_\alpha(x_1,x_2),\quad (x_1,x_2)\in\overline{\Sigma},
\end{aligned} \tag{5.9.18}$$

where $v_\alpha^0, w^0, \psi_\alpha^0, T^0, \nu_\alpha, \omega$ and χ_α are prescribed functions. We consider the boundary conditions

$$\begin{aligned}
&M_{\beta\alpha}n_\beta=\widetilde{M}_\alpha,\quad \tau_{\alpha3}n_\alpha=\widetilde{\tau},\\
&\mu_{\beta\alpha}n_\beta=\widetilde{\mu}_\alpha,\qquad Q_\alpha n_\alpha=\widetilde{Q}\quad \text{on } \Gamma\times\mathcal{T},
\end{aligned} \tag{5.9.19}$$

where the functions $\widetilde{M}_\alpha, \widetilde{\tau}, \widetilde{\mu}_\alpha$ and $\widetilde{Q}$ are given.

In what follows we extend the uniqueness result given by Eringen [105] to the case of thermoelastic plates. We suppose for the purpose of this result that the underlying time interval is $[0,\infty)$. Let f be a continuous function on $\Sigma\times\mathcal{T}$. We denote

$$\widehat{f}(x_1,x_2,t)=\int_0^t f(x_1,x_2,\tau)d\tau,\quad (x_1,x_2)\in\Sigma,\ t\in\mathcal{T}.$$

We introduce the notation

$$H=\widehat{s}+\rho_0T_0\sigma^0, \tag{5.9.20}$$

where

$$\rho_0\sigma^0=I(bv^0_{\rho,\rho}+aT^0). \tag{5.9.21}$$

In view of (5.9.10), (5.9.16), (5.9.18)-(5.9.21) we get

Lemma 5.9.1. *The functions $\sigma\in C^{0,1}$, $Q_\alpha\in C^{1,0}$ and $R\in C^0$ satisfy the equations* (5.9.10) *and the initial condition $\sigma(x_1,x_2,0)=\sigma^0(x_1,x_2)$, $(x_1,x_2)\in\overline{\Sigma}$, if and only if*

$$\rho_0T_0\sigma=\widehat{Q}_{\alpha,\alpha}-2h\widehat{R}+H\quad \textit{on } \Sigma\times[0,\infty). \tag{5.9.22}$$

The proof is immediate.

Let us consider two external data systems $Z^{(\alpha)}=\{F^{(\alpha)}, G_\rho^{(\alpha)}, L_\rho^{(\alpha)}, s^{(\alpha)}, \widetilde{M}_\rho^{(\alpha)}, \widetilde{\tau}^{(\alpha)}, \widetilde{\mu}_\rho^{(\alpha)}, \widetilde{Q}^{(\alpha)}, v_\rho^{0(\alpha)}, w^{0(\alpha)}, .., \chi_\rho^{(\alpha)}\}$, $(\alpha=1,2)$. Let $A^{(\alpha)}=$

$\{w^{(\alpha)},\ \psi_\rho^{(\alpha)},\ v_\rho^{(\alpha)},\ T^{(\alpha)},\ \varepsilon_{ij}^{(\alpha)},\ \eta_{\nu\rho}^{(\alpha)},\ \tau_{ij}^{(\alpha)},\ \mu_{\kappa\nu}^{(\alpha)},\ \sigma^{(\alpha)},\ M_{\nu\kappa}^{(\alpha)},\ Q_\nu^{(\alpha)},\ R^{(\alpha)}\}$ be a solution corresponding to $Z^{(\alpha)}$. We denote

$$M_\alpha^{(\rho)} = M_{\beta\alpha}^{(\rho)} n_\beta,\ \tau^{(\rho)} = \tau_{\alpha 3}^{(\rho)} n_\alpha,\ \mu_\alpha^{(\rho)} = \mu_{\beta\alpha}^{(\rho)} n_\beta,$$
$$Q^{(\rho)} = Q_\alpha^{(\rho)} n_\alpha,\ H^{(\alpha)} = \widehat{s}^{(\alpha)} + \rho_0 T_0 \sigma^{0(\alpha)} \tag{5.9.23}$$

and introduce the notations

$$P_{\kappa\nu}(p,q) = \int_\Gamma \big[M_\alpha^{(\kappa)}(p) v_\alpha^{(\nu)}(q) + 2h\tau^{(\kappa)}(p) w^{(\nu)}(q) + 2h\mu_\alpha^{(\kappa)}(p)\psi_\alpha^{(\nu)}(q) -$$
$$-\frac{1}{T_0}\widehat{Q}^{(\kappa)}(p) T^{(\nu)}(q)\big] ds + \int_\Sigma \big[2hF^{(\kappa)}(p) w^{(\nu)}(q) + 2hG_\alpha^{(\kappa)}(p)\psi_\alpha^{(\nu)}(q) +$$
$$+L_\alpha^{(\kappa)}(p) v_\alpha^{(\nu)}(q) - \frac{1}{T_0} H^{(\kappa)}(p) T^{(\nu)}(q)\big] dA,\ p,q \in \mathcal{T},$$

$$K_{\kappa\nu}(p,q) = \int_\Sigma \{\rho_0 I \ddot{v}_\alpha^{(\kappa)}(p) v_\alpha^{(\nu)}(q) + 2h\rho_0 \ddot{w}^{(\kappa)}(p) w^{(\nu)}(q) +$$
$$+2hJ\ddot{\psi}_\alpha^{(\kappa)}(p)\psi_\alpha^{(\nu)}(q) - \frac{1}{T_0} k\big[I\widehat{T}_{,\alpha}^{(\kappa)}(p) T_{,\alpha}^{(\nu)}(q) + 2h\widehat{T}^{(\kappa)}(p) T^{(\nu)}(q)\big]\} dA, \tag{5.9.24}$$

where, for convenience, we have suppressed the arguments x_1 and x_2.

Theorem 5.9.1. *Let*

$$E_{\kappa\nu}(p,q) = P_{\kappa\nu}(p,q) - K_{\kappa\nu}(p,q), \tag{5.9.25}$$

for all $p,q \in \mathcal{T}$. *Then*

$$E_{\kappa\nu}(p,q) = E_{\nu\kappa}(q,p). \tag{5.9.26}$$

Proof. Let

$$I_{\kappa\nu}(p,q) = M_{\alpha\beta}^{(\kappa)}(p)\varepsilon_{\alpha\beta}^{(\nu)}(q) - \rho_0\sigma^{(\kappa)}(p) T^{(\nu)}(q) +$$
$$+2h[\tau_{3\alpha}^{(\kappa)}(p)\varepsilon_{3\alpha}^{(\nu)}(q) + \tau_{\alpha 3}^{(\kappa)}(p)\varepsilon_{\alpha 3}^{(\nu)}(q) + \mu_{\alpha\beta}^{(\kappa)}(p)\eta_{\alpha\beta}^{(\nu)}(q)].$$

In view of (5.9.16) we obtain

$$I_{\kappa\nu}(p,q) = I\{\lambda\varepsilon_{\rho\rho}^{(\kappa)}(p)\varepsilon_{\eta\eta}^{(\nu)}(q) + (\mu+\kappa)\varepsilon_{\alpha\beta}^{(\kappa)}(p)\varepsilon_{\alpha\beta}^{(\nu)}(q) +$$
$$+\mu\varepsilon_{\beta\alpha}^{(\kappa)}(p)\varepsilon_{\alpha\beta}^{(\nu)}(q) - b\big[\varepsilon_{\rho\rho}^{(\nu)}(q) T^{(\kappa)}(p) + \varepsilon_{\rho\rho}^{(\kappa)}(p) T^{(\nu)}(q)\big]\} +$$
$$+2h\{(\mu+\kappa)\varepsilon_{3\alpha}^{(\kappa)}(p)\varepsilon_{3\alpha}^{(\nu)}(q) + \mu\big[\varepsilon_{\alpha 3}^{(\kappa)}(p)\varepsilon_{3\alpha}^{(\nu)}(q) + \varepsilon_{\alpha 3}^{(\nu)}(q)\varepsilon_{3\alpha}^{(\kappa)}(p)\big] + \tag{5.9.27}$$
$$+(\mu+k)\varepsilon_{\alpha 3}^{(\kappa)}(p)\varepsilon_{\alpha 3}^{(\nu)}(q) + \alpha\eta_{\rho\rho}^{(\kappa)}(p)\eta_{\beta\beta}^{(\nu)}(q) + \beta\eta_{\rho\alpha}^{(\kappa)}(p)\eta_{\alpha\rho}^{(\nu)}(q) +$$
$$+\gamma\eta_{\alpha\beta}^{(\kappa)}(p)\eta_{\alpha\beta}^{(\nu)}(q)]\}.$$

It follows from (5.9.27) that

$$I_{\kappa\nu}(p,q) = I_{\nu\kappa}(q,p). \tag{5.9.28}$$

On the other hand, in view of (5.9.6), (5.9.8), (5.9.22) and (5.9.14) we obtain

$$\begin{aligned}
I_{\kappa\nu}(p,q) = & \{M_{\alpha\beta}^{(\kappa)}(p)v_\beta^{(\nu)}(q) - \frac{1}{T_0}\,\dot{Q}_\alpha^{(\kappa)}(p)T^{(\nu)}(q) + \\
& +2h\big[\mu_{\alpha\beta}^{(\kappa)}(p)\psi_\beta^{(\nu)}(q) + \tau_{\alpha 3}^{(\kappa)}(p)w^{(\nu)}(q)\big]\}_{,\alpha} + 2h\big[F^{(\kappa)}(p)w^{(\nu)}(q) + \\
& +G_\alpha^{(\kappa)}(p)\psi_\alpha^{(\nu)}(q)\big] + L_\alpha^{(\kappa)}(p)v_\alpha^{(\nu)}(q) - \frac{1}{T_0}\,H^{(\kappa)}(p)T^{(\nu)}(q) - \\
& -\rho_0 I\ddot{v}_\alpha^{(\kappa)}(p)v_\alpha^{(\nu)}(q) - \rho_0\ddot{u}^{(\kappa)}(p)u^{(\nu)}(q) - 2h\big[\rho_0\ddot{w}^{(\kappa)}(p)w^{(\nu)}(q) + \\
& +J\ddot{\psi}_\alpha^{(\kappa)}(p)\psi_\alpha^{(\nu)}(q)\big] + \frac{1}{T_0}\,k\big[I\widehat{T}_{,\alpha}^{(\kappa)}(p)T_{,\alpha}^{(\nu)}(q) + 2h\widehat{T}^{(\kappa)}(p)T^{(\nu)}(q)\big].
\end{aligned} \tag{5.9.29}$$

By using the divergence theorem, from (5.9.24) and (5.9.25) we find that

$$\int_\Sigma I_{\kappa\nu}(p,q)dA = E_{\kappa\nu}(p,q).$$

In view of (5.9.28) we conclude that (5.9.26) holds. □

A consequence of Theorem 5.9.1 is the following proposition.

Corollary 5.9.1. *Let*

$$\begin{aligned}
P(p,q) = & \int_\Gamma \{M_\alpha(p)v_\alpha(q) + 2h\tau(p)w(q) + \\
& +2h\mu_\alpha(p)\psi_\alpha(q) - \frac{1}{T_0}\,\widehat{Q}(p)T(q)\}ds + \\
& + \int_\Sigma [L_\alpha(p)v_\alpha(q) + 2hF(p)w(q) + 2hG_a(p)\psi_\alpha(q) - \frac{1}{T_0}\,H(p)T(q)]dA,
\end{aligned} \tag{5.9.30}$$

for all $p,q \in \mathcal{T}$, *where* $(w,\psi_\alpha,v_\alpha,T)$ *is a solution of the boundary-initial-value problem* (5.9.17)-(5.9.19). *Then*

$$\frac{d}{dt}\left\{\int_\Sigma [\rho_o I v_\alpha v_\alpha + 2h(\rho_0 w^2 + J\psi_\alpha\psi_\alpha)]dA + \right.$$

$$+\frac{1}{T_0}k\int_0^t\int_\Sigma(I\widehat{T}_{,\alpha}\widehat{T}_{,\alpha}+2h\widehat{T}^2)dt\,dA\Big\}=\int_0^t[P(t-\tau,t+\tau)-$$
$$-P(t+\tau,t-\tau)]d\tau+\int_\Sigma\{\rho_0 I[\dot{v}_\alpha(0)v_\alpha(2t)+\dot{v}_\alpha(2t)v_\alpha(0)]+ \tag{5.9.31}$$
$$+2hJ[\dot{\psi}_\alpha(0)\psi_\alpha(2t)+\dot{\psi}_\alpha(2t)\psi_\alpha(0)]+2h\rho_0[\dot{w}(0)w(2t)+$$
$$+\dot{w}(2t)w(0)]\}dA.$$

Proof. On the basis of (5.9.26) we can write

$$\int_0^t E_{11}(t+\tau,t-\tau)d\tau=\int_0^t E_{11}(t-\tau,t+\tau)d\tau. \tag{5.9.32}$$

We apply this relation for $w^{(1)}=w$, $\psi_\alpha^{(1)}=\psi_\alpha$, $v_\alpha^{(1)}=v_\alpha$ and $T^{(1)}=T$. In view of (5.9.24), (5.9.25) and (5.9.30) we obtain

$$\int_0^t E_{11}(t+\tau,t-\tau)d\tau=\int_0^t P(t+\tau,t-\tau)d\tau-\int_\Sigma[\rho_0 I\ddot{v}_\alpha(t+\tau)v_\alpha(t-\tau)+$$
$$+2hJ\ddot{\psi}_\alpha(t+\tau)\psi_\alpha(t-\tau)+2h\rho_0\ddot{w}(t+\tau)w(t-\tau)]dA+$$
$$+\frac{1}{T_0}k\int_\Sigma[I\widehat{T}_{,\alpha}(t+\tau)T_{,\alpha}(t-\tau)+2h\widehat{T}(t+\tau)T(t-\tau)]dA \tag{5.9.33}$$

and

$$\int_0^t E_{11}(t-\tau,t+\tau)d\tau=\int_0^t P(t-\tau,t+\tau)d\tau-\int_\Sigma[\rho_0 I\ddot{v}_\alpha(t-\tau)v_\alpha(t+\tau)+$$
$$+2hJ\ddot{\psi}_\alpha(t-\tau)\psi_\alpha(t+\tau)+2h\rho_0\ddot{w}_\alpha(t-\tau)w(t+\tau)]dA+$$
$$+\frac{1}{T_0}k\int_\Sigma[I\widehat{T}_{,\alpha}(t-\tau)T_{,\alpha}(t+\tau)+2h\widehat{T}(t-\tau)T(t+\tau)]dA. \tag{5.9.34}$$

If we substitute (5.9.33) and (5.9.34) in (5.9.32) and take into account the identities (5.3.17), then we conclude that (5.9.31) holds. □

Corollary 5.9.1 forms the basis of the following uniqueness result.

Theorem 5.9.2. *Assume that*

(i) ρ_0 *and* J *are strictly positive;*

(ii) k *is strictly positive.*

Then the boundary-initial-value problem has at most one solution.

Proof. If there are two solutions, then their difference $(w,\psi_\alpha,v_\alpha,T)$ corresponds to null data. In view of (5.9.31) and the initial conditions we get

$$\int_{\Sigma}[\rho_0 I v_\alpha v_\alpha + 2h(\rho_0 w^2 + J\psi_\alpha\psi_\alpha)]dA+ \\ +\frac{1}{T_0}k\int_0^t\int_{\Sigma}(I\widehat{T}_{,\alpha}\widehat{T}_{,\alpha}+2h\widehat{T}^2)d\tau\, dA=0. \tag{5.9.35}$$

By the hypotheses of theorem and (5.9.35) we conclude that $w = 0$, $\psi_\alpha = 0$, $v_\alpha = 0$ and $T = 0$ on $\Sigma\times\mathcal{T}$. □

Theorem 5.9.1 also implies the following reciprocal theorem.

Theorem 5.9.3. *Let $A^{(\alpha)}$ be a solution corresponding to the external data system $Z^{(\alpha)}$ $(\alpha = 1, 2)$. Then*

$$\begin{aligned}&\int_{\Sigma}[\mathcal{M}_\alpha^{(1)}*v_\alpha^{(2)}+2h\mathcal{H}^{(1)}*w^{(2)}+2h\mathcal{N}_\alpha^{(1)}*\psi_\alpha^{(2)}-\frac{1}{T_0}\,j*H^{(1)}*T^{(2)}]dA+\\ &+\int_{\Gamma}j*[M_\alpha^{(1)}*v_\alpha^{(2)}+2h\tau^{(1)}*w^{(2)}+2h\mu_\alpha^{(1)}*\psi_\alpha^{(2)}-\frac{1}{T_0}\,\widehat{Q}^{(1)}*T^{(2)}]ds=\\ &=\int_{\Sigma}[\mathcal{M}_\alpha^{(2)}*v_\alpha^{(1)}+2h\mathcal{H}^{(2)}*w^{(1)}+2h\mathcal{N}_\alpha^{(2)}*\psi_\alpha^{(1)}-\frac{1}{T_0}\,j*H^{(2)}*T^{(1)}]dA+\\ &+\int_{\Gamma}j*[M_\alpha^{(2)}*v_\alpha^{(1)}+2h\tau^{(2)}*w^{(1)}+2h\mu_\alpha^{(2)}*\psi_\alpha^{(1)}-\frac{1}{T_0}\,\widehat{Q}^{(2)}*T^{(1)}]ds,\end{aligned} \tag{5.9.36}$$

where

$$\mathcal{M}_\alpha^{(\kappa)}=j*L_\alpha^{(\kappa)}+\rho_0 I(t\nu_\alpha^{(\kappa)}+v_\alpha^{0(\kappa)}),\;\; \mathcal{N}_\alpha^{(\kappa)}=j*G_\alpha^{(\kappa)}+J(t\chi_\alpha^{(\kappa)}+\psi_\alpha^{0(\kappa)}),$$

$$\mathcal{H}^{(\kappa)}=j*F^{(\kappa)}+\rho_0(t\omega^{(\kappa)}+w^{0(\kappa)}),\;\; j(t)=t,\; t\in\mathcal{T}. \tag{5.9.37}$$

Proof. If we take in (5.9.26), $p = y$ and $q = t - y$, and integrate with respect to y from 0 to t, we arrive at

$$\begin{aligned}&\int_{\Gamma}\left[M_\alpha^{(1)}*v_\alpha^{(2)}+2h\tau^{(1)}*w^{(2)}+2h\mu_\alpha^{(1)}*\psi_\alpha^{(2)}-\frac{1}{T_0}\,\widehat{Q}^{(1)}*T^{(2)}\right]ds+\\ &+\int_{\Sigma}\left[2hF^{(1)}*w^{(2)}+2hG_\alpha^{(1)}*\psi_\alpha^{(2)}+L_\alpha^{(1)}*v_\alpha^{(2)}-\frac{1}{T_0}\,H^{(1)}*T^{(2)}\right]dA-\\ &-\int_{\Sigma}\Big\{\rho_0 I\ddot{v}_\alpha^{(1)}*v_\alpha^{(2)}+2h\rho_0\ddot{w}^{(1)}*w^{(2)}+2hJ\ddot{\psi}_\alpha^{(1)}*\psi_\alpha^{(2)}-\end{aligned}$$

$$-\frac{1}{T_0}k\left[I\widehat{T}^{(1)}_{,\alpha}*T^{(2)}_{,\alpha}+2h\widehat{T}^{(1)}*T^{(2)}\right]\Big\}dA=\int_\Gamma\Big[M^{(2)}_\alpha*v^{(1)}_\alpha+$$

$$+2h\tau^{(2)}*w^{(1)}+2h\mu^{(2)}_\alpha*\psi^{(1)}_\alpha-\frac{1}{T_0}\widehat{Q}^{(2)}*T^{(1)}\Big]ds+\int_\Sigma\Big[2hF^{(2)}*w^{(1)}+$$

$$+2hG^{(2)}_\alpha*\psi^{(1)}_\alpha+L^{(2)}_\alpha*v^{(1)}_\alpha-\frac{1}{T_0}H^{(2)}*T^{(1)}\Big]dA-\int_\Sigma\{\rho_0 I\ddot{v}^{(2)}_\alpha*v^{(1)}_\alpha+$$

$$+2h\rho_0\ddot{w}^{(2)}*w^{(1)}+2hJ\ddot{\psi}^{(2)}_\alpha*\psi^{(1)}_\alpha-\frac{1}{T_0}k\big[I\widehat{T}^{(2)}_{,\alpha}*T^{(1)}_{,\alpha}+2h\widehat{T}^{(2)}*T^{(1)}\big]\}dA.\tag{5.9.38}$$

If we take the convolution of the relation (5.9.38) with j and use (5.9.37) then we obtain (5.9.36). □

As in Section 2.2 we can obtain some applications of the above reciprocal theorem.

5.10 A Generalized Theory of Micropolar Thermoelasticity

In recent years there has been some interest in a theory of thermomechanics which will allow for "second sound" effects. Some authors have discussed generalized theories on the basis of a modified law of heat conduction. A large number of works are devoted to theories of thermomechanics which are based on a generalized entropy inequality. An account of the historical development of the subject as well as references to various contributions may be found in the works of Nowacki [289], Suhubi [355], Chandrasekharaiah [44], [48], Ignaczak [195], [200], Jou et al. [212], and Müller and Ruggeri [268]. In [130], Green and Lindsay have presented a theory of thermoelasticity by discussing restrictions on the constitutive equations with the help of an entropy production inequality proposed by Green and Laws [129]. The theory admits the possibility of "second sound" and leads to the result that the linear heat conduction tensor is symmetric. The result was generalized to the theory of micropolar thermoelastic bodies by Boschi and Ieşan [27]. In this section we present the linear theory of micropolar thermoelasticity established in [27] and derive uniqueness results.

a) **Basic equations.** Throughout this section we consider the linear theory of micropolar bodies and we shall use the notations of Section 5.1. We postulate the energy balance in the form (cf. (5.1.8))

$$\int_P (\rho_0 \dot{u}_i \ddot{u}_i + I_{ij} \ddot{\varphi}_j \dot{\varphi}_i + \rho_0 \dot{e}) dV = \int_P \rho_0 (f_i \dot{u}_i + g_i \dot{\varphi}_i + S) dV + \\ + \int_{\partial P} (t_i \dot{u}_i + m_i \dot{\varphi}_i + q) dA, \tag{5.10.1}$$

for every part P of B and every time. Following Green and Rivlin [125], we consider a second motion which differs from the given motion of the continuum by a constant rigid body translational velocity. Under this transformation the functions $\rho_0, e, \ddot{u}_i, \varphi_i, I_{ij}, f_i, g_i, S, t_i, m_i$ and q are not affected. From (5.10.1) we obtain

$$\int_P \rho_0 \ddot{u}_i dV = \int_P \rho_0 f_i dV + \int_{\partial P} t_i dA.$$

By the well-known method we find that

$$t_i = t_{ji} n_j \tag{5.10.2}$$

and

$$t_{ji,j} + \rho_0 f_i = \rho_0 \ddot{u}_i. \tag{5.10.3}$$

In view of (5.10.2) and (5.10.3) the equation (5.10.1) reduces to

$$\int_P (I_{ij} \ddot{\varphi}_j \dot{\varphi}_i + \rho_0 \dot{e}) dV = \int_P [t_{ji} \dot{u}_{i,j} + \rho_0 (g_i \dot{\varphi}_i + S)] dV + \\ + \int_{\partial P} (m_i \dot{\varphi}_i + q) dA. \tag{5.10.4}$$

Let us now consider a motion of the continuum which differs from the given motion only by a superposed uniform rigid body angular velocity. Suppose that at time t, the body is rotated back into original orientation. We also assume that $\rho_0 \dot{e}$, S, $\rho_0 g_i - I_{ij} \ddot{\varphi}_j$, t_{ij}, m_i and q are unaltered by such motion so that, in the linear theory, from (5.10.4) we obtain

$$\int_P [I_{ij} \ddot{\varphi}_j (\dot{\varphi}_i + b_i) + \rho_0 \dot{e}] dV = \int_P [t_{ji} (\dot{u}_{i,j} + \varepsilon_{jik} b_k) + \\ + \rho_0 g_i (\dot{\varphi}_i + b_i) + \rho_0 S] dV + \int_{\partial P} [m_i (\dot{\varphi}_i + b_i) + q] dA, \tag{5.10.5}$$

where b_i are arbitrary constants. It follows from (5.10.4) and (5.10.5) that

$$\left[\int_P (I_{ij} \ddot{\varphi}_j - \rho_0 g_i - \varepsilon_{irs} t_{rs}) dV - \int_{\partial P} m_i dA \right] b_i = 0, \tag{5.10.6}$$

for all arbitrary constants b_k. Since the quantities in the square brackets in (5.10.6) are independent of b_s, we get

$$\int_P I_{ij}\ddot{\varphi}_j dV = \int_P (\rho_0 g_i + \varepsilon_{irs} t_{rs}) dV + \int_{\partial P} m_i dA, \tag{5.10.7}$$

for any P of B. By the well-known method the relation (5.10.7) leads to

$$m_i = m_{ji} n_j, \tag{5.10.8}$$

where m_{ij} is the couple stress tensor. The local form of the balance equation (5.10.7) is

$$m_{ji,j} + \varepsilon_{irs} t_{rs} + \rho_0 g_i = I_{ij}\ddot{\varphi}_j. \tag{5.10.9}$$

With the help of (5.10.8) and (5.10.9), the relation (5.10.4) reduces to

$$\int_P \rho_0 \dot{e} dV = \int_P (t_{ij}\dot{e}_{ij} + m_{ij}\dot{\kappa}_{ij} + \rho_0 S) dV + \int_{\partial P} q\, dA, \tag{5.10.10}$$

where e_{ij} and κ_{ij} are defined in (5.1.28). If we apply this equation to a region which in the reference state was a tetrahedron bounded by coordinate planes through the point X and by a plane whose unit normal is n_k, we obtain

$$q = q_j n_j, \tag{5.10.11}$$

where q_s is the heat flux across the surface that was originally the coordinate plane perpendicular to the x_s-axis. In view of (5.10.11) we obtain the local form of the energy balance

$$\rho_o \dot{e} = t_{ij}\dot{e}_{ij} + m_{ij}\dot{\kappa}_{ij} + \rho_0 S + q_{j,j}. \tag{5.10.12}$$

In what follows we shall restrict our attention to the linear theory of homogeneous bodies. Following [130], we adopt the entropy production inequality proposed by Green and Laws [129],

$$\int_P \rho_0 \dot{\eta}\, dV - \int_P \rho_0 \frac{1}{\Phi} S\, dV - \int_{\partial P} \frac{1}{\Phi} q\, dA \geq 0, \tag{5.10.13}$$

for all regions P of B and every time. Here, η is the entropy per unit mass and Φ is a strictly positive function. With the help of (5.10.11), the inequality (5.10.13) leads to the local form

$$\rho_0 \dot{\eta} - \frac{1}{\Phi} \rho_0 S - \left(\frac{1}{\Phi} q_i \right)_{,i} \geq 0. \tag{5.10.14}$$

If we denote

$$A = e - \eta\Phi, \tag{5.10.15}$$

then from (5.10.12) and (5.10.14) we get

$$-\rho_0(\dot{A} + \eta\dot{\Phi}) + t_{ij}\dot{e}_{ij} + m_{ij}\dot{\kappa}_{ij} + \frac{1}{\Phi}\, q_i\Phi_{,i} \geq 0. \tag{5.10.16}$$

Let θ be the temperature measured from the constant absolute temperature T_0 of the body in its reference configuration. We define a micropolar thermoelastic material by the following constitutive equations

$$\begin{aligned} &A = \widehat{A}(e_{ij}, \kappa_{ij}, \theta, \theta_{,i}, \dot{\theta}), && \Phi = \widehat{\Phi}(e_{ij}, \kappa_{ij}, \theta, \theta_{,i}, \dot{\theta}),\\ &t_{rs} = \widehat{t}_{rs}(e_{ij}, \kappa_{ij}, \theta, \theta_{,i}, \dot{\theta}), && m_{rs} = \widehat{m}_{rs}(e_{ij}, \kappa_{ij}, \theta, \theta_{,i}, \dot{\theta}),\\ &\eta = \widehat{\eta}(e_{ij}, \kappa_{ij}, \theta, \theta_{,i}, \dot{\theta}), && q_k = \widehat{q}_k(e_{ij}, \kappa_{ij}, \theta, \theta_{,i}, \dot{\theta}), \end{aligned} \tag{5.10.17}$$

the response functions being of class C^2 and consistent with the assumption of the linear theory. It follows from (5.10.16) and (5.10.17) that

$$\Phi = \widehat{\Phi}(\theta, \dot{\theta}),\ A = \widehat{A}(e_{ij}, \kappa_{ij}, \theta, \theta_{,i}, \dot{\theta}), \tag{5.10.18}$$

$$\frac{\partial A}{\partial \dot{\theta}} + \eta\,\frac{\partial \Phi}{\partial \dot{\theta}} = 0, \tag{5.10.19}$$

$$q_i\,\frac{\partial \Phi}{\partial \dot{\theta}} = \Phi\,\frac{\partial \sigma}{\partial \theta_{,i}}, \tag{5.10.20}$$

$$\begin{aligned} &\left(t_{ij} - \frac{\partial \sigma}{\partial e_{ij}}\right)\dot{e}_{ij} + \left(m_{ij} - \frac{\partial \sigma}{\partial \kappa_{ij}}\right)\dot{\kappa}_{ij} -\\ &-\left(\frac{\partial \sigma}{\partial \theta} + \rho_0\eta\,\frac{\partial \Phi}{\partial \theta}\right)\dot{\theta} + \frac{1}{\Phi}\, q_i\,\frac{\partial \Phi}{\partial \theta}\,\theta_{,i} \geq 0, \end{aligned} \tag{5.10.21}$$

where

$$\sigma = \rho_0 A. \tag{5.10.22}$$

Following Green and Lindsay [130] we assume that

$$\Phi(\theta, 0) = T_0 + \theta,\ \frac{\partial \Phi}{\partial \dot{\theta}} \neq 0. \tag{5.10.23}$$

From (5.10.19)-(5.10.22) we obtain

$$\rho_0\eta = -\frac{\partial \sigma}{\partial \dot{\theta}} \Big/ \frac{\partial \Phi}{\partial \dot{\theta}}, \tag{5.10.24}$$

$$q_i = \Phi\,\frac{\partial \sigma}{\partial \theta_{,i}} \Big/ \frac{\partial \Phi}{\partial \dot{\theta}}, \tag{5.10.25}$$

$$t_{ij} = \frac{\partial \sigma}{\partial e_{ij}}, \quad m_{ij} = \frac{\partial \sigma}{\partial \kappa_{ij}}\cdot \tag{5.10.26}$$

The inequality (5.10.21) reduces to

$$\frac{1}{\Phi}\, q_i \,\frac{\partial \Phi}{\partial \theta}\,\theta_{,i} - \left(\frac{\partial \sigma}{\partial \theta} + \rho_0 \eta \,\frac{\partial \Phi}{\partial \theta}\right)\dot{\theta} \geq 0. \tag{5.10.27}$$

In view of (5.10.15), (5.10.22), (5.10.18), (5.10.24) and (5.10.26), the energy equation (5.10.12) becomes

$$\rho_0 \Phi \dot{\eta} + \left(\frac{\partial \sigma}{\partial \theta} + \rho_0 \eta \,\frac{\partial \Phi}{\partial \theta}\right)\dot{\theta} + \frac{\partial \sigma}{\partial \theta_{,i}}\,\dot{\theta}_{,i} = q_{i,i} + \rho_0 S. \tag{5.10.28}$$

Assuming that the initial body is free from stress, couple stress and heat flux, in the context of linear theory we have

$$\begin{aligned}
\sigma \;&= \sigma_0 - a\theta - b_0\dot{\theta} - \frac{1}{2}\, d\theta^2 - e\theta\dot{\theta} - \frac{1}{2}\, f\dot{\theta}^2 + a_i\theta\theta_{,i} + \\
&+ \alpha_0 b_i \dot{\theta}\theta_{,i} + \frac{1}{2}\,\alpha_0 K_{ij}\theta_{,i}\theta_{,j} + a_{ij}e_{ij}\theta + a_{ijk}e_{ij}\theta_{,k} + \\
&+ b_{ij}e_{ij}\dot{\theta} + b_{ijk}\kappa_{ij}\theta_{,k} + c_{ij}\kappa_{ij}\theta + d_{ij}\kappa_{ij}\dot{\theta} + \\
&+ \frac{1}{2}\, A_{ijrs}e_{ij}e_{rs} + B_{ijrs}e_{ij}\kappa_{rs} + \frac{1}{2}\, C_{ijrs}\kappa_{ij}\kappa_{rs}, \\
\Phi \;&= T_0 + \theta + \alpha_0\dot{\theta} + \beta_0\theta\dot{\theta} + \frac{1}{2}\,\gamma_0\dot{\theta}^2.
\end{aligned} \tag{5.10.29}$$

From (5.10.24)-(5.10.26) and (5.10.29) we obtain

$$\begin{aligned}
t_{ij} &= A_{ijrs}e_{rs} + B_{ijrs}\kappa_{rs} + a_{ij}\theta + b_{ij}\dot{\theta} + a_{ijk}\theta_{,k}, \\
m_{ij} &= B_{rsij}e_{rs} + C_{ijrs}\kappa_{rs} + c_{ij}\theta + d_{ij}\dot{\theta} + b_{ijk}\theta_k, \\
q_i &= \frac{1}{\alpha_0}\, T_0(a_i\theta + \alpha_0 b_i\dot{\theta} + a_{rsi}e_{rs} + b_{rsi}\kappa_{rs} + \alpha_0 K_{ij}\theta_{,j}), \\
\rho_0\eta &= \frac{1}{\alpha_0}\big[b + (e - \frac{1}{\alpha_0}\, b_0\beta_0) + (f - \frac{1}{\alpha_0}\, b_0\gamma_0)\dot{\theta} - \\
&\qquad\qquad - \alpha_0 b_i\theta_{,i} - b_{ij}e_{ij} - d_{ij}\kappa_{ij}\big].
\end{aligned} \tag{5.10.30}$$

The entropy inequality (5.10.27) implies the following restrictions on the constitutive coefficients

$$\begin{aligned}
&a_i = 0, \quad a_{ijk} = 0, \quad b_{ijk} = 0, \quad b_0 = \alpha_0 a, \quad b_{ij} = \alpha_0 a_{ij}, \\
&d_{ij} = \alpha_0 c_{ij}, \quad e - a\beta_0 = d\alpha_0
\end{aligned} \tag{5.10.31}$$

and

$$(d\alpha_0 - h)\dot{\theta}^2 + 2b_i\dot{\theta}\theta_{,i} + K_{ij}\theta_{,i}\theta_{,j} \geq 0, \tag{5.10.32}$$

where

$$h = \frac{1}{\alpha_0}(f - a\gamma_0). \tag{5.10.33}$$

Thus, we obtain the following constitutive equations

$$\begin{aligned}
t_{ij} &= A_{ijrs}e_{rs} + B_{ijrs}\kappa_{rs} + a_{ij}(\theta + \alpha_0\dot{\theta}),\\
m_{ij} &= B_{rsij}e_{rs} + C_{ijrs}\kappa_{rs} + c_{ij}(\theta + \alpha_0\dot{\theta}),\\
\rho_0\eta &= a + d\theta + h\dot{\theta} - b_i\theta_{,i} - a_{ij}e_{ij} - c_{ij}\kappa_{ij},\\
q_i &= T_0(b_i\dot{\theta} + K_{ij}\theta_{,j}).
\end{aligned} \tag{5.10.34}$$

The constitutive coefficients have the following symmetries

$$A_{ijrs} = A_{rsij}, \quad C_{ijrs} = C_{rsij}, \quad K_{ij} = K_{ji}. \tag{5.10.35}$$

In the context of the linear theory, the energy equation (5.10.28) reduces to

$$\rho_0 T_0 \dot{\eta} = q_{i,i} + \rho_0 S. \tag{5.10.36}$$

The basic equations of the theory are the equations of motion (5.10.3), (5.10.9), the equation of energy (5.10.36), the constitutive equations (5.10.34) and the geometrical equations (5.1.28).

If the material is isotropic with a centre of symmetry, then the constitutive equations take the form

$$\begin{aligned}
t_{ij} &= \lambda e_{rr}\delta_{ij} + (\mu + \kappa)e_{ij} + \mu e_{ji} - b(\theta + \alpha_0\dot{\theta})\delta_{ij},\\
m_{ij} &= \alpha\kappa_{rr}\delta_{ij} + \beta\kappa_{ji} + \gamma\kappa_{ij},\\
\rho_0\eta &= a + d\theta + h\dot{\theta} + be_{rr},\\
q_i &= k\theta_{,i}.
\end{aligned} \tag{5.10.37}$$

The field equations for isotropic and homogeneous bodies follow from (5.10.3), (5.10.9), (5.10.36), (5.10.37) and (5.1.28),

$$\begin{aligned}
&(\mu + \kappa)\Delta u_i + (\lambda + \mu)u_{j,ji} + \kappa\varepsilon_{irs}\varphi_{s,r} - b(\theta_{,i} +\\
&\qquad + \alpha_0\dot{\theta}_{,i}) + \rho_0 f_i = \rho_0\ddot{u}_i,\\
&\gamma\Delta\varphi_i + (\alpha + \beta)\varphi_{j,ji} + \kappa\varepsilon_{irs}u_{s,r} - 2\kappa\varphi_i + \rho_0 g_i = J\ddot{\varphi}_i,\\
&k\Delta\theta - T_0(d\dot{\theta} + h\ddot{\theta} + b\dot{u}_{i,i}) = -\rho_0 S \text{ on } B\times I.
\end{aligned} \tag{5.10.38}$$

To the basic equations we must adjoin initial conditions and boundary conditions. The initial conditions are

$$\begin{aligned}
&u_i(\boldsymbol{x},0)=u_i^0(\boldsymbol{x}),\ \dot{u}_i(\boldsymbol{x},0)=v_i^0(\boldsymbol{x}),\ \varphi_i(\boldsymbol{x},0)=\varphi_i^0(\boldsymbol{x}),\\
&\dot{\varphi}_i(\boldsymbol{x},0)=\nu_i^0(\boldsymbol{x}),\ \eta(\boldsymbol{x},0)=\eta^0(\boldsymbol{x}),\ \theta(\boldsymbol{x},0)=\theta^0(\boldsymbol{x}),\quad \boldsymbol{x}\in\overline{B},
\end{aligned} \tag{5.10.39}$$

where $u_i^0, v_i^0, \varphi_i^0, \nu_i^0, \eta^0$ and θ^0 are prescribed functions. In the case of the mixed boundary value problem the boundary conditions are (5.2.1).

b) **Uniqueness theorem.** In what follows we shall use the notation (5.1.35).

Theorem 5.10.1. *Assume that*

(i) ρ_0, α_0 *and* h *are strictly positive;*
(ii) I_{ij} *is symmetric and positive definite;*
(iii) W *is a positive semi-definite quadratic form;*
(iv) *the inequality* (5.10.32) *holds.*

Then the mixed problem of thermoelastodynamics has at most one solution.

Proof. We introduce the function G by

$$G=t_{ij}\dot{e}_{ij}+m_{ij}\dot{\kappa}_{ij}+\rho_0\dot{\eta}(\theta+\alpha_0\dot{\theta}).$$

It follows from (5.10.34) and (5.1.35) that

$$G=\dot{W}+(d\dot{\theta}+h\ddot{\theta}-b_i\dot{\theta}_{,i})(\theta+\alpha_0\dot{\theta}). \tag{5.10.40}$$

On the other hand, from (5.10.3), (5.10.9), (5.1.28) and (5.10.36) we obtain

$$\begin{aligned}
G\ =\ &\left[t_{ji}\dot{u}_i+m_{ji}\dot{\varphi}_i+\frac{1}{T_0}\,q_j(\theta+\alpha_0\dot{\theta})\right]_{,j}+\\
&+\rho_0\left[f_i\dot{u}_i+g_i\dot{\varphi}_i+\frac{1}{T_0}\,S(\theta+\alpha_0\dot{\theta})\right]-\\
&-\rho_0\ddot{u}_i\dot{u}_i-I_{ij}\ddot{\varphi}_j\dot{\varphi}_i-\frac{1}{T_0}\,q_i(\theta+\alpha_0\dot{\theta})_{,i}.
\end{aligned} \tag{5.10.41}$$

From (5.10.40) and (5.10.41) we get

$$\begin{aligned}
&\dot{W}+\rho_0\ddot{u}_i\dot{u}_i+I_{ij}\ddot{\varphi}_j\dot{\varphi}_i+(d\dot{\theta}+h\ddot{\theta}-b_i\dot{\theta}_{,i})(\theta+\alpha_0\dot{\theta})=\\
&\quad=\left[t_{ji}\dot{u}_i+m_{ji}\dot{\varphi}_i+\frac{1}{T_0}\,q_j(\theta+\alpha_0\dot{\theta})\right]_{,j}+\\
&\quad+\rho_0\left[f_i\dot{u}_i+g_i\dot{\varphi}_i+\frac{1}{T_0}\,S(\theta+\alpha_0\dot{\theta})\right].
\end{aligned} \tag{5.10.42}$$

We define the function U on I by

$$
\begin{aligned}
U = \int_B \Big\{ W &+ \frac{1}{2}\left[\rho_0 \dot{u}_i \dot{u}_i + I_{ij}\dot{\varphi}_j\dot{\varphi}_i + \frac{1}{\alpha_0} h(\theta + \alpha_0\dot{\theta})^2\right] + \\
&+ \alpha_0 \left[(\alpha_0 d - h)\left(\frac{1}{\alpha_0}\theta\right)^2 - 2b_i\left(\frac{1}{\alpha_0}\theta\right)\theta_{,i} + K_{ij}\theta_{,j}\theta_{,i}\right]\Big\} dV.
\end{aligned}
\tag{5.10.43}
$$

We note that the inequality (5.10.32) implies that

$$
(\alpha_0 d - h)x^2 + 2b_i x y_i + K_{ij} y_i y_j \geq 0,
$$

for any x and y_i, so that from the hypotheses (i)-(iii) and (5.10.43) we conclude that $U \geq 0$ on I.

If we integrate the relation (5.10.42) over B and use the divergence theorem, then we obtain

$$
\begin{aligned}
\dot{U} = \int_B \rho_0 &\left[f_i \dot{u}_i + g_i \dot{\varphi}_i + \frac{1}{T_0} S(\theta + \alpha_0 \dot{\theta})\right] dV + \\
+ \int_{\partial B} &\left[t_i \dot{u}_i + m_i \dot{\varphi}_i + \frac{1}{T_0} q(\theta + \alpha_0 \dot{\theta})\right] dA.
\end{aligned}
\tag{5.10.44}
$$

We now suppose that there are two solutions of the mixed problem. Then their difference $(u_i^*, \varphi_i^*, \theta^*)$ corresponds to null data. From (5.10.43) we obtain $\dot{U}^* = 0$ on I. Taking into account the initial conditions and the hypotheses of the theorem, we conclude that $u_i^* = 0$, $\varphi_i^* = 0$ and $\theta^* = 0$ on $B \times I$. □

We note that for non-polar theory the uniqueness theorem has been established by Green and Lindsay [130].

c) **Reciprocity.** We now use the method established in [184] to derive a reciprocal theorem. Throughout this section we restrict our attention to centrosymmetric bodies,

$$
B_{ijrs} = 0, \quad b_i = 0, \quad c_{ij} = 0. \tag{5.10.45}
$$

We shall show that a reciprocity relation implies another uniqueness result. In what follows we use the notation (5.3.1) and assume that $I = [0, \infty)$.

Lemma 5.10.1. *The functions $\eta \in C^{0,1}$ and $q_i \in C^{1,0}$ satisfy the equation (5.10.36) and the initial condition $\eta(\boldsymbol{x}, 0) = \eta_0(\boldsymbol{x})$, $\boldsymbol{x} \in \overline{B}$, if and only if*

$$
T_0(\rho_0 \eta - a) = \widehat{q}_{i,i} + R \quad \text{on } B \times [0, \infty), \tag{5.10.46}
$$

where

$$R = \rho_0 \widehat{S} + T_0(\rho_0 \eta_0 - a). \tag{5.10.47}$$

The proof of this lemma is immediate.

In view of Lemma 5.10.1, the boundary-initial-value problem can be characterized by the equations (5.10.3), (5.10.9), (5.10.36), (5.10.34), (5.1.28) on $B \times I$, the boundary conditions (5.2.1) and the initial conditions

$$\begin{aligned} &u_i(\boldsymbol{x},0) = u_i^0(\boldsymbol{x}), \quad \dot{u}_i(\boldsymbol{x},0) = v_i^0(\boldsymbol{x}), \quad \varphi_i(\boldsymbol{x},0) = \varphi_i^0(\boldsymbol{x}), \\ &\dot{\varphi}_i(\boldsymbol{x},0) = \nu_i^0(\boldsymbol{x}), \quad \theta(\boldsymbol{x},0) = \theta^0(\boldsymbol{x}), \qquad \boldsymbol{x} \in \overline{B}. \end{aligned} \tag{5.10.48}$$

We introduce the notation

$$Nf = f + \alpha_0 \dot{f}. \tag{5.10.49}$$

Consider now two external data systems $L^{(\alpha)} = (f_i^{(\alpha)}, g_i^{(\alpha)}, S^{(\alpha)}, \widetilde{u}_i^{(\alpha)}, \widetilde{\varphi}_i^{(\alpha)}, \widetilde{\theta}^{(\alpha)}, \widetilde{t}_i^{(\alpha)}, \widetilde{m}_i^{(\alpha)}, \widetilde{q}^{(\alpha)}, u_i^{0(\alpha)}, v_i^{0(\alpha)}, \varphi_i^{0(\alpha)}, \nu_i^{0(\alpha)}, \eta^{0(\alpha)}, \theta^{0(\alpha)})$, $(\alpha = 1,2)$. Let $p^{(\alpha)} = (u_i^{(\alpha)}, \varphi_i^{(\alpha)}, \theta^{(\alpha)}, e_{ij}^{(\alpha)}, \kappa_{ij}^{(\alpha)}, t_{ij}^{(\alpha)}, m_{ij}^{(\alpha)}, \eta^{(\alpha)}, q_i^{(\alpha)})$ be a solution corresponding to $L^{(\alpha)}$.

Theorem 5.10.2. *Let*

$$\begin{aligned} \Gamma_{\kappa\nu}(r,s) = &\int_{\partial B} \Big[t_i^{(\kappa)}(\boldsymbol{x},r) u_i^{(\nu)}(\boldsymbol{x},s) + m_i^{(\kappa)}(\boldsymbol{x},r)\varphi_i^{(\nu)}(\boldsymbol{x},s) - \\ &- \frac{1}{T_0}\,\widehat{q}^{(\kappa)}(\boldsymbol{x},r) N\theta^{(\nu)}(\boldsymbol{x},s)\Big] dA + \int_B \Big[\rho_0 f_i^{(\kappa)}(\boldsymbol{x},r) u_i^{(\beta)}(\boldsymbol{x},s) + \\ &+ \rho_0 g_i^{(\kappa)}(\boldsymbol{x},r)\varphi_i^{(\nu)}(\boldsymbol{x},s) - \frac{1}{T_0}\, R^{(\kappa)}(\boldsymbol{x},r) N\theta^{(\nu)}(\boldsymbol{x},s)\Big] dV - \\ &- \int_B \Big[\rho_0 \ddot{u}_i^{(\kappa)}(\boldsymbol{x},r) u_i^{(\nu)}(\boldsymbol{x},s) + I_{ij}\ddot{\varphi}_j^{(\kappa)}(\boldsymbol{x},r)\varphi_i^{(\nu)}(\boldsymbol{x},s) - \\ &- h\dot{\theta}^{(\kappa)}(\boldsymbol{x},r)\theta^{(\nu)}(\boldsymbol{x},s) - \alpha_0 d\theta^{(\kappa)}(\boldsymbol{x},r)\dot{\theta}^{(\nu)}(\boldsymbol{x},s)\Big] dV + \\ &+ \frac{1}{T_0}\int_B \widehat{q}_i^{(\kappa)}(\boldsymbol{x},r) N\theta_{,i}^{(\nu)}(\boldsymbol{x},s) dV, \end{aligned} \tag{5.10.50}$$

for all $r, s \in I$. *If the relations* (5.10.45) *hold, then*

$$\Gamma_{\kappa\nu}(r,s) = \Gamma_{\nu\kappa}(s,r), \quad (\kappa,\nu = 1,2), \tag{5.10.51}$$

for any $r, s \in I$.

Proof. If we denote

$$\begin{aligned}\Pi_{\kappa\nu}(r,s) = t_{ij}^{(\kappa)}(r)e_{ij}^{(\nu)}(s) + m_{ij}^{(\kappa)}(r)\kappa_{ij}^{(\nu)}(s)- \\ -(\rho_0\eta^{(\kappa)}(r) - a)N\theta^{(\nu)}(s),\end{aligned} \tag{5.10.52}$$

then, on the basis of (5.10.34), we obtain

$$\begin{aligned}\Pi_{\kappa\nu}(r,s) = A_{ijmn}e_{mn}^{(\kappa)}(r)e_{ij}^{(\nu)}(s) + C_{ijmn}\kappa_{mn}^{(\kappa)}(r)\kappa_{ij}^{(\nu)}(s)+ \\ +a_{ij}\left[e_{ij}^{(\nu)}(s)N\dot{\theta}^{(\kappa)}(r) + e_{ij}^{(\kappa)}(r)N\dot{\theta}^{(\nu)}(s)\right] - \alpha_0 h\dot{\theta}^{(\kappa)}(r)\dot{\theta}^{(\nu)}(s)- \\ -d\theta^{(\kappa)}(r)\theta^{(\nu)}(s) - h\dot{\theta}^{(\kappa)}(r)\theta^{(\nu)}(s) - \alpha_0 d\theta^{(\kappa)}(r)\dot{\theta}^{(\nu)}(s).\end{aligned} \tag{5.10.53}$$

We introduce the notations

$$\Pi_{\kappa\nu}^{*}(r,s) = \Pi_{\kappa\nu}(r,s) + h\dot{\theta}^{(\kappa)}(r)\theta^{(\nu)}(s) + \alpha_0 d\theta^{(\kappa)}(r)\dot{\theta}^{(\nu)}(s). \tag{5.10.54}$$

From (5.10.53), (5.10.54) and (5.10.35) we get

$$\Pi_{\kappa\nu}^{*}(r,s) = \Pi_{\nu\kappa}^{*}(s,r), \quad r,s \in I. \tag{5.10.55}$$

On the other hand, in view of the equations (5.1.28), (5.10.3), (5.10.9) and (5.10.46), we arrive at

$$\begin{aligned}\Pi_{\kappa\nu}(r,s) = \left[t_{ji}^{(\kappa)}(r)u_i^{(\nu)}(s) + m_{ji}^{(\kappa)}(r)\varphi_i^{(\nu)}(s) - \frac{1}{T_0}\widehat{q}_j^{(\kappa)}(r)N\theta^{(\nu)}(s)\right]_{,j} + \\ +\rho_0\left(f_i^{(\kappa)}(r)u_i^{(\nu)}(s) + g_i^{(\kappa)}(r)\varphi_i^{(\nu)}(s)\right) - \frac{1}{T_0}R^{(\kappa)}(r)N\theta^{(\nu)}(s)- \\ -\rho_0\ddot{u}_i^{(\kappa)}(r)u_i^{(\nu)}(s) - I_{ij}\ddot{\varphi}_j^{(\kappa)}(r)\varphi^{(\nu)}(s) + \frac{1}{T_0}\widehat{q}_i^{(\kappa)}(r)N\theta_{,i}^{(\nu)}(s).\end{aligned} \tag{5.10.56}$$

It follows from (5.10.54), (5.10.56), (5.10.50) and the divergence theorem, that

$$\int_B \Pi_{\kappa\nu}^{*}(r,s)dV = \Gamma_{\kappa\nu}(r,s). \tag{5.10.57}$$

The relations (5.10.55) and (5.10.57) imply (5.10.51). □

If we take in (5.10.51) $r = \tau$ and $s = t - \tau$ and integrate from 0 to t, then we obtain the following reciprocal theorem.

Theorem 5.10.3. *Assume that I_{ij} is symmetric and that the relations (5.10.45) hold. Let* $p^{(\alpha)} = (u_i^{(\alpha)}, \varphi_i^{(\alpha)}, \theta^{(\alpha)}, e_{ij}^{(\alpha)}, \kappa_{ij}^{(\alpha)}, t_{ij}^{(\alpha)}, m_{ij}^{(\alpha)}, \eta^{(\alpha)}, q_i^{(\alpha)})$

be a solution corresponding to the external data system $L^{(\alpha)}$, $(\alpha = 1, 2)$. Then

$$\begin{aligned}&\int_B \left[F_i^{(1)} * u_i^{(2)} + G_i^{(1)} * \varphi_i^{(2)} - \frac{1}{T_0} j * R^{(1)} * N\theta^{(2)}\right] dV + \\ &+ \int_{\partial B} j * \left[t_i^{(1)} * u_i^{(2)} + m_i^{(1)} * \varphi_i^{(2)} - \frac{1}{T_0} \widehat{q}^{(1)} * N\theta^{(2)}\right] dA = \\ &= \int_B \left[F_i^{(2)} * u_i^{(1)} + G_i^{(2)} * \varphi_i^{(1)} - \frac{1}{T_0} j * R^{(2)} * N\theta^{(1)}\right] dV + \\ &+ \int_{\partial B} j * \left[t_i^{(2)} * u_i^{(1)} + m_i^{(2)} * \varphi_i^{(1)} - \frac{1}{T_0} \widehat{q}^{(2)} * N\theta^{(1)}\right] dA.\end{aligned} \tag{5.10.58}$$

Here we have used the notations (5.3.10). The relation (5.10.58) has been established in [178].

Theorem 5.10.2 implies the following result.

Theorem 5.10.4. *Let*

$$\begin{aligned}\Omega(r, s) = &\int_B \left[\rho_0 f_i(r) u_i(s) + \rho_0 g_i(r) \varphi_i(s) - \frac{1}{T_0} R(r) N\theta(s)\right] dV + \\ &+ \int_{\partial B} \left[t_i(r) u_i(s) + m_i(r) \varphi_i(s) - \frac{1}{T_0} \widehat{q}(r) N\theta(s)\right] dA,\end{aligned} \tag{5.10.59}$$

for all $r, s \in I$. Assume that (5.10.45) hold. Then

$$\begin{aligned}&\frac{d}{dt}\Big\{\int_B (\rho_0 u_i u_i + I_{ij}\varphi_i\varphi_j + \alpha_0 K_{ij}\widehat{\theta}_{,j}\widehat{\theta}_{,i}) dV + \\ &+ \int_0^t \int_B [(\alpha_0 d - h)\theta^2 + K_{ij}\widehat{\theta}_{,j}\widehat{\theta}_{,i}]\, dt\, dV\Big\} = \\ &= \int_0^t [\Omega(t-s, t+s) - \Omega(t+s, t-s)] ds + \int_B \{\rho_0 [\dot{u}_i(2t) u_i(0) + \\ &+ \dot{u}_i(0) u_i(2t)] + I_{ij}[\dot{\varphi}_i(2t)\varphi_j(0) + \varphi_i(2t)\dot{\varphi}_j(0)] + \\ &+ (\alpha_0 d - h)\theta(0)\theta(2t) + \alpha_0 K_{ij}\widehat{\theta}_{,j}(2t)\theta_{,i}(0)\} dV.\end{aligned} \tag{5.10.60}$$

Proof. In view of (5.10.51), we have

$$\int_0^t \Gamma_{11}(t+s, t-s) ds = \int_0^t \Gamma_{11}(t-s, t+s) ds. \tag{5.10.61}$$

From (5.10.50) and (5.10.59) we obtain

$$\int_0^t \Gamma_{11}(t+s,t-s)ds = \int_0^t \Omega(t+s,t-s)ds - \int_0^t\int_B [\rho_0 \ddot{u}_i(t+s)u_i(t-s)+$$
$$+I_{ij}\ddot{\varphi}_j(t+s)\varphi_i(t-s) - h\dot{\theta}(t+s)\theta(t-s) - \alpha_0 d\theta(t+s)\dot{\theta}(t-s)]ds\, dV+$$
$$+\int_0^t\int_B K_{ij}\widehat{\theta}_{,j}(t+s)[\dot{\widehat{\theta}}_{,i}(t-s)+\alpha_0\ddot{\widehat{\theta}}_{,i}(t-s)]ds\, dV,$$
$$\int_0^t \Gamma_{11}(t-s,t+s)ds = \int_0^t \Omega(t-s,t+s)ds - \int_0^t\int_B [\rho_0 \ddot{u}_i(t-s)u_i(t+s)+$$
$$+I_{ij}\ddot{\varphi}_j(t-s)\varphi_i(t+s) - h\dot{\theta}(t-s)\theta(t+s) - \alpha_0 d\theta(t-s)\dot{\theta}(t+s)]ds\, dV+$$
$$+\int_0^t\int_B K_{ij}\widehat{\theta}_{,j}(t-s)[\dot{\widehat{\theta}}_{,i}(t+s)+\alpha_0\ddot{\widehat{\theta}}_{,i}(t+s)]ds\, dV. \tag{5.10.62}$$

If we use the relations (5.3.17), then from (5.10.62) we find that

$$\int_0^t \Gamma_{11}(t+s,t-s)ds = \int_0^t \Omega(t+s,t-s)ds - \int_B \{\rho_0[\dot{u}_i(2t)u_i(0)-$$
$$-\dot{u}_i(t)u_i(t)] + I_{ij}[\dot{\varphi}_j(2t)\varphi_i(0) - \dot{\varphi}_j(t)\varphi_i(t)] - h[\theta(2t)\theta(0)-$$
$$-\theta^2(t)] - \alpha_0 d[\theta^2(t) - \theta(0)\theta(2t)]\}dV - \int_0^t\int_B [\rho_0\dot{u}_i(t-s)\dot{u}_i(t+s)+$$
$$+I_{ij}\dot{\varphi}_j(t-s)\dot{\varphi}_i(t+s) - h\theta(t+s)\dot{\theta}(t-s) - \alpha_0 d\dot{\theta}(t+s)\theta(t-s)]ds\, dV+$$
$$+\int_0^t\int_B K_{ij}[\dot{\widehat{\theta}}_{,j}(t+s)\widehat{\theta}_{,i}(t-s) + \alpha_0\theta_{,j}(t+s)\theta_{,i}(t-s)]ds\, dV+$$
$$+\int_B K_{ij}[\widehat{\theta}_{,j}(\widehat{\theta}_{,i} + \alpha_0\dot{\widehat{\theta}}_{,i}) - \alpha_0\widehat{\theta}_{,j}(2t)\theta_{,i}(0)]dV,$$

$$\int_0^t \Gamma_{11}(t-s,t+s)ds = \int_0^t \Omega(t-s,t+s)ds - \int_B \{\rho_0[\dot{u}_i(t)u_i(t)-$$
$$-\dot{u}_i(0)u_i(2t)] + I_{ij}[\dot{\varphi}_j(t)\varphi_i(t) - \dot{\varphi}_j(0)\varphi_i(2t)]-$$
$$-h\dot{\theta}(t-s)\theta(t+s) - \alpha_0 d\theta(t-s)\dot{\theta}(t+s)]ds\, dV+$$
$$+\int_0^t\int_B K_{ij}\widehat{\theta}_{,j}(t-s)\dot{\widehat{\theta}}_{,i}(t+s)ds\, dV+ \tag{5.10.63}$$
$$+\int_0^t\int_B \alpha_0 K_{ij}\theta_{,i}(t+s)\theta_{,j}(t-s)ds\, dV - \int_B \alpha_0 K_{ij}\dot{\widehat{\theta}}_{,i}\widehat{\theta}_{,j}dV.$$

The desired result follows from (5.10.61) and (5.10.63). □

An immediate consequence of Theorem 5.10.4 is the following uniqueness result.

Theorem 5.10.5. *Assume that*

(i) ρ_0 *and* $\alpha_0 d - h$ *are strictly positive;*

(ii) $\alpha_0 \geq 0$ *and* K_{ij} *is positive semi-definite;*

(iii) I_{ij} *is symmetric and positive definite;*

(iv) *the relations* (5.10.45) *hold.*

Then the mixed problem of thermoelastodynamics has at most one solution.

Proof. Assume that there are two solutions. Then their difference (u_i, φ_i, θ) corresponds to null data and from (5.10.60) we obtain

$$\begin{aligned}\int_B [\rho_0 u_i u_i + I_{ij}\varphi_i\varphi_j + \alpha_0 K_{ij}\widehat{\theta}_{,i}\widehat{\theta}_{,j}]dV + \\ + \int_0^t \int_B [(\alpha_0 d - h)\theta^2 + K_{ij}\widehat{\theta}_{,j}\widehat{\theta}_{,i}]ds\, dV = 0.\end{aligned} \tag{5.10.64}$$

By using the hypotheses (i)-(iv) and (5.10.64) we obtain $u_i = 0$, $\varphi_i = 0$ and $\theta = 0$ on $B \times I$. □

The generalized micropolar thermoelasticity has been investigated by various authors (see, e.g., Dost and Tabbarok [90], Chandrasekharaiah [43], [47], Vieru [371], De Cicco and Nappa [87]).

Chapter 6

THERMOELASTOSTATICS OF MICROPOLAR BODIES

6.1 Boundary Value Problems. Basic Theorems

In the equilibrium theory, the basic equations reduce to the equations of equilibrium

$$\begin{aligned} &t_{ji,j} + \rho_0 f_i = 0, \\ &m_{ji,j} + \varepsilon_{irs} t_{rs} + \rho_0 g_i = 0, \end{aligned} \tag{6.1.1}$$

the equilibrium energy equation

$$q_{i,i} + \rho_0 S = 0, \tag{6.1.2}$$

the constitutive equations

$$\begin{aligned} &t_{ij} = A_{ijrs} e_{rs} + B_{ijrs} \kappa_{rs} - \beta_{ij} \theta, \\ &m_{ij} = B_{rsij} e_{rs} + C_{ijrs} \kappa_{rs} - \xi_{ij} \theta, \\ &q_i = k_{ij} \theta_{,j} \end{aligned} \tag{6.1.3}$$

and the geometrical equations

$$e_{ij} = u_{j,i} + \varepsilon_{jik} \varphi_k, \quad \kappa_{ij} = \varphi_{j,i}, \tag{6.1.4}$$

on B. The constitutive coefficients satisfy the relations (5.1.33).

We consider the boundary conditions

$$\begin{aligned} &u_i = \widetilde{u}_i \ \text{on} \ \overline{S}_1, \quad t_{ji} n_j = \widetilde{t}_i \ \text{on} \ S_2, \\ &\varphi_i = \widetilde{\varphi}_i \ \text{on} \ \overline{S}_3, \quad m_{ji} n_j = \widetilde{m}_i \ \text{on} \ S_4, \\ &\theta = \widetilde{\theta} \ \text{on} \ \overline{S}_5, \quad q_i n_i = \widetilde{q} \ \text{on} \ S_6, \end{aligned} \tag{6.1.5}$$

where S_k, $(k = 1, 2, \ldots, 6)$, are parts of ∂B such that $\overline{S}_1 \cup S_2 = \overline{S}_3 \cup S_4 = \overline{S}_5 \cup S_6 = \partial B$, $S_1 \cap S_2 = S_3 \cap S_4 = S_5 \cap S_6 = \emptyset$, and $\widetilde{u}_i, \widetilde{t}_i, \widetilde{\varphi}_i, \widetilde{m}_i, \widetilde{\theta}$

and $\widetilde{q}$ are given. We assume that: (α) the functions ρ_0, f_i, g_i and S are continuous on $\overline{B}$; (β) the constitutive coefficients are continuously differentiable on $\overline{B}$ and satisfy (5.1.32), (5.1.33); (γ) $\widetilde{u}_i$ are continuous on $\overline{S}_1, \widetilde{\varphi}_i$ are continuous on $\overline{S}_3, \widetilde{\theta}$ is continuous on $\overline{S}_5, \widetilde{t}_i$ are piecewise regular on $S_2, \widetilde{m}_i$ are piecewise regular on S_4 and $\widetilde{q}$ is piecewise regular on S_6.

As in classical theory, the temperature θ is independent of the displacement and microrotation fields. The function θ is the solution of a boundary value problem which consists in the equations (6.1.2), $(6.1.3)_3$ on B and the boundary conditions $(6.1.5)_{5,6}$. In discussion of the general theory we shall treat the temperature field as a known function.

We say that the ordered array of functions $(u_i, \varphi_i, e_{ij}, \kappa_{ij}, t_{ij}, m_{ij})$ is an admissible state provided: (i) $u_i, \varphi_i \in C^2(B) \cap C^1(\overline{B})$; (ii) e_{ij} and κ_{ij} are continuous tensor fields on $\overline{B}$; (iii) $t_{ij}, m_{ij}, t_{ji,j}, m_{ji,j} \in C^0(\overline{B})$. Clearly, the set of all admissible states is a vector space provided we define addition and scalar multiplication by $(u_i, \varphi_i, \ldots, m_{ij}) + (u'_i, \varphi'_i, \ldots, m'_{ij}) = (u_i + u'_i, \varphi_i + \varphi'_i, \ldots, m_{ij} + m'_{ij}), \lambda(u_i, \varphi_i, \ldots, m_{ij}) = (\lambda u_i, \lambda\varphi_i, \ldots, \lambda m_{ij})$.

By a thermoelastic state corresponding to the body loads f_i, g_i and the temperature field θ we mean an admissible state that satisfies the equations (6.1.1), $(6.1.3)_{1,2}$ and (6.1.4) on B. After the determination of the temperature θ, we have to find a thermoelastic state corresponding to the body loads f_i, g_i and to the temperature θ, which satisfies the boundary conditions $(6.1.5)_{1-4}$. If such a thermoelastic state exists, it is called a solution of the mixed problem. The ordered array of functions $(f_i, g_i, \widetilde{u}_i, \widetilde{\varphi}_i, \widetilde{t}_i, \widetilde{m}_i, \theta)$ is called external data system on B for the mixed problem. When $S_2 = S_4 = S_6 = \partial B$, the resulting problem is called the Neumann problem.

With a view toward deriving a reciprocal theorem we assume that the body is subjected to two external data systems $L^{(\alpha)} = (f_i^{(\alpha)}, g_i^{(\alpha)}, \widetilde{u}_i^{(\alpha)}, \widetilde{\varphi}_i^{(\alpha)}, \widetilde{t}_i^{(\alpha)}, \widetilde{m}_i^{(\alpha)}, \theta^{(\alpha)}), (\alpha = 1,2)$. Let $s^{(\alpha)} = (u_i^{(\alpha)}, \varphi_i^{(\alpha)}, e_{ij}^{(\alpha)}, \kappa_{ij}^{(\alpha)}, t_{ij}^{(\alpha)}, m_{ij}^{(\alpha)})$ be a solution corresponding to $L^{(\alpha)}, (\alpha = 1,2)$.

Theorem 6.1.1 *If $s^{(\alpha)}$ are thermoelastic states corresponding to the external data systems $L^{(\alpha)}, (\alpha = 1,2)$, then*

$$\int_B \{[\rho_0 f_i^{(1)} - (\beta_{ji}\theta^{(1)})_{,j}]u_i^{(2)} + [\rho_0 g_i^{(1)} - (\xi_{ji}\theta^{(1)})_{,j} - \varepsilon_{irs}\beta_{rs}\theta^{(1)}]\varphi_i^{(2)}\}dV +$$

$$+ \int_{\partial B} [(t_i^{(1)} + \beta_{ji}\theta^{(1)}n_j)u_i^{(2)} + (m_i^{(1)} + \xi_{ji}\theta^{(1)}n_j)\varphi_i^{(2)}]dA =$$

$$= \int_B \{[\rho_0 f_i^{(2)} - (\beta_{ji}\theta^{(2)})_{,j}]u_i^{(1)} + [\rho_0 g_i^{(2)} - (\xi_{ji}\theta^{(2)})_{,j} - \varepsilon_{irs}\beta_{rs}\theta^{(2)}]\varphi_i^{(1)}\}dV +$$
$$+ \int_B [(t_i^{(2)} + \beta_{ji}\theta^{(2)}n_j)u_i^{(1)} + (m_i^{(2)} + \xi_{ji}\theta^{(2)}n_j)\varphi_i^{(1)}]dA, \tag{6.1.6}$$

where

$$t_i^{(\alpha)} = t_{ji}^{(\alpha)}n_j, \quad m_i^{(\alpha)} = m_{ji}^{(\alpha)}n_j. \tag{6.1.7}$$

Proof. We introduce the following notation

$$2E_{\alpha\beta} = (t_{ij}^{(\alpha)} + \beta_{ij}\theta^{(\alpha)})e_{ij}^{(\beta)} + (m_{ij}^{(\alpha)} + \xi_{ij}\theta^{(\alpha)})\kappa_{ij}^{(\beta)}, \quad (\alpha, \beta = 1, 2). \tag{6.1.8}$$

By (6.1.3),

$$2E_{\alpha\beta} = A_{ijrs}e_{rs}^{(\alpha)}e_{ij}^{(\beta)} + B_{ijrs}(e_{ij}^{(\beta)}\kappa_{rs}^{(\alpha)} + e_{ij}^{(\alpha)}\kappa_{rs}^{(\beta)}) + C_{ijrs}\kappa_{rs}^{(\alpha)}\kappa_{ij}^{(\beta)}. \tag{6.1.9}$$

Since the constitutive coefficients satisfy (5.1.3) we obtain

$$E_{\alpha\beta} = E_{\beta\alpha}. \tag{6.1.10}$$

On the other hand, by (6.1.4) and (6.1.1), we get

$$\begin{aligned} 2E_{\alpha\beta} &= [(t_{ij}^{(\alpha)} + \beta_{ij}\theta^{(\alpha)})u_j^{(\beta)} + (m_{ij}^{(\alpha)} + \xi_{ij}\theta^{(\alpha)})\varphi_j^{(\beta)}]_{,i} + \\ &+ [\rho_0 f_j^{(\alpha)} - (\beta_{ij}\theta^{(\alpha)})_{,i}]u_j^{(\beta)} + \\ &+ [\rho_0 g_j^{(\alpha)} - (\xi_{ij}\theta^{(\alpha)})_{,i} + \varepsilon_{jsr}\beta_{rs}\theta^{(\alpha)}]\varphi_j^{(\beta)}. \end{aligned} \tag{6.1.11}$$

By the divergence theorem and (6.1.10),

$$\begin{aligned} 2\int_B E_{\alpha\beta}dv &= \int_{\partial B} [(t_j^{(\alpha)} + \beta_{ij}\theta^{(\alpha)}n_i)u_j^{(\beta)} + (m_j^{(\alpha)} + \\ &+ \xi_{ij}\theta^{(\alpha)}n_i)\varphi_j^{(\beta)}]dA + \int_B \{[\rho_0 f_j^{(\alpha)} - (\beta_{ij}\theta^{(\alpha)})_{,i}]u_j^{(\beta)} + \\ &+ [\rho_0 g_j^{(\alpha)} - (\xi_{ij}\theta^{(\alpha)})_{,i} + \varepsilon_{jsr}\beta_{rs}\theta^{(\alpha)}]\varphi_j^{(\beta)}\}dV. \end{aligned} \tag{6.1.12}$$

The relations (6.1.10) and (6.1.12) imply the desired result. □

We note that the reciprocity relation can be written in the form

$$\begin{aligned} &\int_B \rho_0(f_i^{(1)}u_i^{(2)} + g_i^{(1)}\varphi_i^{(2)})dV + \int_{\partial B}(t_i^{(1)}u_i^{(2)} + m_i^{(1)}\varphi_i^{(2)})dA + \\ &+ \int_B \theta^{(1)}(\beta_{ji}e_{ij}^{(2)} + \xi_{ji}\kappa_{ij}^{(2)})dV = \int_B \rho_0(f_i^{(2)}u_i^{(1)} + g_i^{(2)}\varphi_i^{(1)})dV + \\ &+ \int_{\partial B}(t_i^{(2)}u_i^{(1)} + m_i^{(2)}\varphi_i^{(1)})dA + \int_B \theta^{(2)}(\beta_{ji}e_{ij}^{(1)} + \xi_{ji}\kappa_{ij}^{(1)})dV. \end{aligned} \tag{6.1.13}$$

Let us assume that

$$L^{(1)}=(f_i^{(1)}=0, g_i^{(1)}=0, \widetilde{u}_i^{(1)}=0, \widetilde{\varphi}_i^{(1)}=0, \widetilde{t}_i^{(1)}=0, \widetilde{m}_i^{(1)}=0, \theta^{(1)}),$$

$$L^2=(f_i^{(2)}=\delta_{ik}\delta(x-y)/\rho_0, g_i^{(2)}=0, \widetilde{u}_i^{(2)}=0, \widetilde{\varphi}_i^{(2)}=0, \widetilde{t}_i^{(2)}=0, \widetilde{m}_i^{(2)}=0, \theta^{(2)}=0),$$

where $\delta(\cdot)$ is the Dirac measure and y is a fixed point. In this case the reciprocity relation (6.1.13) leads to

$$u_k^{(1)}(\mathbf{y}) = \int_B \theta^{(1)}(\mathbf{x})[\beta_{ji}(\mathbf{x})e_{ij}^{(2k)}(\mathbf{x};\mathbf{y}) + \xi_{ji}(\mathbf{x})\kappa_{ij}^{(2k)}(\mathbf{x};\mathbf{y})]dV_x. \quad (6.1.14)$$

We now consider the external data system $L^{(3)} = (f_i^{(3)} = 0, g_i^{(3)} = \delta_{is}\delta(x-y)/\rho_0, \widetilde{u}_i^{(3)} = 0, \widetilde{\varphi}_i^{(3)} = 0, \widetilde{t}_i^{(3)} = 0, \widetilde{m}_i^{(3)} = 0, \theta^{(3)} = 0)$ and denote by $(u_i^{(3s)}, \varphi_i^{(3s)}, e_{ij}^{(3s)}, \kappa_{ij}^{(3s)}, t_{ij}^{(3s)}, m_{ij}^{(3s)})$ the solution of the mixed problem corresponding to $L^{(3)}$. If we apply the reciprocity relation (6.1.13) to the external data systems $L^{(1)}$ and $L^{(3)}$, then we obtain

$$\varphi_s^{(1)}(\mathbf{y}) = \int_B \theta^{(1)}(\mathbf{x})[\beta_{ji}(\mathbf{x})e_{ij}^{(3s)}(\mathbf{x};\mathbf{y}) + \xi_{ji}(\mathbf{x})\kappa_{ij}^{(3s)}(\mathbf{x};\mathbf{y})]dV_x. \quad (6.1.15)$$

The relations (6.1.14) and (6.1.15) generalize Maysel formulae in classical thermoelastostatics.

Now, we assume that $L^{(1)} = L^{(2)} = (f_i, g_i, \widetilde{u}_i, \widetilde{\varphi}_i, \widetilde{t}_i, \widetilde{m}_i, \theta), s^{(1)} = s^{(2)} = (u_i, \varphi_i, e_{ij}, \kappa_{ij}, t_{ij}, m_{ij})$. It follows from (6.1.9) and (6.1.12), that for $\alpha = \beta = 1$ we obtain

$$\begin{aligned} 2\int_B W dV = &\int_{\partial B} \{[\rho_0 f_i - (\beta_{ji}\theta)_{,j}]u_i + [\rho_0 g_i - (\xi_{ji}\theta)_{,j} + \\ &+ \varepsilon_{isr}\beta_{rs}\theta]\varphi_i\} dV + \int_{\partial B} [(t_i + \beta_{ji}\theta n_j)u_i + (m_i + \xi_{ji}\theta n_j)\varphi_i] dA, \end{aligned} \quad (6.1.16)$$

where W is given by (5.1.35). The relation (6.1.16) leads to the following uniqueness result.

Theorem 6.1.2. *If W is a positive definite quadratic form, then any two solutions of the mixed problem are equal modulo a rigid displacement. Moreover, if S_1 is non-empty, then the mixed problem has at most one solution.*

Proof. Let $s^0 = (u_i^0, \varphi_i^0, e_{ij}^0, \kappa_{ij}^0, t_{ij}^0, m_{ij}^0)$ be the difference of two solutions of the mixed problem. According to the linearity of the problem, s^0 is a solution corresponding to null external data. From (6.1.16) we obtain

$$\int_B W^0 dV = 0, \quad (6.1.17)$$

where

$$2W^0 = A_{ijrs}e^0_{ij}e^0_{rs} + 2B_{ijrs}e^0_{ij}\kappa^0_{rs} + C_{ijrs}\kappa^0_{ij}\kappa^0_{rs}.$$

Since W^0 is positive definite, from (6.1.17) we conclude that $e^0_{ij}=0$, $\kappa^0_{ij} = 0$ and therefore

$$u^0_{j,i} + \varepsilon_{jis}\varphi^0_s = 0, \varphi^0_{j,i} = 0. \tag{6.1.18}$$

From (6.1.18) we get

$$u^0_i = a_i + \varepsilon_{ijk}b_j x_k, \quad \varphi^0_i = b_i, \tag{6.1.19}$$

where a_i and b_i are arbitrary constants. If S_1 is non-empty, then $a_i = 0$ and $b_i = 0$. □

We now present a minimum principle in linear thermoelastostatics. By a kinematically admissible state we mean an admissible state that satisfies the constitutive equations (6.1.3), the geometrical equations (6.1.4) and the boundary conditions

$$u_i = \widetilde{u}_i \text{ on } \overline{S}_1, \quad \varphi_i = \widetilde{\varphi}_i \text{ on } \overline{S}_3. \tag{6.1.20}$$

Theorem 6.1.3. *Assume that the internal energy density W is positive definite. Let $\mathcal{K}$ denote the set of all kinematically admissible states, and let Λ be the functional on $\mathcal{K}$ defined by*

$$\begin{aligned}\Lambda(s) = &\int_B \{\frac{1}{2}A_{ijrs}e_{ij}e_{rs} + B_{ijrs}e_{ij}\kappa_{rs} + \frac{1}{2}C_{ijrs}\kappa_{ij}\kappa_{rs} - \\ &-[\rho_0 f_i - (\beta_{ji}\theta)_{,j}]u_i - [\rho_0 g_i - (\xi_{ji}\theta)_{,j} - \varepsilon_{irs}\beta_{rs}\theta]\varphi_i\}dV - \\ &-\int_{S_2}(\widetilde{t}_i + \beta_{ji}\theta n_j)u_i dA - \int_{S_4}(\widetilde{m}_i + \xi_{ji}\theta n_j)\varphi_i dA,\end{aligned} \tag{6.1.21}$$

for every $s = (u_i, \varphi_i, e_{ij}, \kappa_{ij}, t_{ij}, m_{ij}) \in \mathcal{K}$. Further, let s be a solution of the mixed problem. Then

$$\Lambda(s) \leq \Lambda(s^*), \tag{6.1.22}$$

for every $s^* \in \mathcal{K}$, and equality holds only if $s^* = s$ modulo a rigid displacement.

Proof. Let $s, s^* \in \mathcal{K}$ and define

$$s' = s^* - s, \quad s' = (u'_i, \varphi'_i, e'_{ij}, \kappa'_{ij}, t'_{ij}, m'_{ij}). \tag{6.1.23}$$

Then s' is an admissible state and

$$\begin{aligned} &e'_{ij} = u'_{j,i} + \varepsilon'_{jik}\varphi'_k, \quad \kappa'_{ij} = \varphi'_{j,i}, \\ &t'_{ij} = A_{ijrs}e'_{rs} + B_{ijrs}\kappa'_{rs}, \\ &m'_{ij} = B_{rsij}e'_{rs} + C_{ijrs}\kappa'_{rs}, \\ &u'_i = 0 \text{ on } \overline{S}_1, \quad \varphi'_i = 0 \text{ on } \overline{S}_3. \end{aligned} \tag{6.1.24}$$

In view of (6.1.21) and (6.1.24) we find that

$$\begin{aligned} \Lambda(s^*) = \Lambda(s) + &\int_B \left(\frac{1}{2}A_{ijrs}e'_{ij}e'_{rs} + B_{ijrs}e'_{ij}\kappa'_{rs} + \frac{1}{2}C_{ijrs}\kappa'_{ij}\kappa'_{rs} \right) dV - \\ &- \int_B [(t_{ji,j} + \rho_0 f_i)u'_i + (m_{ji,j} + \varepsilon_{irs}t_{rs} + \rho_0 g_i)\varphi'_i]dV + \\ &+ \int_{S_2} (t_{ji}n_j - \widetilde{t}_i)u'_i dA + \int_{S_4} (m_{ji}n_j - \widetilde{m}_i)\varphi'_i dA. \end{aligned} \tag{6.1.25}$$

Since s is a solution of the mixed problem and the internal energy density is positive, (6.1.25) implies that $\Lambda(s) \leq \Lambda(s^*)$ for any $s^* \in \mathcal{K}$. Moreover, $\Lambda(s^*) = \Lambda(s)$ only if $e'_{ij} = 0$ and $\kappa'_{ij} = 0$, and we conclude that s and s^* must be equal modulo a rigid displacement. □

Other variational theorems have been presented by Hlavácek and Hlavácek [152] and Nowacki [289]. The existence question appropriate to basic boundary value problems has been discussed by Hlavácek and Hlavácek [152], Hlavácek and Nečas [153], Ieşan [164], Kupradze, Gegelia, Bashelishvili and Burchuladze [232].

6.2 Special Results for Homogeneous and Isotropic Bodies

In this section, we assume that the body is homogeneous and isotropic so that the constitutive equations (6.1.3) are replaced by (5.1.34) where $\lambda, \mu, \kappa, \alpha, \beta, \gamma$ and k are prescribed constants. The boundary value problem for the temperature field reduces to

$$k\Delta\theta = -\rho_0 S \text{ on } B, \tag{6.2.1}$$

$$\theta = \widetilde{\theta} \text{ on } \overline{S}_5, \quad k\frac{\partial\theta}{\partial n} = \widetilde{q} \text{ on } S_6. \tag{6.2.2}$$

After the determination of the function θ we have to study the equations of equilibrium (6.1.1), the geometrical equations (6.1.4) and the constitutive equations

$$
\begin{aligned}
t_{ij} &= \lambda e_{rr}\delta_{ij} + (\mu+\kappa)e_{ij} + \mu e_{ji} - b\theta\delta_{ij},\\
m_{ij} &= \alpha\kappa_{rr}\delta_{ij} + \beta\kappa_{ji} + \gamma\kappa_{ij},
\end{aligned}
\tag{6.2.3}
$$

on B, with the boundary conditions $(6.1.5)_{1-4}$. From (6.1.1), (6.2.3) and (6.1.4) we obtain the field equations expressed in terms of the displacement and microrotation vector fields

$$
\begin{aligned}
&(\mu+\kappa)\Delta u_i + (\lambda+\mu)u_{j,ji} + \kappa\varepsilon_{irs}\varphi_{s,r} = b\theta_{,i} - \rho_0 f_i,\\
&\gamma\Delta\varphi_i + (\alpha+\beta)\varphi_{j,ji} + \kappa\varepsilon_{irs}u_{s,r} - 2\varphi_i = -\rho_0 g_i.
\end{aligned}
\tag{6.2.4}
$$

The constitutive equations can be written in the form

$$
\begin{aligned}
e_{ij} &= \frac{1}{2\mu+\kappa}t_{(ij)} + \frac{1}{\kappa}t_{<ij>} - \frac{\lambda t_{ss}\delta_{ij}}{(2\mu+\kappa)(3\lambda+2\mu+\kappa)} + \frac{\beta}{3\lambda+2\mu+\kappa}\theta\delta_{ij},\\
\kappa_{ij} &= \frac{1}{\gamma+\beta}m_{(ij)} + \frac{1}{\gamma-\beta}m_{<ij>} - \frac{\alpha m_{ss}\delta_{ij}}{(\gamma+\beta)(3\alpha+\beta+\gamma)},
\end{aligned}
\tag{6.2.5}
$$

where $A_{(ij)}$ and $A_{<ij>}$ denote here the symmetric and anti-symmetric parts of A_{ij}, respectively.

i) *Temperature fields that induces stress and couple-stress free states.* Let B be simply-connected, and let e_{ij} and κ_{ij} be a class C^1 tensors fields that satisfy the equations

$$
e_{ij,k} - e_{kj,i} - \varepsilon_{jis}\kappa_{ks} + \varepsilon_{jks}\kappa_{is} = 0, \quad \kappa_{ij,k} = \kappa_{kj,i}. \tag{6.2.6}
$$

Then there exist the displacements u_i and the microrotations φ_i which satisfy the geometrical equations (6.1.4). The restrictions (6.2.6) are the conditions of compatibility.

We now consider a simply-connected body B and seek a temperature field that induce a state in B with the property that $t_{ij}=0$ and $m_{ij}=0$. In this case, from (6.2.5) we obtain

$$
e_{ij} = \frac{\beta}{3\lambda+2\mu+\kappa}\theta\delta_{ij}, \quad \kappa_{ij} = 0. \tag{6.2.7}
$$

The equations (6.2.6) reduce to $\theta_{,i}=0$ on B. We conclude that $t_{ij}=0$ and $m_{ij}=0$ on B only if θ is constant.

ii) *Thermoelastic potential displacement.* We assume that the body forces and body couples are absent. In this case the equations (6.2.4) reduce to

$$
\begin{aligned}
&(\mu+\kappa)\Delta u_i + (\lambda+\mu)u_{j,ji} + \kappa\varepsilon_{irs}\varphi_{s,r} = b\theta_{,i},\\
&\gamma\Delta\varphi_i + (\alpha+\beta)\varphi_{j,ji} + \kappa\varepsilon_{irs}u_{s,r} - 2\varphi_i = 0.
\end{aligned}
\tag{6.2.8}
$$

The problem can be reduced to one for an unheated body, with zero body loads. We seek a particular solution of the system (6.2.8) in the form

$$u_i' = \Phi_{,i}, \quad \varphi_i' = 0, \tag{6.2.9}$$

where the thermoelastic potential displacement Φ is of class C^3 on B. The equations (6.2.8) are satisfied if Φ is a solution of the equation

$$\Delta\Phi = \frac{b}{\lambda + 2\mu + \kappa}\theta \ \text{ on } \ B. \tag{6.2.10}$$

It follows from (6.2.3) and (6.2.9) that the stresses and couple-stresses corresponding to the functions (6.2.9) are given by

$$t_{ij}' = (2\mu + \kappa)(\Phi_{,ij} - \delta_{ij}\Delta\Phi), m_{ij}' = 0. \tag{6.2.11}$$

If we take

$$u_i = u_i' + u_i'', \quad \varphi_i = \varphi_i' + \varphi_i'',$$

then from (6.2.3) we obtain

$$t_{ij} = t_{ij}' + t_{ij}'', \quad m_{ij} = m_{ij}'', \tag{6.2.12}$$

where

$$\begin{aligned} t_{ij}'' &= \lambda e_{rr}''\delta_{ij} + (\mu + \kappa)e_{ij}'' + \mu e_{ji}'', \\ m_{ij}'' &= \alpha\kappa_{rr}''\delta_{ij} + \beta\kappa_{ji}'' + \gamma\kappa_{ij}'', \end{aligned} \tag{6.2.13}$$

and $e_{ij}'' = u_{j,i}'' + \varepsilon_{jik}\varphi''$, $\kappa_{ij}'' = \varphi_{j,i}''$. Clearly, the functions t_{ij}'' and m_{ij}'' satisfy the equations

$$t_{ji,j}'' = 0, \quad m_{ji,j}'' + \varepsilon_{irs}t_{rs}'' = 0, \quad \text{on } B. \tag{6.2.14}$$

The boundary conditions reduce to

$$\begin{aligned} &u_i'' = \widetilde{u}_i - u_i' \ \text{ on } \ \overline{S}_1, \quad t_{ji}''n_j = \widetilde{t}_i - t_{ji}'n_j \ \text{ on } \ S_2, \\ &\varphi_i'' = \widetilde{\varphi}_i \ \text{ on } \ \overline{S}_3, \quad m_{ji}''n_j = \widetilde{m}_i \ \text{ on } \ S_4. \end{aligned} \tag{6.2.15}$$

In the case of the unbounded space the solution of the problem is given by (6.2.9). For example, if a homogeneous and isotropic space is subjected to the concentrated heat source $\rho_0 S = A\delta(x - y), (A = const.)$, then θ satisfies the equation

$$\Delta\theta = -\frac{A}{k}\delta(x - y).$$

The temperature field which vanishes at infinity is

$$\theta = \frac{A}{4\pi k r},$$

where $r = |\mathbf{x} - \mathbf{y}|$. From (6.2.10) we obtain

$$\Delta\Phi = \frac{bA}{4\pi k(\lambda + 2\mu + \kappa)r},$$

so that

$$\Phi = \frac{bAr}{8\pi k(\lambda + 2\mu + \kappa)}. \tag{6.2.16}$$

It follows from (6.2.9) and (6.2.16) that the solution of the problem is given by

$$u_i = \frac{bA(x_i - y_i)}{8\pi k(\lambda + 2\mu + \kappa)r}, \quad \varphi_i = 0.$$

iii) *A solution of Galerkin type.* We now present a solution of the field equations (6.2.4) established by Sandru [326]. We introduce the notations

$$\begin{aligned} &L_1 = (\lambda + 2\mu + \kappa)\Delta, \; L_2 = (\mu + \kappa)\Delta, \; L_3 = (\alpha + \beta + \gamma)\Delta - 2\kappa, \\ &L_4 = \gamma\Delta - 2\kappa, \; \delta = \left[\frac{\kappa(2\mu + \kappa)}{\gamma(\mu + \kappa)}\right]^{1/2}, \; l = \left(\frac{2\kappa}{\alpha + \beta + \gamma}\right)^{1/2}, \\ &\tau_1^{-1} = \gamma(\mu + \kappa)(\lambda + 2\mu + \kappa), \; \tau_2^{-1} = \gamma(\mu + \kappa)(\alpha + \beta + \gamma). \end{aligned} \tag{6.2.17}$$

Theorem 6.2.1. *Let*

$$\begin{aligned} u_i &= L_1 L_4 P_i - [(\lambda + \mu)L_4 - \kappa^2]P_{j,ji} - \kappa\varepsilon_{ijk}L_3 Q_{k,j}, \\ \varphi_i &= L_2 L_3 Q_i - [(\alpha + \beta)L_2 - \kappa^2]Q_{j,ji} - \kappa\varepsilon_{ijk}L_1 P_{k,j}, \end{aligned} \tag{6.2.18}$$

where P_i and Q_i are class C^6 fields on B that satisfy the equations

$$\begin{aligned} \Delta\Delta(\Delta - \delta^2)P_i &= -\tau_1(\rho_0 f_i - b\theta_{,i}), \\ \Delta\Delta(\Delta - l^2)Q_i &= -\tau_2\rho_0 g_i. \end{aligned} \tag{6.2.19}$$

Then u_i and φ_i satisfy equations (6.2.4).

Proof. By (6.2.18),

$$\begin{aligned}(\mu+\kappa)\Delta u_i+(\lambda+\mu)u_{j,ji}+\kappa\varepsilon_{irs}\varphi_{s,r}=\\=(\mu+\kappa)\Delta L_1L_4P_i-(\mu+\kappa)[(\lambda+\mu)L_4-\kappa^2]\Delta P_{j,ji}-\\-\kappa(\mu+\kappa)\varepsilon_{ijk}L_3\Delta Q_{k,j}+(\lambda+\mu)L_1L_4P_{j,ji}-\\-(\lambda+\mu)[(\lambda+\mu)L_4-\kappa^2]\Delta P_{s,si}+\kappa\varepsilon_{irs}L_2L_3Q_{s,r}-\\-\kappa^2L_1P_{j,ji}+\kappa^2\Delta L_1P_i=(\lambda+2\mu+\kappa)\Delta[(\mu+\kappa)\Delta(\gamma\Delta-2\kappa)+\\+\kappa^2\Delta]=\gamma(\mu+\kappa)(\lambda+2\mu+\kappa)\Delta\Delta(\Delta-\delta^2)P_i.\end{aligned}\tag{6.2.20}$$

Similarly,

$$\begin{aligned}\gamma\Delta\varphi_i+(\alpha+\beta)\varphi_{j,ji}+\kappa\varepsilon_{irs}u_{s,r}-2\varphi_i=\\=\gamma(\mu+\kappa)(\alpha+\beta+\gamma)\Delta\Delta(\Delta-l^2).\end{aligned}\tag{6.2.21}$$

The desired result follows from (6.2.20), (6.2.21) and (6.2.19).□

The solution for special types of temperature fields have been presented by Nowacki [289] and Dyszlewicz [96].

6.3 The Equilibrium Plane Problems

In this section we study the plane strain problem in the framework of the equilibrium theory of micropolar thermoelastic bodies. The plane strain problem in the dynamic theory has been defined in Section 5.8. In the absence of time dependence, a state of plane strain of homogeneous and isotropic bodies, parallel to the x_1Ox_2-plane is characterized by

$$\begin{aligned}&u_\alpha=u_\alpha(x_1,x_2),\ \varphi_3=\varphi(x_1,x_2),\ \theta=\theta(x_1,x_2),\\&u_3=0,\ \ \varphi_\alpha=0,\ \ (x_1,x_2)\in\Sigma.\end{aligned}\tag{6.3.1}$$

Throughout this section we consider boundary conditions of Neumann type. It follows from (5.8.3), (5.8.5) and (5.8.6) that the temperature field satisfies the boundary value problem

$$k\Delta\theta=-\rho_0S\ \text{ on }\Sigma,\ \ k\frac{\partial\theta}{\partial n}=\widetilde{q}\ \text{ on }L.\tag{6.3.2}$$

After the determination of the function θ we have to find the functions u_α and φ from the equilibrium equations

$$t_{\beta\alpha,\beta}+\rho_0f_\alpha=0,\quad m_{\alpha3,\alpha}+\varepsilon_{3\alpha\beta}t_{\alpha\beta}+\rho_0g_3=0,\tag{6.3.3}$$

the constitutive equations

$$t_{\alpha\beta} = \lambda e_{\rho\rho}\delta_{\alpha\beta} + (\mu+\kappa)e_{\alpha\beta} + \mu e_{\beta\alpha} - b\theta\delta_{\alpha\beta},$$
$$m_{\alpha 3} = \gamma\kappa_{\alpha 3} \tag{6.3.4}$$

and the geometrical equations

$$e_{\alpha\beta} = u_{\beta,\alpha} + \varepsilon_{3\beta\alpha}\varphi, \quad \kappa_{\alpha 3} = \varphi_{,\alpha}, \tag{6.3.5}$$

on Σ, and the boundary conditions

$$t_{\beta\alpha}n_\beta = \widetilde{t}_\alpha, \quad m_{\alpha 3}n_\alpha = \widetilde{m}_3 \text{ on } L. \tag{6.3.6}$$

We now state, without proof, the following result [164], [110].

Theorem 6.3.1. *The boundary value problem (6.3.3.)-(6.3.6) has solutions belonging to $C^\infty(\overline{\Sigma})$ if and only if the C^∞ functions $f_\alpha, g_3, \theta, \widetilde{t}_\alpha$ and $\widetilde{m}_3$ satisfy the conditions*

$$\int_\Sigma \rho_0 f_\alpha dA + \int_L \widetilde{t}_\alpha ds = 0,$$
$$\int_\Sigma \rho_0(\varepsilon_{3\alpha\beta}x_\alpha f_\beta + g_3)dA + \int_L (\varepsilon_{3\alpha\beta}x_\alpha\widetilde{t}_\beta + \widetilde{m}_3)ds = 0. \tag{6.3.7}$$

The equations of equilibrium can be expressed in terms of the functions u_α and φ,

$$(\mu+\kappa)\Delta u_\alpha + (\lambda+\mu)u_{\beta,\beta\alpha} + \kappa\varepsilon_{\alpha\beta 3}\varphi_{,\beta} = b\theta_{,\alpha} - \rho_0 f_\alpha,$$
$$\gamma\Delta\varphi + \kappa\varepsilon_{3\alpha\beta}u_{\beta,\alpha} - 2\kappa\varphi = -\rho_0 g_3, \tag{6.3.8}$$

on Σ.

a) **Polar coordinates.** In the solution of various boundary value problems it is convenient to employ the polar coordinates (r,ϑ), such that $x_1 = r\cos\vartheta, x_2 = r\sin\vartheta$. The equations of equilibrium can be written in the form

$$\frac{\partial t_{rr}}{\partial r} + \frac{1}{r}\frac{\partial t_{\vartheta r}}{\partial\vartheta} + \frac{1}{r}(t_{rr} - t_{\vartheta\vartheta}) + \rho_0 f_r = 0,$$
$$\frac{\partial t_{r\vartheta}}{\partial r} + \frac{1}{r}\frac{\partial t_{\vartheta\vartheta}}{\partial\vartheta} + \frac{1}{r}(t_{r\vartheta} + t_{\vartheta r}) + \rho_0 f_\vartheta = 0, \tag{6.3.9}$$
$$\frac{\partial m_{rz}}{\partial r} + \frac{1}{r}\frac{\partial m_{\vartheta z}}{\partial\vartheta} + \frac{1}{r}m_{rz} + t_{r\vartheta} - t_{\vartheta r} + \rho_0 g_3 = 0,$$

on Σ, where $t_{rr}, t_{\vartheta\vartheta}, t_{r\vartheta}$ and $t_{\vartheta r}$ are the physical components of the stress tensor, m_{rz} and $m_{\vartheta z}$ are physical components of couple-stress tensor, and f_r and f_ϑ are the physical components of the body force. The constitutive equations become

$$\begin{aligned}
&t_{rr}=(\lambda+2\mu+\kappa)e_{rr}+\lambda e_{\vartheta\vartheta}-b\theta,\ t_{r\vartheta}=(\mu+\kappa)e_{r\vartheta}+\mu e_{\vartheta r},\\
&t_{\vartheta\vartheta}=\lambda e_{rr}+(\lambda+2\mu+\kappa)e_{\vartheta\vartheta}-b\theta,\ t_{\vartheta r}=(\mu+\kappa)e_{\vartheta r}+\mu e_{r\vartheta},\\
&m_{rz}=\gamma\kappa_{rz},\ m_{\vartheta z}=\gamma\kappa_{\vartheta z},
\end{aligned} \tag{6.3.10}$$

where

$$\begin{aligned}
&e_{rr}=\frac{\partial u_r}{\partial r},\quad e_{\vartheta\vartheta}=\frac{1}{r}\left(\frac{\partial u_\vartheta}{\partial\vartheta}+u_r\right),\quad e_{r\vartheta}=\frac{\partial u_\vartheta}{\partial r}-\varphi,\\
&e_{\vartheta r}=\frac{1}{r}\left(\frac{\partial u_r}{\partial\vartheta}-u_\vartheta\right)+\varphi,\quad \kappa_{\vartheta z}=\frac{1}{r}\frac{\partial\varphi}{\partial\vartheta},\quad \kappa_{rz}=\frac{\partial\varphi}{\partial r}.
\end{aligned} \tag{6.3.11}$$

Here, u_r and u_ϑ are the physical components of the displacement vector field. The equations of equilibrium (6.3.9) can be expressed in terms of the functions u_r, u_ϑ and φ. Thus we obtain

$$\begin{aligned}
&(\lambda+2\mu+\kappa)\left(\frac{\partial^2 u_r}{\partial r^2}+\frac{1}{r}\frac{\partial u_r}{\partial r}-\frac{1}{r^2}u_r+\frac{1}{r}\frac{\partial^2 u_\vartheta}{\partial r\partial\vartheta}-\frac{1}{r^2}\frac{\partial u_\vartheta}{\partial\vartheta}\right)-\\
&-(\mu+\kappa)\frac{1}{r}\left(\frac{\partial^2 u_\vartheta}{\partial r\partial\vartheta}+\frac{1}{r}\frac{\partial u_\vartheta}{\partial\vartheta}-\frac{1}{r}\frac{\partial^2 u_r}{\partial\vartheta^2}\right)+\kappa\frac{1}{r}\frac{\partial\varphi}{\partial\vartheta}+\rho_0 f_r-b\frac{\partial\theta}{\partial r}=0,\\
&(\lambda+2\mu+\kappa)\left(\frac{1}{r^2}\frac{\partial^2 u_\vartheta}{\partial\theta^2}+\frac{1}{r}\frac{\partial^2 u_r}{\partial r\partial\vartheta}+\frac{1}{r^2}\frac{\partial u_r}{\partial\vartheta}\right)-(\mu+\kappa)\left(\frac{1}{r}\frac{\partial^2 u_r}{\partial r\partial\vartheta}-\frac{1}{r^2}\frac{\partial u_r}{\partial\vartheta}-\right.\\
&\left.-\frac{\partial^2 u_\vartheta}{\partial r^2}-\frac{1}{r}\frac{\partial u_\vartheta}{\partial r}+\frac{1}{r^2}u_\vartheta\right)-\kappa\frac{\partial\varphi}{\partial r}+\rho_0 f_\vartheta-b\frac{1}{r}\frac{\partial\theta}{\partial\vartheta}=0,\\
&\gamma\left(\frac{\partial^2\varphi}{\partial r^2}+\frac{1}{r}\frac{\partial\varphi}{\partial r}+\frac{1}{r^2}\frac{\partial^2\varphi}{\partial\vartheta^2}\right)+\kappa\left(\frac{\partial u_\vartheta}{\partial r}+\frac{1}{r}u_\vartheta-\frac{1}{r}\frac{\partial u_r}{\partial\vartheta}-2\varphi\right)+\rho_0 g_3=0.
\end{aligned} \tag{6.3.12}$$

Let $\widetilde{t}_r$ and $\widetilde{t}_\vartheta$ be physical components of the given traction vector. Then the conditions (6.3.6) can be expressed in the form

$$\begin{aligned}
&t_{rr}n_r+t_{\vartheta r}n_\vartheta=\widetilde{t}_r,\quad t_{r\vartheta}n_r+t_{\vartheta\vartheta}n_\vartheta=\widetilde{t}_\vartheta,\\
&m_{rz}n_r+m_{\vartheta z}n_\vartheta=\widetilde{m}_3 \text{ on } L,
\end{aligned} \tag{6.3.13}$$

where n_r and n_ϑ are physical components of the vector $\mathbf{n}$.

b) **Potential functions of displacement and microrotation.** We now give a representation of solutions of the equations (6.3.8). We introduce the operator

$$\mathcal{D}=\frac{1}{\tau_1}\Delta\Delta(\Delta-\delta^2), \tag{6.3.14}$$

where τ_1 and δ are defined in (6.2.17).

Theorem 6.3.2. *Let*

$$\begin{aligned} u_\alpha &= (\lambda+2\mu+\kappa)\Delta(\gamma\Delta-2\kappa)\Gamma_\alpha-[\gamma(\lambda+\mu)\Delta- \\ &-\kappa(2\lambda+2\mu+\kappa)]\Gamma_{\beta,\beta\alpha}+\kappa(\lambda+2\mu+\kappa)\varepsilon_{3\beta\alpha}\Delta\Gamma_{3,\beta}, \\ \varphi &= (\lambda+2\mu+\kappa)[\kappa\varepsilon_{3\alpha\beta}\Delta\Gamma_{\alpha,\beta}+(\mu+\kappa)\Delta\Delta\Gamma_3], \end{aligned} \tag{6.3.15}$$

where Γ_j are class C^6 fields on Σ that satisfy the equations

$$\mathcal{D}\Gamma_\alpha = b\theta_{,\alpha}-\rho_0 f_\alpha, \quad \mathcal{D}\Gamma_3 = -\rho_0 g_3. \tag{6.3.16}$$

Then u_α and φ satisfy the equations (6.3.8) on Σ.

The proof of this theorem, which we omit, is analogous to that given for Theorem 6.2.1.

c) **Generalization of Muskhelishvili's method.** We assume that the heat source is absent. In this case, and in the framework of classical thermoelasticity, Muskhelishvili [270] established a method of reduction of thermoelastic plane problem to an isothermal one. The result has been extended to micropolar thermoelastic bodies in [162]. In what follows we present this method. From (6.3.2) we obtain $\Delta\theta = 0$ on Σ. Thus, we can introduce the analytic function

$$F(z) = \theta + iP, \tag{6.3.17}$$

of the complex variable $z = x_1 + ix_2$. If we define the functions w_α by

$$w_1 + iw_2 = \int F(z)dz, \tag{6.3.18}$$

then we have

$$w_{\alpha,\beta} = \theta\delta_{\alpha\beta} + \varepsilon_{3\beta\alpha}P. \tag{6.3.19}$$

We introduce the functions u_α^0 and φ^0 on Σ by

$$u_\alpha = u_\alpha^0 + Hw_\alpha, \quad \varphi = \varphi^0 + HP, \tag{6.3.20}$$

where

$$H = \frac{b}{2\lambda+2\mu+\kappa}. \tag{6.3.21}$$

Let us denote

$$e_{\alpha\beta}^0 = u_{\beta,\alpha}^0 + \varepsilon_{3\beta\alpha}\varphi^0, \quad \kappa_{\alpha 3}^0 = \varphi_{,\alpha}^0. \tag{6.3.22}$$

From (6.3.5), (6.3.19), (6.3.20) and (6.3.22) we obtain

$$e_{\alpha\beta} = e^0_{\alpha\beta} + H\theta\delta_{\alpha\beta}, \quad \kappa_{\alpha 3} = \kappa^0_{\alpha 3} + HP_{,\alpha}. \tag{6.3.23}$$

We introduce the notations

$$t^0_{\alpha\beta} = \lambda e^0_{\rho\rho}\delta_{\alpha\beta} + (\mu+\kappa)e^0_{\alpha\beta} + \mu e^0_{\beta\alpha}, \quad m^0_{\alpha 3} = \gamma\kappa^0_{\alpha 3}. \tag{6.3.24}$$

With the help of (6.3.4), (6.3.23), (6.3.24) and

$$\theta_{,\alpha} = \varepsilon_{3\alpha\beta}P_{,\beta}, \quad P_{,\alpha} = \varepsilon_{3\alpha\beta}\theta_{,\beta},$$

we obtain

$$t_{\alpha\beta} = t^0_{\alpha\beta}, \quad m_{\alpha 3} = m^0_{\alpha 3} + \gamma H\varepsilon_{3\rho\alpha}\theta_{,\rho}. \tag{6.3.25}$$

The equations (6.3.24) are the constitutive equations in the isothermal theory corresponding to the plane deformation characterized by the displacements u^0_α and the microrotation φ^0. We assume that $f_\alpha = 0$ and $g_3 = 0$. Upon substituting $t_{\alpha\beta}$ and $m_{\alpha 3}$ given by (6.3.25) into the equations of equilibrium we obtain

$$t^0_{\beta\alpha,\beta} = 0, \quad m^0_{\alpha 3,\alpha} + \varepsilon_{3\rho\nu}t^0_{\rho\nu} = 0, \tag{6.3.26}$$

on Σ. We assume that L is a piecewise smooth curve parameterized by its arc length s. The mechanical boundary conditions (6.3.6) reduce to

$$t^0_{\beta\alpha}n_\beta = \widetilde{t}_\alpha, \quad m^0_{\alpha 3}n_\alpha = \widetilde{m}_3 + \gamma H\frac{d\theta}{ds}, \text{ on } L. \tag{6.3.27}$$

We conclude that the thermoelastic plane problem associated to the simply-connected domain Σ is reduced to the plane strain problem of isothermal elastostatics ($\theta = 0$) characterized by the field equations (6.3.22), (6.3.24), (6.3.26) on Σ and the boundary conditions (6.3.27) on L.

Remark 1. Assume that: (α) the domain Σ is simply-connected; (β) the heat source is absent; (γ) the mechanical loads are absent. Then the thermoelastic plane strain problem associated to Σ reduces to solving of the isothermal elastic plane strain problem corresponding to zero body loads, zero boundary traction and the boundary couple $\gamma H d\theta/ds$.

Let us suppose that Σ is a bounded multiply-connected domain. We assume that inside Σ there are m contours $L_k (k = 1, 2, \ldots, m)$ which have no common points. Let L_{m+1} be the outer contour. In this case the function F has the form (3.2.19), and w_1 and w_2 satisfy (3.2.21).

Since the functions u_α and φ are single-valued, it follows from (6.3.20), (3.2.19) and (3.2.21) that the functions u^0_α and φ^0 are multi-valued,

$$\begin{aligned}
\left[u_1^0\right]_{L_k} &= 2\pi H(B_k x_2 + b_k),\\
\left[u_2^0\right]_{L_k} &= -2\pi H(B_k x_1 + a_k),\\
\left[\varphi^0\right]_{L_k} &= -2\pi H B_k.
\end{aligned}$$

In this way, the thermoelastic problem associated to the multiply-connected domain Σ reduces to a plane problem of mechanical dislocations defined by the field equations (6.3.22), (6.3.24), (6.3.26) on Σ, the boundary conditions (6.3.27) on L and the dislocation characteristics

$$\varepsilon_k^0 = -2\pi H B_k, \quad \alpha_k^0 = 2\pi H b_k, \quad \beta_k^0 = -2\pi H a_k. \tag{6.3.28}$$

Remark 2. Assume that: (i) the domain Σ is multiply-connected; (ii) the heat source is absent; (iii) the mechanical loads are absent. Then the thermoelastic plane strain problem associated to Σ reduces to solving of the isothermal plane problem of dislocations corresponding to zero body loads, zero boundary traction, the boundary couple $\gamma H d\theta/ds$, and to the dislocation characteristics (6.3.28).

d) **Application.** We now use the method presented in Section 6.3 (c) to study the thermoelastic plane strain of a tube. Let Σ be defined by $\Sigma = \{\mathbf{x} : R_1^2 < x_1^2 + x_2^2 < R_2^2, x_3 = 0\}, (R_1, R_2 > 0)$. We assume that the tube is in a state of plane strain in the absence of mechanical loads and heat sources. Let T_1 be the constant temperature of the inner surface and T_2 the constant temperature on the outer surface. In this case the temperature θ is given by $\theta = C_1 \ln r + C_2$, where $r = (x_1^2 + x_2^2)^{1/2}$ and C_α are constants defined by

$$C_1 \ln R_1 + C_2 = T_1, \quad C_1 \ln R_2 + C_2 = T_2.$$

The function F introduced by (6.3.17) can be expressed in the form $F(z) = C_1 \log z + C_2$. According to Remark 2, Section 6.3 (c), in this case the thermoelastic problem reduces to the problem of mechanical dislocations defined by the equations (6.3.22), (6.3.24), (6.3.26) on Σ, the dislocation characteristics

$$\varepsilon_1^0 = -2\pi H B_1, \quad \alpha_1^0 = 0, \quad \beta_1^0 = 0 \tag{6.3.29}$$

and the boundary conditions

$$t^0_{\beta\alpha} n_\beta = 0, \quad m^0_{\alpha 3} n_\alpha = 0 \text{ for } r = R_1 \text{ and } r = R_2. \tag{6.3.30}$$

We shall study this problem by using the polar coordinates and the equations from Section 6.3(a). Thus we have

$$u_1^0 + iu_2^0 = (u_r + iu_\vartheta)e^{i\vartheta}. \tag{6.3.31}$$

The boundary conditions (6.3.30) reduce to

$$t_{rr} = 0, \quad t_{r\vartheta} = 0, \quad m_{rz} = 0 \text{ for } r = R_1 \text{ and } r = R_2. \tag{6.3.32}$$

We seek the solution $(u_r, u_\vartheta, \varphi^0)$ in the form

$$u_r = \Lambda r \ln r + rF(r), \; u_\vartheta = \frac{1}{2\pi}\varepsilon_1^0 r\vartheta, \; \varphi^0 = \frac{1}{2\pi}\varepsilon_1^0\vartheta, \tag{6.3.33}$$

where F is an unknown function of class C^2 and

$$\Lambda = \frac{(2\mu+\kappa)}{4\pi(\lambda+2\mu+\kappa)}\varepsilon_1^0. \tag{6.3.34}$$

We note that the functions u_r, u_ϑ and φ^0 given by (6.3.33) correspond to the dislocation characteristics (6.3.29). It follows from (6.3.33), (6.3.11) and (6.3.10), (for isothermal theory), that

$$\begin{aligned}
t_{rr}&=(\lambda+2\mu+\kappa)(\Lambda\ln r+\Lambda+F+rF')+\lambda(\Lambda\ln r+F+\frac{1}{2\pi}\varepsilon_1^0),\\
t_{\vartheta\vartheta}&=\lambda(\Lambda\ln r+\Lambda+F+rF')+(\lambda+2\mu+\kappa)(\Lambda\ln r+F+\frac{1}{2\pi}\varepsilon_1^0),\\
t_{r\vartheta}&=t_{\vartheta r}=0,\; m_{rz}=0,\; m_{\vartheta z}=\frac{1}{2\pi r}\gamma\varepsilon_1^0,
\end{aligned} \tag{6.3.35}$$

where $f'(r) = df(r)/dr$. The equations of equilibrium reduce to

$$\frac{d}{dr}\left[\frac{1}{r}\frac{d}{dr}(r^2F)\right] = 0.$$

Thus, we obtain

$$F = D_1 + D_2 r^{-2}, \tag{6.3.36}$$

where D_1 and D_2 are arbitrary constants. From (6.3.35) and (6.3.36) we find

$$t_{rr} = (2\lambda+2\mu+\kappa)(\Lambda\ln r + \frac{1}{4\pi}\varepsilon_1^0 + D_1) - (2\mu+\kappa)D_2 r^{-2}.$$

The boundary conditions (6.3.32) are satisfied if and only if

$$(2\lambda+2\mu+\kappa)D_1-(2\mu+\kappa)R_1^{-2}D_2 = -(2\lambda+2\mu+\kappa)(\Lambda\ln R_1+\frac{1}{4\pi}\varepsilon_1^0),$$

$$(2\lambda+2\mu+\kappa)D_1-(2\mu+\kappa)R_2^{-2}D_2= - (2\lambda+2\mu+\kappa)(\Lambda \ln R_2+\frac{1}{4\pi}\varepsilon_1^0).$$

Clearly, these equations determine the constants D_1 and D_2. In this way the function F is determined by (6.3.36). Thus, the solution of the isothermal plane strain problem is given by (6.3.33) and the solution of thermoelastic problem follows from (6.3.20), (6.3.31) and (6.3.33).

e)**Generalized plane strain problem.** We now assume that the cylinder B is occupied by a nonhomogeneous and anisotropic material. We suppose that the constitutive coefficients are independent of the axial coordinate,

$$A_{ijrs}=A_{ijrs}(x_1,x_2), B_{ijrs}=B_{ijrs}(x_1,x_2), C_{ijrs}=C_{ijrs}(x_1,x_2),$$
$$\beta_{ij}=\beta_{ij}(x_1,x_2), \xi_{ij}=\xi_{ij}(x_1,x_2), k_{ij}=k_{ij}(x_1,x_2),\ (x_1,x_2)\in\Sigma. \tag{6.3.37}$$

We assume that the domain Σ is C^∞-smooth and that the constitutive coefficients belong to C^∞. We consider a C^∞-theory but it is possible to get a classical solution for more general assumptions of regularity (cf. Fichera [110]).

The state of generalized plane strain of the cylinder B, parallel to the plane x_1Ox_2, is characterized by

$$u_i = u_i(x_1,x_2),\ \varphi_i = \varphi_i(x_1,x_2),\ \ \theta = \theta(x_1,x_2),\quad (x_1,x_2)\in\Sigma. \tag{6.3.38}$$

This restriction, in conjunction with the relations (6.1.4), (6.1.3), implies that $e_{ij}, \kappa_{ij}, t_{ij}, m_{ij}$ and q_i are independent of x_3. From (6.1.4) and (6.3.38) we get

$$e_{\alpha i} = u_{i,\alpha} + \varepsilon_{i\alpha k}\varphi_k,\ e_{3i} = \varepsilon_{i3k}\varphi_k,\ \ \kappa_{\alpha i} = \varphi_{i,\alpha} \tag{6.3.39}$$

and $\kappa_{i3} = 0$. The constitutive equations (6.1.3) reduce to

$$\begin{aligned} t_{\alpha i} &= A_{\alpha ijk}e_{jk} + B_{\alpha i\beta j}\kappa_{\beta j} - \beta_{\alpha i}\theta,\\ t_{3\alpha} &= A_{3\alpha jk}e_{jk} + B_{3\alpha\beta j}\kappa_{\beta j} - \beta_{3\alpha}\theta,\\ m_{\alpha i} &= B_{jk\alpha i}e_{jk} + C_{\alpha i\beta j}\kappa_{\beta j} - \xi_{\alpha i}\theta, \end{aligned} \tag{6.3.40}$$

$$q_\alpha = k_{\alpha\beta}\theta_{,\beta} \tag{6.3.41}$$

and

$$\begin{aligned} t_{33} &= A_{33jk}e_{jk} + B_{33\beta j}\kappa_{\beta j} - \beta_{33}\theta,\\ m_{3i} &= B_{jk3i}e_{jk} + C_{3i\beta j}\kappa_{\beta j} - \xi_{3i}\theta,\ q_3 = k_{3\beta}\theta_{,\beta}. \end{aligned}$$

We assume that the body loads $\mathbf{f}, \mathbf{g}$ and S and the boundary data $\widetilde{\mathbf{t}}, \widetilde{\mathbf{m}}$ and $\widetilde{q}$ are independent of x_3. The equations of equilibrium (6.1.1) reduce to

$$t_{\alpha i,\alpha} + \rho_0 f_i = 0, \quad m_{\alpha i,\alpha} + \varepsilon_{ijk} t_{jk} + \rho_0 g_i = 0 \text{ on } \Sigma. \tag{6.3.42}$$

The energy equation (6.1.2) becomes

$$q_{\alpha,\alpha} = -\rho_0 S \quad \text{on } \Sigma. \tag{6.3.43}$$

The boundary conditions take the form

$$t_{\alpha i} n_\alpha = \widetilde{t}_i, \quad m_{\alpha i} n_\alpha = \widetilde{m}_i \text{ on } L \tag{6.3.44}$$

and

$$q_\alpha n_\alpha = \widetilde{q} \quad \text{on } L. \tag{6.3.45}$$

The temperature field θ is the solution of the boundary value problem characterized by the equations (6.3.41), (6.3.43) on Σ and the boundary conditions (6.3.45) on L.

The generalized plane strain problem consists in solving the equations (6.3.39), (6.3.40) and (6.3.42) on Σ with the boundary conditions (6.3.44) on L.

The functions t_{33}, m_{3i} and q_3 can be found after the displacement, microrotation and temperature fields have been determined.

We note that the conditions of equilibrium for the cylinder B can be written in the form

$$\begin{aligned} &\int_\Sigma \rho_0 f_i dA + \int_L \widetilde{t}_i ds = 0, \\ &\int_\Sigma \rho_0 (\varepsilon_{3\alpha\beta} x_\alpha f_\beta + g_3) dA + \int_L (\varepsilon_{3\alpha\beta} x_\alpha \widetilde{t}_\beta + \widetilde{m}_3) ds = 0, \end{aligned} \tag{6.3.46}$$

$$\int_\Sigma t_{3\alpha} dA = \int_\Sigma \rho_0 (x_\alpha f_3 - \varepsilon_{3\alpha\beta} g_\beta) dA + \int_L (x_\alpha \widetilde{t}_3 - \varepsilon_{3\alpha\beta} \widetilde{m}_\beta) ds. \tag{6.3.47}$$

The conditions (6.3.47) are identically satisfied on the basis of the relations (6.3.42) and (6.3.44). Thus,

$$\begin{aligned} \int_\Sigma t_{32} dA &= \int_\Sigma [t_{23} + x_2 (t_{\alpha 3,\alpha} + \rho_0 f_3) + m_{\alpha 1,\alpha} + \rho_0 g_1] dA = \\ &= \int_\Sigma [(x_2 t_{\alpha 3})_{,\alpha} + m_{\alpha 1,\alpha} + \rho_0 x_2 f_3 + \rho_0 g_1] dA = \\ &= \int_\Sigma \rho_0 (x_2 f_3 + g_1) dA + \int_L (x_2 \widetilde{t}_3 + \widetilde{m}_1) ds. \end{aligned}$$

In a similar way we can prove that the second condition from (6.3.47) is satisfied. Following Fichera [110], p. 385 we obtain

Theorem 6.3.3 *The boundary value problem (6.3.39), (6.3.40), (6.3.42), (6.3.44) has a solution belonging to C^∞ if and only if the functions $f_i, g_3, \widetilde{t}_i$ and $\widetilde{m}_3$ satisfy the conditions (6.3.46).*

6.4 Exercises

1. *Investigate the thermoelastic deformation of a circular solid cylinder subjected to a uniform heat flux vector.*

Solution. We consider a homogeneous and isotropic cylinder with the cross section $\Sigma = \{\mathbf{x} : x_1^2 + x_2^2 < a^2, x_3 = 0\}, (a > 0)$. Suppose that the cylinder is in equilibrium in the absence of mechanical loads and heat sources. We consider that the heat flux vector has the form $\mathbf{q} = kQ\mathbf{e}_1$ where Q is a given constant. In this case the temperature field satisfies the equation $\Delta\theta = 0$ on Σ and the boundary condition $\partial\theta/\partial n = Qn_1$ for $r = a$. The solution of this boundary value problem is $\theta = Qx_1$ on Σ. In what follows we shall use the method presented in Section 6.3(c). The function F defined by (6.3.17) has the form $F(z) = Qz$. From (6.3.18) we obtain

$$w_1 = \frac{1}{2}Qr^2 \cos 2\vartheta, \quad w_2 = \frac{1}{2}Qr^2 \sin 2\vartheta.$$

On the basis of Remark 1, Section 6.3(c), the thermoelastic problem reduces to the solving of the isothermal elastic plane strain problem defined by the (6.3.22), (6.3.24), (6.3.26) on Σ and the boundary conditions

$$t^0_{\beta\alpha} n_\beta = 0, \quad m^0_{\alpha 3} n_\alpha = -\gamma H Q n_2 \quad \text{for } r = a. \tag{6.4.1}$$

The boundary conditions (6.4.1) can be expressed in polar coordinates in the form

$$t_{rr} = 0, t_{r\vartheta} = 0, \quad m_{rz} = -\gamma H Q \sin\vartheta \quad \text{for } r = a. \tag{6.4.2}$$

We seek the solution of the isothermal problem assuming that

$$u_r = U(r)\cos\vartheta, \quad u_\vartheta = V(r)\sin\vartheta, \quad \varphi^0 = W(r)\sin\vartheta, \tag{6.4.3}$$

where U, V and W are functions only on r. From the constitutive equations and (6.4.3) we obtain

$$\begin{aligned}
t_{rr} &= [(\lambda + 2\mu + \kappa)U' + \frac{1}{r}\lambda(U + V)]\cos\vartheta,\\
t_{\vartheta\vartheta} &= [\lambda U' + \frac{1}{r}(\lambda + 2\mu + \kappa)(U + V)]\cos\vartheta,\\
t_{r\vartheta} &= [(\mu + \kappa)V' - \frac{1}{r}\mu(U + V) - \kappa W]\sin\vartheta,\\
t_{\vartheta r} &= [\mu V' - \frac{1}{r}(\mu + \kappa)(U + V) + \kappa W]\sin\vartheta,\\
m_{rz} &= \gamma W' \sin\vartheta, \quad m_{\vartheta z} = \frac{1}{r}\gamma W \cos\vartheta.
\end{aligned} \tag{6.4.4}$$

It follows from the equilibrium equations and (6.4.4) that the functions U, V and W satisfy the following system of equations

$$\begin{aligned}
&r^2\frac{d^2U}{dr^2} + r\frac{dU}{dr} - (1 + d_1)U + (1 - d_1)r\frac{dV}{dr} - (1 + d_1)V + d_2 rW = 0,\\
&d_1(r^2\frac{d^2V}{dr^2} + r\frac{dV}{dr}) - (1 + d_1)V - (1 - d_1)r\frac{dU}{dr} - (1 + d_1)U - d_2 r^2\frac{dW}{dr} = 0,\\
&r^2\frac{d^2W}{dr^2} + r\frac{dW}{dr} - (1 + 2d_3 r^2)W + d_3 r^2\frac{dV}{dr} + d_3 rV + d_3 rU = 0,
\end{aligned} \tag{6.4.5}$$

where

$$d_1 = \frac{\mu + \kappa}{\lambda + 2\mu + \kappa}, \quad d_2 = \frac{\kappa}{\lambda + 2\mu + \kappa}, \quad d_3 = \frac{\kappa}{\gamma}. \tag{6.4.6}$$

Let us introduce the independent variable t through the relation $t = \ln r$. If we denote $D = d/dt$ then the first two equations from (6.4.5) can be written in the form

$$\begin{aligned}
&\left[D^2 - (1 + d_1)\right]U + [(1 - d_1)D - (1 + d_1)]V = -e^t d_2 W,\\
&[(d_1 - 1)D - (1 + d_1)]U + \left[d_1 D^2 - (1 + d_1)\right]V = e^t d_2 DW.
\end{aligned} \tag{6.4.7}$$

The general solution of the homogeneous system (6.4.7) which correspond to a non-rigid displacement is given by

$$\begin{aligned}
U_0 &= E_1 t + E_2 e^{2t} + E_3 e^{-2t},\\
V_0 &= \frac{d_1 - 1}{d_1 + 1}E_1 - E_1 t - \frac{3 - d_1}{1 - 3d_1}E_2 e^{2t} + E_3 e^{-2t},
\end{aligned} \tag{6.4.8}$$

where E_j are arbitrary constants. A particular solution of nonhomogeneous system is

$$U^* = -\frac{1}{2d_1}d_2(S_1 - S_2 e^{-2t}), \quad V^* = \frac{1}{2d_1}d_2(S_1 + S_2 e^{-2t}), \tag{6.4.9}$$

where

$$S_1(t) = \int^t e^x W(x)dx, \quad S_2(t) = \int^t e^{3x} W(x)dx. \tag{6.4.10}$$

In view of (6.4.8)-(6.4.10) we find that

$$\begin{aligned}
U=&E_1 \ln r + E_2 r^2 + E_3 r^{-2} - \frac{1}{2d_1} d_2 \left(\int^r W(x)dx - r^{-2} \int^r x^2 W(x)dx \right), \\
V =& \frac{d_1 - 1}{d_1 + 1} E_1 - E_1 \ln r - \frac{3 - d_1}{1 - 3d_1} E_2 r^2 + E_3 r^{-2} + \\
&+ \frac{1}{2d_1} d_2 \left(\int^r W(x)dx + r^{-2} \int^r x^2 W(x)dx \right).
\end{aligned} \tag{6.4.11}$$

Since U and V must be finite for $r = 0$ we conclude that $E_1 = 0$ and $E_3 = 0$. If we substitute U and V from (6.4.11) into the third equation of (6.4.5) then we obtain

$$r^2 \frac{d^2 W}{dr^2} + r \frac{dW}{dr} - (1 + \delta^2 r^2)W = \frac{8}{1 - 3d_1} d_3 E_2 r^3,$$

where δ is defined in (6.2.17). The solution of this equation is

$$W = E_4 K_1(\delta r) + E_5 I_1(\delta r) - \frac{8(\mu + \kappa)}{(2\mu + \kappa)(1 - 3d_1)} E_2 r, \tag{6.4.12}$$

where K_n and I_n are the modified Bessel functions of order n and E_4 and E_5 are arbitrary constants. Since W must be finite for $r = 0$ we have $E_4 = 0$. It follows from (6.4.11) and (6.4.12) that the solution of the system (6.4.5) can be expressed in the form

$$\begin{aligned}
U &= \Lambda_1 E_2 r^2 - \frac{1}{2d_1 \delta} d_2 E_5 [I_0(\delta r) - I_2(\delta r)], \\
V &= -\Lambda_2 E_2 r^2 + \frac{1}{2d_1 \delta} d_2 E_5 [I_0(\delta r) + I_2(\delta r)], \\
W &= E_5 I_1(\delta r) - \Lambda_3 E_2 r,
\end{aligned} \tag{6.4.13}$$

where

$$\begin{aligned}
\Lambda_1 &= 1 + \frac{\kappa}{(2\mu + \kappa)(1 - 3d_1)}, \quad \Lambda_2 = \frac{1}{1 - 3d_1}(3 - d_1 + \frac{3\kappa}{2\mu + \kappa}), \\
\Lambda_3 &= \frac{8(\mu + \kappa)}{(2\mu + \kappa)(1 - 3d_1)}.
\end{aligned} \tag{6.4.14}$$

In view of (6.4.4) and (6.4.13) we obtain

$$t_{rr} = [\Gamma_1 E_2 r - \frac{1}{r}\gamma\delta E_5 I_2(\delta r)]\cos\vartheta,$$
$$t_{r\vartheta} = -[\Gamma_2 E_2 r + \frac{1}{r}\gamma\delta E_5 I_2(\delta r)]\sin\vartheta,$$
$$m_{rz} = [\gamma\delta E_5 I_1'(\delta r) - \gamma\Lambda_3 E_2]\sin\vartheta,$$

where

$$\Gamma_1 = (3\lambda + 4\mu + 2\kappa)\Lambda_1 - \lambda\Lambda_2, \ \ \Gamma_2 = \mu\Lambda_1 + (\mu + 2\kappa)\Lambda_2 - \kappa\Lambda_3.$$

We can see that $\Gamma_1 = -\Gamma_2$ so that the first two boundary conditions (6.4.2) imply that

$$E_2 = \frac{1}{a^2\Gamma_1}\gamma\delta E_5 I_2(\delta a). \tag{6.4.15}$$

The remaining boundary condition leads to

$$E_5 = -HQ[\delta I_1'(\delta a) - \frac{1}{a^2\Gamma_1}\delta\gamma\Lambda_3 I_2(\delta a)]^{-1}. \tag{6.4.16}$$

Thus, the solution of the isothermal plane strain problem is given by (6.4.3), (6.4.13) where the constants E_2 and E_5 are given by (6.4.15) and (6.4.16). The solution of the thermoelastic problem follows from (6.3.20) and (6.4.3).

2. *A homogeneous and isotropic thermoelastic material occupies a right hollow cylinder with the generic cross-section* $\Sigma = \{\mathbf{x} : R_1^2 < x_1^2 + x_2^2 < R_2^2, x_3 = \ell\}$. *The body is in equilibrium in absence of mechanical loads. Investigate the deformation of the cylinder when the temperature field is independent of the axial coordinate and the surface of the body is free of tractions and couple-stresses.*

Solution. The cylinder is in a state of plane strain. We assume that the solution $\theta = \theta(x_1, x_2), (x_1, x_2) \in \Sigma$, of the boundary value problem (6.3.2) is known. The problem consists in finding of the functions u_1, u_2 and φ on Σ which satisfy the equations (6.3.3)-(6.3.5) on Σ and the boundary conditions (6.3.6) on L, where $f_\alpha = 0, g_3 = 0, \widetilde{t}_\alpha = 0$ and $\widetilde{m}_3 = 0$. We study this boundary value problem in polar coordinates. Thus, we have to study the system of equations (6.3.12), with $f_r = f_\vartheta = 0, g_3 = 0$, in the circular ring-shaped region $r \in [R_1, R_2], \vartheta \in [0, 2\pi)$ (Fig. 3). The boundary conditions (6.3.13) reduce to

$$t_{rr} = 0, t_{r\vartheta} = 0, m_{rz} = 0 \text{ for } r = R_1 \text{ and } r = R_2. \tag{6.4.17}$$

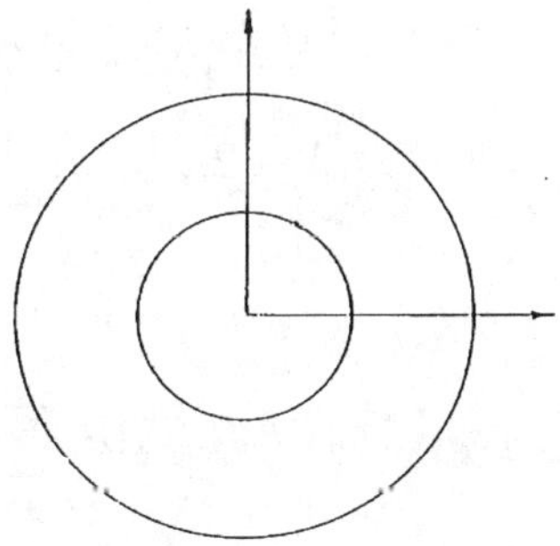

Figure 3

Since the solution is periodic in ϑ with period of 2π, we may expand u_r, u_ϑ and φ as follows

$$u_r = \mathrm{Re}\sum_{n=-\infty}^{\infty} U^{(n)}(r)\exp(in\vartheta),\ u_\vartheta = \mathrm{Re}\sum_{n=-\infty}^{\infty} iV^{(n)}(r)\exp(in\vartheta),$$
$$\varphi = \mathrm{Re}\sum_{n=-\infty}^{\infty} iW^{(n)}(r)\exp(in\vartheta). \tag{6.4.18}$$

We can write the temperature θ in the form

$$\theta = \mathrm{Re}\sum_{n=-\infty}^{\infty} T^{(n)}(r)\exp(in\vartheta), \tag{6.4.19}$$

where $T^{(n)}$ are given. If we substitute (6.4.18) and (6.4.19) into equations (6.3.12), (with $f_r = 0, f_\vartheta = 0, g_3 = 0$), then we obtain the following equations for the Fourier coefficients $U^{(n)}, V^{(n)}$ and $W^{(n)}$

$$r^2\frac{d^2U^{(n)}}{dr^2} + r\frac{dU^{(n)}}{dr} - (1+n^2d_1)U^{(n)} - n(1-d_1)r\frac{dV^{(n)}}{dr} +$$
$$+ n(1+d_1)V^{(n)} - nd_2rW^{(n)} = b_0r^2\frac{dT^{(n)}}{dr},$$
$$n(1-d_1)r\frac{dU^{(n)}}{dr} + n(1+d_1)U^{(n)} + d_1(r^2\frac{d^2V^{(n)}}{dr^2} + r\frac{dV^{(n)}}{dr}) -$$
$$-(n^2+d_1)V^{(n)} - d_2r^2\frac{dW^{(n)}}{dr} = b_0nrT^{(n)},$$
$$r^2\frac{d^2W^{(n)}}{dr^2} + r\frac{dW^{(n)}}{dr} - (n^2+2d_3r^2)W^{(n)} + \tag{6.4.20}$$
$$+d_3r(r\frac{dV^{(n)}}{dr} + V^{(n)} - nU^{(n)}) = 0,$$

where we have used the notations from (6.4.6) and

$$b_0 = b/(\lambda + 2\mu + \kappa). \tag{6.4.21}$$

We can write

$$\begin{aligned}
t_{rr} &= \text{Re} \sum_{n=-\infty}^{\infty} T_{rr}^{(n)}(r)\exp(in\vartheta), \; t_{r\vartheta} = \text{Re} \sum_{n=-\infty}^{\infty} T_{r\vartheta}^{(n)}(r)\exp(in\vartheta),\\
t_{\vartheta r} &= \text{Re} \sum_{n=-\infty}^{\infty} T_{\vartheta r}^{(n)}(r)\exp(in\vartheta), \; t_{\vartheta\vartheta} = \text{Re} \sum_{n=-\infty}^{\infty} T_{\vartheta\vartheta}^{(n)}(r)\exp(in\vartheta),\\
m_{rz} &= \text{Re} \sum_{n=-\infty}^{\infty} M_{rz}^{(n)}(r)\exp(in\vartheta), m_{\vartheta z} = \text{Re} \sum_{n=-\infty}^{\infty} M_{\vartheta z}^{(n)}(r)\exp(in\vartheta),
\end{aligned} \tag{6.4.22}$$

It follows from (6.3.10), (6.3.11), (6.4.18) and (6.4.21) that

$$\begin{aligned}
T_{rr}^{(n)} &= \lambda(\frac{dU^{(n)}}{dr} + \frac{1}{r}U^{(n)} - \frac{n}{r}V^{(n)}) + (2\mu+\kappa)\frac{dU^{(n)}}{dr} - bT^{(n)},\\
T_{\vartheta\vartheta}^{(n)} &= \lambda(\frac{dU^{(n)}}{dr} + \frac{1}{r}U^{(n)} - \frac{n}{r}V^{(n)}) + (2\mu+\kappa)\frac{1}{r}(U^{(n)} - nV^{(n)}) - bT^{(n)},\\
T_{r\vartheta}^{(n)} &= i\left[\mu\left(\frac{dV^{(n)}}{dr} + \frac{n}{r}U^{(n)} - \frac{1}{r}V^{(n)}\right) + \kappa\left(\frac{dV^{(n)}}{dr} - W^{(n)}\right)\right],\\
T_{\vartheta r}^{(n)} &= i\left[\mu\left(\frac{dV^{(n)}}{dr} - \frac{1}{r}V^{(n)} + \frac{n}{r}U^{(n)}\right) + \kappa\left(\frac{n}{r}U^{(n)} - \frac{1}{r}V^{(n)} + W^{(n)}\right)\right],\\
M_{rz}^{(n)} &= i\gamma\frac{dW^{(n)}}{dr}, \quad M_{\vartheta z}^{(n)} = -\frac{1}{r}n\gamma W^{(n)}.
\end{aligned} \tag{6.4.23}$$

The general solution of the system of equations (6.4.20) can be determined as in the previous exercise. We present this solution in three cases: (i) $n=0$; (ii) $n=\pm 1$; (iii) $|n| \geq 2$. In the case $n=0$ the general solution has the form

$$\begin{aligned}
U^{(0)} &= \frac{1}{E}(1+\nu)[(1-2\nu)A_1^{(0)}r + A_2^{(0)}r^{-1}] + b_0 r^{-1}\int_{R_2}^{r} xT^{(0)}(x)dx,\\
V^{(0)} &= \frac{1}{E}(1+\nu)\{A_3^{(0)}r + A_4^{(0)}r^{-1} + \frac{1}{\delta}e[A_6^0 I_1(\delta r) - A_5^{(0)}K_1(\delta r)]\},\\
W^{(0)} &= \frac{1}{E}(1+\nu)[A_3^{(0)} + A_5^{(0)}K_0(\delta r) + A_6^{(0)}I_0(\delta r)],
\end{aligned} \tag{6.4.24}$$

where $A_s^{(0)}, (s = 1, 2, \dots, 6)$, are arbitrary constants and

$$\nu = \frac{\lambda}{2\lambda + 2\mu + \kappa}, \quad E = (2\mu + \kappa)(1 + \nu), \quad e = \frac{\kappa}{\mu + \kappa}. \tag{6.4.25}$$

It follows from (6.4.23) and (6.4.24) that

$$\begin{aligned}
T_{rr}^{(0)} &= A_1^{(0)} - A_2^{(0)} r^{-2} - (2\mu + \kappa) b_0 r^{-2} \int_{R_1}^{r} x T^{(0)}(x) dx, \\
T_{\vartheta\vartheta}^{(0)} &= A_1^0 + A_2^{(0)} r^{-2} - (2\mu + \kappa) b_0 [T^{(0)}(r) - r^{-2} \int_{R_1}^{r} x T^{(0)}(x) dx], \\
T_{r\vartheta}^{(0)} &= i \left\{ \frac{1}{\delta r} e [A_5^0 K_1(\delta r) - A_6^{(0)} I_1(\delta r)] - A_4^{(0)} r^{-2} \right\}, \\
T_{\vartheta r}^{(0)} &= i \left\{ e [A_6^{(0)} I_1'(\delta r) - A_5^{(0)} K_1'(\delta r)] - A_4^{(0)} r^{-2} \right\}, \\
M_{rz}^{(0)} &= \frac{1}{E} (1 + \nu) i \delta \gamma [A_6^{(0)} I_1(\delta r) - A_5^{(0)} K_1(\delta r)], M_{\vartheta z}^{(0)} = 0,
\end{aligned} \tag{6.4.26}$$

where $f'(x) = df(x)/dx$.

In the case (ii) the solution has the form

$$\begin{aligned}
U^{(n)} &= \frac{1}{E}(1+\nu)\{A_1^{(n)} + \frac{1}{2} e(\nu-1) A_2^{(n)} + (3-4\nu) A_2^{(n)} \ln r - \\
&-(1 - 4\nu) A_3^{(n)} r^2 + A_4^{(n)} r^{-2} + n \frac{e}{\delta^2 r} [A_5^{(n)} K_1(\delta r) + \\
&+A_6^{(n)} I_1(\delta r)]\} + \frac{1}{2} b_0 [\int_{R_1}^{r} T^{(n)}(x) dx + r^{-2} \int_{R_1}^{r} x^2 T^{(n)}(x) dx], \\
V^{(n)} &= \frac{n}{E}(1+\nu)\{A_1^{(n)} + [1 + \frac{1}{2} e(\nu-1)] A_2^{(n)} + (3-4\nu) A_2^{(n)} \ln r + \\
&+(5-4\nu) A_3^{(n)} r^2 - A_4^{(n)} r^{-2} + \frac{n}{\delta} e [A_5^{(n)} K_1'(\delta r) + A_6^{(n)} I_1'(\delta r)]\} + \\
&+\frac{n}{2} b_0 [\int_{R_1}^{r} T^{(n)}(x) dx - r^{-2} \int_{R_1}^{r} x^2 T^{(n)}(x) dx]\}, \\
W^{(n)} &= \frac{1}{E}(1+\nu)\{n(1-\nu)[2A_2^{(n)} r^{-1} + 8A_3^{(n)} r] + A_5^{(n)} K_1(\delta r) + \\
&+A_6^{(n)} I_1(\delta r)\},
\end{aligned} \tag{6.4.27}$$

where $n = \pm 1$ and $A_s^{(n)}, (s = 1, 2, \dots, 6)$, are arbitrary constants. We now obtain

$$\begin{aligned}
T_{rr}^{(n)} &= (3 - 2\nu) A_2^{(n)} r^{-1} - 2A_3^{(n)} r - 2A_4^{(n)} r^{-3} - \frac{n}{\delta r} e [A_5^{(n)} K_2(\delta r) - \\
&\quad -A_6^{(n)} I_2(\delta r)] - (2\mu + \kappa) b_0 r^{-3} \int_{R_1}^{r} x^2 T^{(n)}(x) dx,
\end{aligned}$$

$$
\begin{aligned}
T_{\vartheta\vartheta}^{(n)} &= (2\nu - 1)A_2^{(n)} r^{-1} - 6A_3^{(n)} r + 2A_4^{(n)} r^{-3} + \frac{n}{\delta r} e[A_5^{(n)} K_2(\delta r) - \\
&\quad - A_6^{(n)} I_2(\delta r)] - (2\mu + \kappa) b_0 [T^{(n)}(r) - r^{-3} \int_{R_1}^{r} x^2 T^{(n)}(x) dx], \\
T_{r\vartheta}^{(n)} &= ni\{(1 - 2\nu)A_2^{(n)} r^{-1} + 2A_3^{(n)} r + 2A_4^{(n)} r^{-3} + \frac{ne}{\delta r}[A_5^{(n)} K_2(\delta r) - \\
&\quad - A_6^{(n)} I_2(\delta r)] + (2\mu + \kappa) b_0 r^{-3} \int_{R_1}^{r} x^2 T^{(n)}(x) dx\}, \\
T_{\vartheta r}^{(n)} &= T_{r\vartheta}^{(n)} + ie[A_5^{(n)} K_1(\delta r) + A_6^{(n)} I_1(\delta r)], \\
M_{rz}^{(n)} &= \frac{1}{E} i\gamma (1 + \nu)\{n(1 - \nu)(8A_3^{(n)} - 2A_2^{(n)} r^{-2}) + \\
&\quad + \delta[A_5^{(n)} K_1'(\delta r) + A_6^{(n)} I_1'(\delta r)]\}, \\
M_{\vartheta z}^{(n)} &= -\frac{1}{E}\gamma(1 + \nu)\{(1 - \nu)(2A_2^{(n)} r^{-2} + 8A_3^{(n)}) + \\
&+ \frac{n}{r}[A_5^{(n)} K_1(\delta r) + A_6^{(n)} I_1(\delta r)\}, \text{ for } n = \pm 1.
\end{aligned} \tag{6.4.28}
$$

For $|n| \geq 2$ the solution is given by

$$
\begin{aligned}
U^{(n)} &= \frac{1}{E}(1 + \nu)\{(n - 2 + 4\nu)A_1^{(n)} r^{n+1} + (n + 2 - 4\nu)A_2^{(n)} r^{1-n} + \\
&\quad + A_3^{(n)} r^{n-1} + A_4^{(n)} r^{-1-n} + \frac{ne}{\delta^2 r}[A_5^{(n)} K_n(\delta r) + A_6^{(n)} I_n(\delta r)]\} + \\
&\quad + \frac{1}{2} b_0 [r^{n-1} \int_{R_1}^{r} x^{1-n} T^{(n)}(x) dx + r^{-1-n} \int_{R_1}^{r} x^{1+n} T^{(n)}(x) dx], \\
V^{(n)} &= \frac{1}{E}(1 + \nu)\{(n + 4 - 4\nu)A_1^{(n)} r^{1+n} - (n - 4 + 4\nu)A_2^{(n)} r^{1-n} + \\
&\quad + A_3^{(n)} r^{n-1} - A_4^{(n)} r^{-1-n} + \frac{e}{\delta}[A_5^{(n)} K_n'(\delta r) + A_6^{(n)} I_n'(\delta r)]\} + \\
&\quad + \frac{1}{2} b_0 [r^{n-1} \int_{R_1}^{r} x^{1-n} T^{(n)}(x) dx - r^{-1-n} \int_{R_1}^{r} x^{1+n} T^{(n)}(x) dx], \\
W^{(n)} &= \frac{1}{E}(1 + \nu)\{4(1 - \nu)[(n + 1)A_1^{(n)} r^{n} - (n - 1)A_2^{(n)} r^{-n}] + \\
&\quad + A_5^{(n)} K_n(\delta r) + A_6^{(n)} I_n(\delta r)\},
\end{aligned} \tag{6.4.29}
$$

where $A_s^{(n)}$, $(s = 1, 2, \ldots, 6)$, are arbitrary constants. The corresponding stresses and couple stresses have the forms

$$
T_{rr}^{(n)} = (n - 2)(n + 1)A_1^{(n)} r^{n} - (n + 2)(n - 1)A_2^{(n)} r^{-n} +
$$

$$+(n-1)A_3^{(n)}r^{n-2}-(n+1)A_4^{(n)}r^{-2-n}-\frac{e}{2\delta r}\{A_5^{(n)}[(n+1)K_{n+1}(\delta r)+$$
$$+(n-1)K_{n-1}(\delta r)]-A_6^{(n)}[(n+1)I_{n+1}(\delta r)+(n-1)I_{n-1}(\delta r)]\}+$$
$$+\frac{1}{2}(2\mu+\kappa)b_0[(n-1)r^{n-2}\int_{R_1}^{r}x^{1-n}T^{(n)}(x)dx-$$
$$-(n+1)r^{-2-n}\int_{R_1}^{r}x^{1+n}T^{(n)}(x)dx],$$
$$-iT_{r\vartheta}^{(n)}=n[(n+1)A_1^{(n)}r^{n}+(n-1)A_2^{(n)}r^{-n}]+(n-1)A_3^{(n)}r^{n-2}+$$
$$+(n+1)A_4^{(n)}r^{-2-n}+\frac{e}{2\delta r}\{A_5^{(n)}[(n+1)K_{n+1}(\delta r)-$$
$$-(n-1)K_{n-1}(\delta r)]-A_6^{(n)}[(n+1)I_{n+1}(\delta r)-(n-1)I_{n-1}(\delta r)]\}+$$
$$+\frac{1}{2}(2\mu+\kappa)b_0[(n-1)r^{n-2}\int_{R_1}^{r}x^{1-n}T^{(n)}(x)dx+$$
$$+(n+1)r^{-2-n}\int_{R_1}^{r}x^{1+n}T^{(n)}(x)dx],$$
$$M_{rz}^{(n)}=\frac{i}{E}\gamma(1+\nu)\{4n(1-\nu)[(n+1)A_1^{(n)}r^{n-1}+$$
$$+(n-1)A_2^{(n)}r^{-1-n}]+\delta[A_5^{(n)}K_n'(\delta r)+A_6^{(n)}I_n'(\delta r)]\}. \tag{6.4.30}$$

The boundary conditions (6.4.17) reduce to

$$T_{rr}^{(n)}=0,\ \ T_{r\vartheta}^{(n)}=0\ \ M_{rz}^{(n)}=0 \text{ for } r=R_1 \text{ and } r=R_2. \tag{6.4.31}$$

The arbitrary constants are determined from (6.4.31). Thus, for $n=0$, from (6.4.26) and (6.4.31) we obtain

$$A_1^{(0)}=\frac{2\mu+\kappa}{R_2^2-R_1^2}b_0\int_{R_1}^{R_2}xT^{(0)}(x)dx,$$
$$A_2^{(0)}=\frac{2\mu+\kappa}{R_2^2-R_1^2}b_0R_1^2\int_{R_1}^{R_2}xT^{(0)}(x)dx,\ \ A_4^{(0)}=A_5^{(0)}=A_6^{(0)}=0. \tag{6.4.32}$$

In the case $n=\pm1$ we obtain

$$A_2^{(n)}=0,\ \ A_3^{(n)}=B_1^{(n)}+B_2^{(n)}A_4^{(n)}+B_3^{(n)}A_5^{(n)},$$
$$A_6^{(n)}=B_4^{(n)}+B_5^{(n)}A_4^{(n)}+B_6^{(n)}A_5^{(n)},$$
$$A_4^{(n)}=P^{(n)}(B_{10}^{(n)}B_{11}^{(n)}-B_8^{(n)}B_{12}^{(n)}), \tag{6.4.33}$$
$$A_5^{(n)}=P^{(n)}(B_7^{(n)}B_{12}^{(n)}-B_9^{(n)}B_{11}^{(n)}),$$

where

$$
\begin{aligned}
B_1^{(n)} &= (2\mu+\kappa)b_0 g\delta R_2^{-3} I_1'(\delta R_2)\int_{R_1}^{R_2} x^2 T^{(n)}(x)dx,\\
B_2^{(n)} &= 2g\delta R_2^{-3} I_1'(\delta R_2),\\
B_3^{(n)} &= ne\ gR_2^{-1}[K_2(\delta R_2)I_1'(\delta R_2) + I_2(\delta R_2)K_1'(\delta R_2)],\\
B_4^{(n)} &= -8n(1-\nu)g(2\mu+\kappa)b_0 R_2^{-3}\int_{R_1}^{R_2} x^2 T^{(n)} dx,\\
B_5^{(n)} &= -16n(1-\nu)gR_2^{-3},\\
B_6^{(n)} &= 2g[\delta R_2 K_1'(\delta R_2) - \frac{4e}{\delta R_2}(1-\nu)K_2(\delta R_2)],\\
B_7^{(n)} &= -2R_1^{-3} - 2R_1 B_2^{(n)} + \frac{ne}{\delta R_1} I_2(\delta R_1) B_5^{(n)},\\
B_8^{(n)} &= -\frac{ne}{\delta R_1} K_2(\delta R_1) - 2R_1 B_3^{(n)} + \frac{ne}{\delta R_1} I_2(\delta R_1) B_6^{(n)},\\
B_9^{(n)} &= 8n(1-\nu)B_2^{(n)} + \delta I_1'(\delta R_1) B_5^{(n)},\\
B_{10}^{(n)} &= \delta K_1'(\delta R_1) + 8n(1-\nu)B_3^{(n)} + \delta I_1'(\delta R_1) B_6^{(n)},\\
B_{11}^{(n)} &= 2R_1 B_1^{(n)} - \frac{ne}{\delta R_1} I_2(\delta R_1) B_4^{(n)},\\
B_{12}^{(n)} &= -8n(1-\nu)B_1^{(n)} - \delta I_1'(\delta R_1) B_4^{(n)},\\
1/P^{(n)} &= B_7^{(n)} B_{10}^{(n)} - B_8^{(n)} B_9^{(n)},\quad g^{-1} = -2[\delta R_2 I_1'(\delta R_2) +\\
&\quad + \frac{4e}{\delta R_2}(1-\nu) I_2(\delta R_2)],\ \text{for } n=\pm 1.
\end{aligned}
$$

It follows from (6.4.30) and (6.4.31) that for $|n| \geq 2$ we have

$$
\begin{aligned}
A_1^{(n)} &= C_1^{(n)} + C_2^{(n)} A_2^{(n)} + C_3^{(n)} A_4^{(n)} + C_5^{(n)} A_5^{(n)},\\
A_2^{(n)} &= C_{13}^{(n)} + C_{14}^{(n)} A_4^{(n)} + C_{15}^{(n)} A_5^{(n)},\\
A_3^{(n)} &= C_9^{(n)} + C_{10}^{(n)} A_2^{(n)} + C_{11}^{(n)} A_4^{(n)} + C_{12}^{(n)} A_5^{(n)},\\
A_4^{(n)} &= (C_{19}^{(n)} C_{20}^{(n)} - C_{17}^{(n)} C_{21}^{(n)})H^{(n)},\\
A_5^{(n)} &= (C_{16}^{(n)} C_{21}^{(n)} - C_{18}^{(n)} C_{20}^{(n)})H^{(n)},\\
A_6^{(n)} &= C_5^{(n)} + C_6^{(n)} A_2^{(n)} + C_7^{(n)} A_4^{(n)} + C_8^{(n)} A_5^{(n)}.
\end{aligned}
$$

Here we have used the notations

$$C_1^{(n)} = -\frac{1}{2}\delta\gamma_1 R_2^{-2-n}(2\mu+\kappa)b_0 I_n'(\delta R_2)\int_{R_1}^{R_2} x^{1+n}T^{(n)}(x)dx,$$

$$C_2^{(n)} = (1-n)\delta\gamma_1 R_2^{-n}[I_n'(\delta R_2) + \frac{2e}{\delta R_2^2}n(1-\nu)I_{n+1}(\delta R_2)],$$

$$C_3^{(n)} = -\delta\gamma_1 R_2^{-2-n} I_n'(\delta R_2),$$

$$C_4^{(n)} = -\tfrac{1}{2}e\gamma_1 R_2^{-1}[I_n'(\delta R_2)K_{1+n}(\delta R_2) + K_n'(\delta R_2)I_{1+n}(\delta R_2)],$$

$$C_5^{(n)} = 2n(n+1)\gamma_1(2\mu+\kappa)(1-\nu)b_0 R_2^{-3}\int_{R_1}^{R_2} x^{1+n}T^{(n)}(x)dx,$$

$$C_6^{(n)} = 4n^2(n-1)(1-\nu)\gamma_1 R_2^{-1},\ \ C_7^{(n)} = 4n(n+1)(1-\nu)\gamma_1 R_2^{-3},$$

$$C_8^{(n)} = \delta^{-2}\gamma_1 R_2^{n-2}[2n(n+1)(1-\nu)eK_{n+1}(\delta R_2) - \delta^2 R_2^2 K_n'(\delta R_2)],$$

$$C_9^{(n)} = -(1+n)R_2^2 C_1^{(n)} - \frac{e}{2\delta}R_2^{-1-n}I_{n-1}(\delta R_2)C_5^{(n)} -$$
$$-\frac{1}{2}(2\mu+\kappa)b_0\int_{R_1}^{R_2} x^{1-n}T^{(n)}(x)dx,$$

$$C_{10}^{(n)} = R_2^{2-2n} - (1+n)R_2^2 C_2^{(n)} - \frac{e}{2\delta}R_2^{1-n}I_{n-1}(\delta R_2)C_6^{(n)},$$

$$C_{11}^{(n)} = -(1+n)R_2^2 C_3^{(n)} - \frac{e}{2\delta}R_2^{1-n}I_{n-1}(\delta R_2)C_7^{(n)},$$

$$C_{12}^{(n)} = \frac{e}{2\delta}R_2^{1-n}K_{n-1}(\delta R_2) - (n+1)R_2^2 C_4^{(n)} - \frac{e}{2\delta}R_2^{1-n}C_8^{(n)}I_{n-1}(\delta R_1),$$

$$C_{13}^{(n)} = \gamma_2[(n+1)R_1^{(n)}C_1^{(n)} + R_1^{n-2}C_9^{(n)} + \frac{e}{2\delta R_1}C_5^{(n)}I_{n-1}(\delta R_1)],$$

$$C_{14}^{(n)} = \gamma_2[(n+1)R_1^n C_3^{(n)} + R_1^{n-2}C_{11}^{(n)} + \frac{e}{2\delta R_1}C_7^{(n)}I_{n-1}(\delta R_1)],$$

$$C_{15}^{(n)} = \gamma_2\{R_1^{n-2}C_{12}^{(n)} + \frac{e}{2\delta R_1}[C_8^{(n)}I_{n-1}(\delta R_1) -$$
$$-K_{n-1}(\delta R_1)] + (1+n)R_1^{(n)}C_4^{(n)}\},$$

$$C_{16}^{(n)} = R_1^{-2-n} + R_1^n C_3^{(n)} + \frac{e}{2\delta R_1}[C_7^{(n)} + C_6^{(n)}C_{14}^{(n)}]I_{n+1}(\delta R_1) +$$
$$+[(n-1)R_1^{-n} + R_1^n C_2^{(n)}]C_{14}^{(n)},$$

$$C_{17}^{(n)} = R_1^n C_4^{(n)} + \frac{e}{2\delta R_1}[(C_8^{(n)} + C_6^{(n)}C_{15}^{(n)})I_{n+1}(\delta R_1) -$$
$$-K_{n+1}(\delta R_1)] + [R_1^n C_2^{(n)} + (n-1)R_1^{-n}]C_{15}^{(n)},$$

$$C_{18}^{(n)} = 4n(1-\nu)[(n+1)R_1^{n-1}(C_3^{(n)} + C_2^{(n)}C_{14}^{(n)}) +$$

$$
\begin{aligned}
&+(n-1)R_1^{-n-1}C_{14}^{(n)}]+(C_7^{(n)}+C_6^{(n)}C_{14}^{(n)})\delta I_n'(\delta R_1),\\
C_{19}^{(n)} &= 4n(1-\nu)[(n+1)R_1^{n-1}(C_4^{(n)}+C_2^{(n)}C_{15}^{(n)})+\\
&+(n-1)R_1^{-n-1}C_{15}^{(n)}]+\delta K_n'(\delta R_1)+(C_8^{(n)}+C_6^{(n)}C_{15}^{(n)})\delta I_n'(\delta R_1),\\
C_{20}^{(n)} &= -R_1^nC_1^{(n)}-\frac{e}{2\delta R_1}[C_5^{(n)}+C_6^{(n)}C_{13}^{(n)}]I_{n+1}(\delta R_1)-\\
&-[R_1^nC_2^{(n)}+(n-1)R_1^{-n}]C_{13}^{(n)},\\
C_{21}^{(n)} &= -\delta C_5^{(n)}I_n'(\delta R_1)-4n(1-\nu)[(n+1)R_1^{n-1}(C_1^{(n)}+\\
&+C_2^{(n)}C_{13}^{(n)})+(n-1)R_2^{-1-n}C_{13}^{(n)}]
\end{aligned}
$$

and

$$
\begin{aligned}
1/H^{(n)} &= C_{16}^{(n)}C_{19}^{(n)}-C_{17}^{(n)}C_{18}^{(n)},\\
\gamma_1^{-1} &= R_2^{n-2}\delta^{-1}[2n(n+1)(1-\nu)eK_{n+1}(\delta R_2)-\delta^2R_2^2K_n'(\delta R_2)],\\
\gamma_2^{-1} &= R_1^{-n}-(n+1)R_1^nC_2^{(n)}-R_1^{n-2}C_{10}^{(n)}-\frac{e}{2\delta R_1}C_6^{(n)}I_{n-1}(\delta R_1).
\end{aligned}
$$

From the conditions that the displacements must be real we find that $A_1^{(-1)}=A_1^{(1)}$. It is easy to see that the constants $A_3^{(0)}$ and $A_1^{(1)}$ correspond to a rigid displacement.

We note that Chiu and Lee [59] have established the general solution for static plane problems of micropolar elasticity in a circular ring-shaped region. This solution was extended to include temperature fields by Chirita [53].

6.5 Thermal Stresses in Beams

In the classical thermoelasticity, the problem of thermal stresses in beams has been intensively studied (see, e.g., Boley and Weiner [23], p. 328). In this section, we study the thermoelastic equilibrium of homogeneous and isotropic micropolar beams. As in the Saint-Venant's problem (see, e.g., Ieşan [181]) the thermoelastic deformation is reduced to the study of plane problems.

Throughout this section the region B refers to the interior of a right cylinder of length h with the generic cross-section Σ and the lateral boundary Π (Fig. 4). The rectangular Cartesian coordinate frame is chosen such that the x_3-axis is parallel to the generators of B and the x_1Ox_2 plane contains one of the terminal cross-sections. We denote by $\Sigma^{(0)}$ and $\Sigma^{(h)}$ the cross-sections located at $x_3=0$ and $x_3=h$,

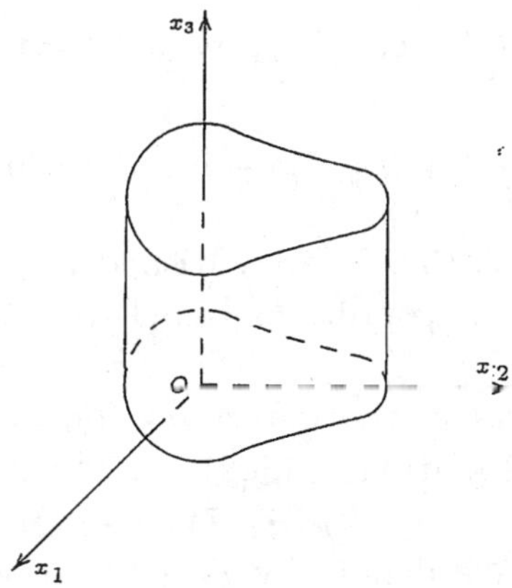

Figure 4

respectively. We assume that the cross-section Σ is a simply-connected regular region, bounded by the closed curve L. We suppose that the body forces and the body couples are absent and the lateral surface is free of applied forces and couples. The cylinder is supposed to be in equilibrium under the action of a given temperature, the loading applied to the ends being statically equivalent to zero. We assume that the temperature θ is a polynomial of degree r in the axial coordinate, namely

$$\theta = \sum_{k=0}^{r} T_k(x_1, x_2)x_3^k, \tag{6.5.1}$$

where T_k are prescribed functions on Σ.

The equations of equilibrium become

$$t_{ji,j} = 0, \quad m_{ji,j} + \varepsilon_{irs}t_{rs} = 0 \text{ on } B. \tag{6.5.2}$$

The constitutive equations are given by (6.2.3) where $\lambda, \mu, \kappa, b, \alpha, \beta$ and γ are prescribed constants. On the lateral boundary of the cylinder we have the conditions

$$t_{\alpha i}n_\alpha = 0, \quad m_{\alpha i}n_\alpha = 0 \text{ on } \Pi. \tag{6.5.3}$$

On the end located at $x_3 = 0$ we have the conditions

$$\int_\Sigma t_{3\alpha}dA = 0, \tag{6.5.4}$$

$$\int_\Sigma t_{33}dA = 0, \tag{6.5.5}$$

$$\int_{\Sigma}(x_\alpha t_{33} - \varepsilon_{3\alpha\beta}m_{3\beta})dA = 0, \tag{6.5.6}$$

$$\int_{\Sigma}(\varepsilon_{3\alpha\beta}x_\alpha t_{3\beta} + m_{33})dA = 0. \tag{6.5.7}$$

If the conditions (6.5.4)-(6.5.7) are satisfied for $x_3 = 0$, then, in view of (6.5.2) and (6.5.3), the conditions (6.5.4)-(6.5.7) are satisfied for any $x_3 \in [0, h]$.

The problem consists in the determination of the functions u_i and φ_i which satisfy the equations (6.5.2), (6.2.3) and (6.1.4) on B, and the boundary conditions (6.5.3)-(6.5.7), assuming that the function θ has the form (6.5.1). We denote by $P^{(n)}$ the problem of thermoelastic equilibrium corresponding to the temperature field

$$\theta = T_n(x_1, x_2)x_3^n, \tag{6.5.8}$$

where n is a positive integer or zero, and T_n is known. Clearly, if we know the solution of the problem $P^{(n)}$ for any n then, according to the linearity of the theory, we can determine the solution when the temperature has the form (6.5.1). To solve the initial problem we use the method of induction. First we establish the solution of the problem $P^{(0)}$. Then, we establish a solution of the problem $P^{(n+1)}$ once a solution of the problem $P^{(n)}$ is known.

a)**The problem** $P^{(0)}$. We first consider two problems, $\mathcal{D}^{(1)}$ and $\mathcal{D}^{(2)}$, of isothermal plane strain (see Section 6.3). The problem $\mathcal{D}^{(1)}$ is characterized by

$$f_\alpha = 0, \quad g_3 = 0, \quad \theta = 0 \text{ on } \Sigma \tag{6.5.9}$$

and the boundary data

$$\widetilde{t}_\alpha = 0, \quad \widetilde{m}_3 = (\beta + \gamma\nu)n_2 \text{ on } L, \tag{6.5.10}$$

where ν is given by (6.4.25). In the problem $\mathcal{D}^{(2)}$ we have

$$f_\alpha = 0, \quad g_3 = 0, \quad \theta = 0 \text{ on } \Sigma \tag{6.5.11}$$

and following boundary data

$$\widetilde{t}_\alpha = 0, \quad \widetilde{m}_3 = -(\beta + \gamma\nu)n_1 \text{ on } L. \tag{6.5.12}$$

We note that the necessary and sufficient conditions (6.3.7) for the existence of the solution are satisfied for each boundary value problem $\mathcal{D}^{(\rho)}$.

We denote by $(u_\alpha^{(\rho)}, \varphi^{(\rho)}, e_{\alpha\beta}^{(\rho)}, \kappa_{\alpha 3}^{(\rho)}, t_{\alpha\beta}^{(\rho)}, m_{\alpha 3}^{(\rho)})$ the solution of the plane strain problem $\mathcal{D}^{(\rho)}, (\rho = 1, 2)$. Thus, the functions $u_\alpha^{(\rho)}, \varphi^{(\rho)}, e_{\alpha\beta}^{(\rho)}, \kappa_{\alpha 3}^{(\rho)}$, $t_{\alpha\beta}^{(\rho)}$ and $m_{\alpha 3}^{(\rho)}$ satisfy the equilibrium equations

$$t_{\beta\alpha,\beta}^{(\rho)} = 0, \quad m_{\alpha 3,\alpha}^{(\rho)} + \varepsilon_{3\alpha\beta} t_{\alpha\beta}^{(\rho)} = 0, \tag{6.5.13}$$

the constitutive equations

$$t_{\alpha\beta}^{(\rho)} = \lambda e_{\eta\eta}^{(\rho)} \delta_{\alpha\beta} + (\mu + \kappa) e_{\alpha\beta}^{(\rho)} + \mu e_{\beta\alpha}^{(\rho)}, \quad m_{\alpha 3}^{(\rho)} = \gamma \kappa_{\alpha 3}^{(\rho)} \tag{6.5.14}$$

and geometrical equations

$$e_{\alpha\beta}^{(\rho)} = u_{\beta,\alpha}^{(\rho)} + \varepsilon_{\beta\alpha 3} \varphi^{(\rho)}, \quad \kappa_{\alpha 3}^{(\rho)} = \varphi_{,\alpha}^{(\rho)}, \tag{6.5.15}$$

on Σ, and the boundary conditions

$$t_{\beta\alpha}^{(\rho)} n_\beta = 0, \quad m_{\alpha 3}^{(\rho)} n_\alpha = (\beta + \gamma\nu) \varepsilon_{3\rho\alpha} n_\alpha \text{ on } L, (\rho = (1, 2). \tag{6.5.16}$$

In what follows we assume that the functions $u_\alpha^{(\rho)}, \varphi^{(\rho)}, e_{\alpha\beta}^{(\rho)}, \kappa_{\alpha 3}^{(\rho)}, t_{\alpha\beta}^{(\rho)}$ and $m_{\alpha 3}^{(\rho)}, (\rho = 1, 2)$, are known.

Let us assume that the considered cylinder is subjected to the temperature field

$$\theta = f(x_1, x_2), \quad (x_1, x_2) \in \Sigma, \tag{6.5.17}$$

where f is a prescribed functions. We seek the solution of the problem $P^{(0)}$ in the form

$$\begin{aligned}
u_1 &= -\frac{1}{2}[x_3^2 + \nu(x_1^2 - x_2^2)] a_1 - a_2 \nu x_1 x_2 - \\
&\quad - a_3 \nu x_1 + a_1 u_1^{(1)} + a_2 u_1^{(2)} + v_1(x_1, x_2), \\
u_2 &= -a_1 \nu x_1 x_2 - \frac{1}{2}[x_3^2 - \nu(x_1^2 - x_2^2)] a_2 - \\
&\quad - a_3 \nu x_2 + a_1 u_2^{(1)} + a_2 u_2^{(2)} + v_2(x_1, x_2), \\
u_3 &= (a_1 x_1 + a_2 x_2 + a_3) x_3, \\
\varphi_1 &= a_2 x_3, \quad \varphi_2 = -a_1 x_3, \\
\varphi_3 &= (a_2 x_1 - a_1 x_2)\nu + a_1 \varphi^{(1)} + a_2 \varphi^{(2)} + w(x_1, x_2),
\end{aligned} \tag{6.5.18}$$

where v_α and w are unknown functions, a_1 and a_2 are unknown constants and $(u_\alpha^{(\rho)}, \varphi^{(\rho)})$ are the displacements and microrotation characterized by the problem $\mathcal{D}^{(\rho)}, (\rho = 1, 2)$.

It follows from (6.1.4), (6.5.15) and (6.5.18) that

$$
\begin{aligned}
&e_{\alpha\beta} = -\nu(a_1x_1 + a_2x_2 + a_3)\delta_{\alpha\beta} + a_1\overset{(1)}{e}_{\alpha\beta} + a_2\overset{(2)}{e}_{\alpha\beta} + \gamma_{\alpha\beta},\\
&e_{33} = a_1x_1 + a_2x_2 + a_3, \quad e_{\alpha 3} = e_{3\alpha} = 0, \quad \kappa_{3\alpha} = \varepsilon_{\alpha\beta 3}a_\beta,\\
&\kappa_{\alpha\beta} = 0, \quad \kappa_{33} = 0, \quad \kappa_{\alpha 3} = \varepsilon_{3\alpha\beta}a_\beta\nu + a_1\overset{(1)}{\kappa}_{\alpha 3} + a_2\overset{(2)}{\kappa}_{\alpha 3} + \zeta_{\alpha 3},
\end{aligned}
\tag{6.5.19}
$$

where

$$
\gamma_{\alpha\beta} = v_{\beta,\alpha} + \varepsilon_{\beta\alpha 3}w, \quad \zeta_{\alpha 3} = w_{,\alpha}. \tag{6.5.20}
$$

We introduce the notations

$$
\begin{aligned}
&s_{\alpha\beta} = \lambda\gamma_{\rho\rho}\delta_{\alpha\beta} + (\mu + \kappa)\gamma_{\alpha\beta} + \mu\gamma_{\beta\alpha} - bf\delta_{\alpha\beta},\\
&\mu_{\alpha 3} = \gamma\zeta_{\alpha 3}.
\end{aligned}
\tag{6.5.21}
$$

In view of (6.2.3), (6.5.17), (6.5.19) and (6.5.21) we obtain

$$
\begin{aligned}
&t_{\alpha\beta} = a_1\overset{(1)}{t}_{\alpha\beta} + a_2\overset{(2)}{t}_{\alpha\beta} + s_{\alpha\beta}, \quad t_{\alpha 3} = t_{3\alpha} = 0,\\
&t_{33}=E(a_1x_1+a_2x_2+a_3)+\lambda(a_1\overset{(1)}{e}_{\alpha\alpha}+a_2\overset{(2)}{e}_{\alpha\alpha} + \gamma_{\alpha\alpha}) - bf,\\
&m_{\alpha\beta}=m_{33}=0, \ m_{\alpha 3}=\varepsilon_{3\alpha\rho}a_\rho(\beta+\gamma\nu)+a_1\overset{(1)}{m}_{\alpha 3}+a_2\overset{(2)}{m}_{\alpha 3}+\mu_{\alpha 3},\\
&m_{3\alpha} = \varepsilon_{3\alpha\rho}a_\rho(\gamma + \beta\nu) + a_1\overset{(1)}{m}_{3\alpha} + a_2\overset{(2)}{m}_{3\alpha} + \beta w_{,\alpha},
\end{aligned}
\tag{6.5.22}
$$

where E is defined in (6.4.25). It follows from (6.5.13) and (6.5.22) that the equations (6.5.2) reduce to

$$
s_{\beta\alpha,\beta} = 0, \quad \mu_{\alpha 3,\alpha} + \varepsilon_{3\alpha\beta}s_{\alpha\beta} = 0 \text{ on } \Sigma. \tag{6.5.23}
$$

With the help of (6.5.16) and (6.5.22), we find that the boundary conditions (6.5.3) take the form

$$
s_{\beta\alpha}n_\beta = 0, \quad \mu_{\alpha 3}n_\alpha = 0 \text{ on } L. \tag{6.5.24}
$$

Thus, we conclude that the functions $v_\alpha, w, \gamma_{\alpha\beta}, \zeta_{\alpha 3}, s_{\alpha\beta}$ and $\mu_{\alpha 3}$ satisfy the equations of equilibrium (6.5.23), the constitutive equations (6.5.21) and the geometrical equations (6.5.20) on Σ, and the boundary conditions (6.5.24) on L. The functions v_α and w are the displacements and microrotation in a thermoelastic plane strain problem corresponding to the temperature field f (cf. Section 6.3). In what follows we assume that this problem is solved, so that we shall treat $v_\alpha, w, \gamma_{\alpha\beta}, \zeta_{\alpha 3}, s_{\alpha\beta}$ and $\mu_{\alpha 3}$ as known functions.

The conditions (6.5.4) and (6.5.7) are satisfied on the basis of the relations (6.5.22). From (6.5.5) and (6.5.22) we obtain

$$a_3 = -a_1 x_1^* - a_2 x_2^* + \frac{1}{EA} P, \tag{6.5.25}$$

where A is the area of Σ and

$$P = \int_\Sigma F dA, F = bf - \lambda \gamma_{\alpha\alpha}, \quad x_\alpha^* = \frac{1}{EA} \int_\Sigma (E x_\alpha + \lambda e_{\rho\rho}^{(\alpha)}) dA. \tag{6.5.26}$$

Therefore, by (6.5.22) and (6.5.25),

$$t_{33} = a_1[E(x_1 - x_1^*) + \lambda e_{\alpha\alpha}^{(1)}] + a_2[E(x_2 - x_2^*) + \lambda e_{\alpha\alpha}^{(2)}] + \frac{1}{A} P - F. \tag{6.5.27}$$

In view of (6.5.22) and (6.5.27), the conditions (6.5.6) become

$$L_{\alpha\beta} a_\beta = M_\alpha, \tag{6.5.28}$$

where

$$\begin{aligned} L_{\alpha\xi} &= \int_\Sigma \{x_\alpha[E(x_\xi - x_\xi^*) + \lambda e_{\eta\eta}^{(\xi)}] + (\gamma + \beta\nu)\delta_{\alpha\xi} - \varepsilon_{\alpha\rho 3} m_{3\rho}^{(\xi)}\} dA, \\ M_\alpha &= \int_\Sigma [x_\alpha(F - \frac{1}{A} P) + \varepsilon_{3\alpha\beta} \mu_{3\beta}] dA. \end{aligned} \tag{6.5.29}$$

Let us prove that $\det(L_{\alpha\beta}) \neq 0$. In view of (6.5.25) the functions u_i and φ_i from (6.5.18) can be written in the form

$$u_\alpha = u_\alpha^0 + v_\alpha, u_3 = u_3^0, \quad \varphi_\alpha = \varphi_\alpha^0, \quad \varphi_3 = \varphi_3^0 + w,$$

where

$$u_i^0 = \sum_{\alpha=1}^2 a_\alpha w_i^{(\alpha)}, \quad \varphi_i^0 = \sum_{\alpha=1}^2 a_\alpha \psi_i^{(\alpha)}. \tag{6.5.30}$$

Clearly,

$$\begin{aligned} & w_1^{(1)} = -\frac{1}{2}[x_3^2 + \nu(x_1^2 - x_2^2)] + \nu x_1 x_1^* + u_1^{(1)}, w_1^{(2)} = \nu x_1(x_2^* - x_2) + u_1^{(2)}, \\ & w_2^{(1)} = \nu x_2(x_1^* - x_1) + u_2^{(1)}, w_2^{(2)} = -\frac{1}{2}[x_3^2 - \nu(x_1^2 - x_2^2)] + \nu x_2^* x_2 + u_2^{(2)}, \\ & w_3^{(\alpha)} = (x_\alpha - x_\alpha^*) x_3, \psi_1^{(1)} = \psi_1^{(3)} = 0, \psi_1^{(2)} = -\psi_2^{(1)} = x_3, \psi_2^{(2)} = \psi_2^{(3)} = 0, \\ & \psi_3^{(1)} = \varphi^{(1)} - x_2 \nu, \psi_3^{(2)} = \varphi^{(2)} + x_1 \nu. \end{aligned} \tag{6.5.31}$$

The components of the stress tensor and couple stress tensor corresponding to the displacements u_i^0 and microrotations φ_i^0 are given by

$$t_{ij}^0 = \sum_{\alpha=1}^{2} a_\alpha \sigma_{ij}^{(\alpha)}, \quad m_{ij}^{(0)} = \sum_{\alpha=1}^{2} a_\alpha \mu_{ij}^{(\alpha)}, \tag{6.5.32}$$

where

$$\begin{aligned} &\sigma_{\alpha\beta}^{(\rho)} = t_{\alpha\beta}^{(\rho)}, \quad \sigma_{\alpha 3}^{(\rho)} = \sigma_{3\alpha}^{(\rho)} = 0, \quad \sigma_{33}^{(\rho)} = E(x_\rho - x_\rho^*) + \lambda e_{\alpha\alpha}^{(\rho)}, \\ &\mu_{\alpha\beta}^{(\rho)} = \mu_{33}^{(\rho)} = 0, \quad \mu_{\alpha 3}^{(\rho)} = \varepsilon_{3\alpha\rho}(\beta + \gamma\nu) + m_{\alpha 3}^{(\rho)}, \\ &\mu_{3\alpha}^{(\rho)} = \varepsilon_{3\alpha\rho}(\gamma + \beta\nu) + m_{3\alpha}^{(\rho)}. \end{aligned} \tag{6.5.33}$$

Clearly, we have

$$t_{ji,j}^0 = 0, \quad m_{ji,j}^0 + \varepsilon_{irs} t_{rs}^0 = 0 \text{ on } B \tag{6.5.34}$$

and

$$t_{\alpha j}^0 n_\alpha = 0, \quad m_{\alpha j}^0 n_\alpha = 0 \text{ on } \Pi. \tag{6.5.35}$$

Moreover,

$$\begin{aligned} t_{ij}^0 &= \lambda e_{rr}^0 \delta_{ij} + (\mu + \kappa) e_{ij}^0 + \mu e_{ji}^0, \\ m_{ij}^0 &= \alpha \kappa_{rr}^0 \delta_{ij} + \beta \kappa_{ji}^0 + \gamma \kappa_{ij}^0, \end{aligned} \tag{6.5.36}$$

where

$$e_{ij}^0 = u_{j,i}^0 + \varepsilon_{jik} \varphi_k^0, \quad \kappa_{ij}^0 = \varphi_{j,i}^0. \tag{6.5.37}$$

By (6.5.30) and (6.5.37)

$$\begin{aligned} &e_{ij}^0 = \sum_{\alpha=1}^{2} a_\alpha \eta_{ij}^{(\alpha)}, \quad \kappa_{ij}^0 = \sum_{\alpha=1}^{2} a_\alpha \xi_{ij}^{(\alpha)}, \\ &\eta_{ij}^\alpha = w_{j,i}^{(\alpha)} + \varepsilon_{jik} \psi_k^{(\alpha)}, \quad \xi_{ij}^{(\alpha)} = \psi_{j,i}^{(\alpha)}. \end{aligned} \tag{6.5.38}$$

It follows from (6.5.34) and (6.5.35) that

$$2 \int_B W^0 dV = \int_{\Sigma^{(0)} \cup \Sigma^{(h)}} (t_{ji}^0 u_i^0 + m_{ji}^0 \varphi_i^0) n_j dA, \tag{6.5.39}$$

where W^0 is the potential energy

$$2W^0 = \lambda e_{rr}^0 e_{ss}^0 + (\mu + \kappa) e_{ij}^0 e_{ij}^0 + \mu e_{ij}^0 e_{ji}^0 + \alpha \kappa_{rr}^0 \kappa_{ss}^0 + \beta \kappa_{ij}^0 \kappa_{ji}^0 + \gamma \kappa_{ij}^0 \kappa_{ij}^0. \tag{6.5.40}$$

On the other hand, in the isothermal theory and in the absence of body loads, the reciprocity relation (6.1.6) reduces to

$$\int_{\partial B}(t'_{ji}u''_i + m'_{ji}u''_i)n_j dA = \int_{\partial B}(t''_{ji}u'_i + m''_{ji}u'_i)n_j dA, \qquad (6.5.41)$$

where t'_{ij} and m'_{ij} correspond to the displacements u'_i and microrotations φ'_i. If we apply the formula (6.5.41) for $u'_i = w_i^{(1)}, \varphi'_i = \psi_i^{(1)}$ and $u''_i = w_i^{(2)}, \varphi''_i = \psi_i^{(2)}$, and use (6.5.31)-(6.5.33) then we get

$$L_{12} = L_{21}. \qquad (6.5.42)$$

From (6.5.39), (6.5.31)-(6.5.33) we obtain

$$2\int_B W^0 dV = h\sum_{\alpha,\beta=1}^{2} L_{\alpha\beta}a_\alpha a_\beta. \qquad (6.5.43)$$

We assume that the potential energy is a positive definite quadratic form (see Theorem 6.1.2). From (6.5.43) we find that

$$\det(L_{\alpha\beta}) \neq 0, \qquad (6.5.44)$$

so that the system (6.5.28) uniquely determines the constants a_1 and a_2. Thus, the solution of the problem $P^{(0)}$ has the form (6.5.18) where $(u_\alpha^{(\rho)}, \varphi^{(\rho)})$ are the solutions of the plane strain problems $\mathcal{D}^{(\rho)}$, $(\rho = 1, 2), v_\alpha, w$ satisfy the thermoelastic plane strain problem (6.5.20), (6.5.21), (6.5.23), (6.5.24) corresponding to the temperature field f, and the constants a_i are given by (6.5.25) and (6.5.28).

b) **Recurrence process.** In what follows we denote by $u_i^*, \varphi_i^*, e_{ij}^*, t_{ij}^*, m_{ij}^*$, respectively, the components of the displacement vector, the components of the microrotation vector, the components of the strain tensor, the components of the stress tensor, the components of the couple-stress tensor from the problem $P^{(n)}$ and by $u_i, \varphi_i, e_{ij}, t_{ij}, m_{ij}$ the analogous functions from the problem $P^{(n+1)}$.

Let us establish the solution of the problem $P^{(n+1)}$ when the solution of the problem $P^{(n)}$ is known. As the solution of the problem $P^{(n)}$ is known for arbitrary $T_n(x_1, x_2)$ we know the solution of the problem $P^{(n)}$ for the temperature distribution $\theta = T_{n+1}(x_1, x_2)x_3^n$. Thus the problem can be presented as follows: to find the functions $u_i, \varphi_i, e_{ij}, \kappa_{ij}, t_{ij}$ and m_{ij} which satisfy the equations (6.5.2), (6.2.3), (6.1.4) and the conditions (6.5.3)-(6.5.7), when the temperature has the form

$$\theta = g(x_1, x_2)x_3^{n+1}, \qquad (6.5.45)$$

assuming that the functions $u_i^*, \varphi_i^*, e_{ij}^*, \kappa_{ij}^*, t_{ij}^*$ and m_{ij}^* which satisfy the equations (6.5.2), (6.2.3), (6.1.4) and the conditions (6.5.3)-(6.5.7), for the temperature distribution

$$\theta = g(x_1, x_2)x_3^n, \tag{6.5.46}$$

are known. We seek the solution of this problem in the form

$$\begin{aligned}
u_1 &= (n+1)\{\int_0^{x_3} u_1^* dx_3 - \frac{1}{2}a_1[x_3^2 + \nu(x_1^2 - x_2^2)] - a_2\nu x_1 x_2 - \\
&\quad - a_3\nu x_1 - \tau x_2 x_3 + a_1 u_1^{(1)} + a_2 u_1^{(2)} + v_1(x_1, x_2)\}, \\
u_2 &= (n+1)\{\int_0^{x_3} u_2^* dx_3 - a_1\nu x_1 x_2 - \frac{1}{2}a_2[x_3^2 - \nu(x_1^2 - x_2^2)] - \\
&\quad - a_3\nu x_2 + \tau x_1 x_3 + a_1 u_2^{(1)} + a_2 u_2^{(2)} + v_2(x_1, x_2)\}, \\
u_3 &= (n+1)\{\int_0^{x_3} u_3^* dx_3 + (a_1 x_1 + a_2 x_2 + a_3)x_3 + F(x_1, x_2)\}, \\
\varphi_1 &= (n+1)\{\int_0^{x_3} \varphi_1^* dx_3 + a_2 x_3 + G_1(x_1, x_2)\}, \\
\varphi_2 &= (n+1)\{\int_0^{x_3} \varphi_2^* dx_3 - a_1 x_3 + G_2(x_1, x_2)\}, \\
\varphi_3 &= (n+1)\{\int_0^{x_3} \varphi_3^* dx_3 + (a_2 x_1 - a_1 x_2)\nu + \tau x_3 + \\
&\quad + a_1\varphi^{(1)} + a_2\varphi^{(2)} + w(x_1, x_2)\},
\end{aligned} \tag{6.5.47}$$

where $(u_\alpha^{(\rho)}, \varphi^{(\rho)})$ are the solutions of the problems $\mathcal{D}^{(\rho)}, (\rho = 1, 2), v_\alpha$, w, F and G_α are unknown functions, and a_i and τ are unknown constants.

From (6.1.4) and (6.5.47) we obtain

$$\begin{aligned}
e_{\alpha\beta} &= (n+1)\{\int_0^{x_3} e_{\alpha\beta}^* dx_3 - \nu(a_1 x_1 + a_2 x_2 + a_3)\delta_{\alpha\beta} + a_1 e_{\alpha\beta}^{(1)} + a_2 e_{\alpha\beta}^{(2)} + \gamma_{\alpha\beta}\}, \\
e_{33} &= (n+1)\{\int_0^{x_3} e_{33}^* dx_3 + a_1 x_1 + a_2 x_2 + a_3 + u_3^*(x_1, x_2, 0)\}, \\
e_{\alpha 3} &= (n+1)\{\int_0^{x_3} e_{\alpha 3}^* dx_3 + F_{,\alpha} + \varepsilon_{\alpha\beta 3} G_\beta\}, \\
e_{3\alpha} &= (n+1)\{\int_0^{x_3} e_{3\alpha}^* dx_3 + \varepsilon_{\beta\alpha 3}(\tau x_\beta + G_\beta) + u_\alpha^*(x_1, x_2, 0)\},
\end{aligned} \tag{6.5.48}$$

where

$$\gamma_{\alpha\beta} = v_{\beta,\alpha} + \varepsilon_{\beta\alpha 3} w. \tag{6.5.49}$$

If we introduce the notations

$$\begin{aligned} \sigma_{\alpha\beta} &= \lambda \gamma_{\rho\rho} \delta_{\alpha\beta} + (\mu + \kappa) \gamma_{\alpha\beta} + \mu \gamma_{\beta\alpha}, \\ \mu_{\alpha 3} &= \gamma w_{,\alpha}, \quad \mu_{3\alpha} = \beta w_{,\alpha}, \end{aligned} \tag{6.5.50}$$

then from (6.2.3), (6.5.45)-(6.5.50) we obtain

$$\begin{aligned}
t_{\alpha\beta} &= (n+1)\{\int_0^{x_3} t^*_{\alpha\beta} dx_3 + a_1 \sigma^{(1)}_{\alpha\beta} + a_2 \sigma^{(2)}_{\alpha\beta} + \sigma_{\alpha\beta} + \lambda \delta_{\alpha\beta} u^*_3(x_1, x_2, 0)\}, \\
t_{33} &= (n+1)\{\int_0^{x_3} t^*_{33} dx_3 + E(a_1 x_1 + a_2 x_2 + a_3) + \\
&\quad + \lambda(a_1 e^{(1)}_{\alpha\alpha} + a_2 e^{(2)}_{\alpha\alpha} + \gamma_{\alpha\alpha}) + (\lambda + 2\mu + \kappa) u^*_3(x_1, x_2, 0)\}, \\
t_{\alpha 3} &= (n+1)\{\int_0^{x_3} t^*_{\alpha 3} dx_3 + (\mu + \kappa) F_{,\alpha} + \\
&\quad + \varepsilon_{\alpha\beta 3}(\kappa G_\beta - \mu \tau x_\beta) + \mu u^*_\alpha(x_1, x_2, 0)\}, \\
t_{3\alpha} &= (n+1)\{\int_0^{x_3} t^*_{3\alpha} dx_3 + \mu F_{,\alpha} + \varepsilon_{\beta\alpha 3}[\kappa G_\beta + (\mu + \kappa) \tau x_\beta] + \\
&\quad + (\mu + \kappa) u^*_\alpha(x_1, x_2, 0)\}, \\
m_{\eta\nu} &= (n+1)\{\int_0^{x_3} m^*_{\eta\nu} dx_3 + \alpha G_{\rho,\rho} \delta_{\eta\nu} + \beta G_{\eta,\nu} + \gamma G_{\nu,\eta} + \\
&\quad + \alpha[\tau + \varphi^*_3(x_1, x_2, 0)] \delta_{\nu\eta}\}, \\
m_{33} &= (n+1)\{\int_0^{x_3} m^*_{33} dx_3 + \alpha G_{\rho,\rho} + (\alpha + \beta + \gamma)[\tau + \varphi^*_3(x_1, x_2, 0)]\}, \\
m_{\alpha 3} &= (n+1)\{\int_0^{x_3} m^*_{\alpha 3} dx_3 + \varepsilon_{\alpha\rho 3} a_\rho (\beta + \gamma\nu) + \\
&\quad + a_1 \mu^{(1)}_{\alpha 3} + a_2 \mu^{(2)}_{\alpha 3} + \mu_{\alpha 3} + \beta \varphi^*_\alpha(x_1, x_2, 0)\}, \\
m_{3\alpha} &= (n+1)\{\int_0^{x_3} m^*_{3\alpha} dx_3 + \varepsilon_{\alpha\rho 3} a_\rho (\gamma + \beta\nu) + \\
&\quad + a_1 \mu^{(1)}_{3\alpha} + a_2 \mu^{(2)}_{3\alpha} + \mu_{3\alpha} + \gamma \varphi^*_\alpha(x_1, x_2, 0)\}.
\end{aligned} \tag{6.5.51}$$

Let (F, F_1, F_2) be an ordered triplet of functions F, F_1, F_2 of class $C^2(\Sigma) \cap C^1(\overline{\Sigma})$. We introduce the operators

$$\begin{aligned}
&\mathcal{L}(F, F_1, F_2) = (\mu + \kappa)\Delta F - \kappa(F_{1,2} - F_{2,1}),\\
&\mathcal{L}_\rho(F, F_1, F_2) = \gamma\Delta F_\rho + (\alpha + \beta)F_{\eta,\eta\rho} + \kappa\varepsilon_{\rho\eta 3}F_{,\eta} - 2\kappa F_\rho,\\
&\Lambda(F, F_1, F_2) = (\mu + \kappa)F_{,\alpha}n_\alpha - \kappa(F_1 n_2 - F_2 n_1),\\
&\Lambda_1(F, F_1, F_2) = [(\alpha+\beta+\gamma)F_{1,1} + \alpha F_{2,2}]n_1 + [\gamma F_{1,2} + \beta F_{2,1}]n_2,\\
&\Lambda_2(F, F_1, F_2) = [\beta F_{1,2} + \gamma F_{2,1}]n_1 + [\alpha F_{1,1} + (\alpha+\beta+\gamma)F_{2,2}]n_2.
\end{aligned} \tag{6.5.52}$$

In view of (6.5.13), (6.5.51) and (6.5.52) the equilibrium equations (6.5.2) reduce to

$$\sigma_{\alpha\beta,\alpha} + P_\beta = 0, \quad \mu_{\alpha 3,\alpha} + \varepsilon_{\alpha\beta 3}\sigma_{\alpha\beta} + L_3 = 0, \tag{6.5.53}$$

$$\mathcal{L}(F, G_1, G_2) = H, \quad \mathcal{L}_\rho(F, G_1, G_2) = H_\rho, \tag{6.5.54}$$

where

$$\begin{aligned}
&P_\beta = \lambda u^*_{3,\beta}(x_1, x_2, 0) + t^*_{3\beta}(x_1, x_2, 0),\\
&L_3 = \beta\varphi^*_{\rho,\rho}(x_1, x_2, 0) + m^*_{33}(x_1, x_2, 0),\\
&H = -\mu u^*_{\alpha,\alpha}(x_1, x_2, 0) - t^*_{33}(x_1, x_2, 0),\\
&H_\rho = -\alpha\varphi^*_{3,\rho}(x_1, x_2, 0) - m^*_{3\rho}(x_1, x_2, 0) +\\
&\qquad + \kappa\varepsilon_{\rho\eta 3}u^*_\eta(x_1, x_2, 0), \ (x_1, x_2) \in \Sigma.
\end{aligned} \tag{6.5.55}$$

By using (6.5.16), (6.5.51) and (6.5.52), the boundary conditions (6.5.3) become

$$\sigma_{\alpha\beta}n_\alpha = p_\beta, \quad \mu_{\alpha 3}n_\alpha = q \ \text{ on } L, \tag{6.5.56}$$

$$\Lambda(F, G_1, G_2) = Q, \quad \Lambda_\alpha(F, G_1, G_2) = Q_\alpha \text{ on } L, \tag{6.5.57}$$

where

$$\begin{aligned}
&p_\beta = -\lambda n_\beta u^*_3(x_1, x_2, 0), \quad q = -\beta n_\alpha\varphi^*_\alpha(x_1, x_2, 0),\\
&Q = \mu\tau(x_2 n_1 - x_1 n_2) - \mu n_\alpha u^*_\alpha(x_1, x_2, 0),\\
&Q_\rho = -\alpha[\tau + \varphi^*_3(x_1, x_2, 0)]n_\rho, \quad (x_1, x_2) \in L.
\end{aligned} \tag{6.5.58}$$

From (6.5.49), (6.5.50), (6.5.53) and (6.5.56) it follows that the functions $v_\alpha, w, \gamma_{\alpha\beta}, \sigma_{\alpha\beta}, \mu_{\alpha 3}$ satisfy the equations of an isothermal plane strain problem. Let us show that this problem has a solution. The necessary and sufficient conditions for the existence of a solution are

$$\begin{aligned}
&\int_L p_\alpha ds + \int_\Sigma P_\alpha dA = 0,\\
&\int_L (x_1 p_2 - x_2 p_1 + q)ds + \int_\Sigma (x_1 P_2 - x_2 P_1 + L_3)dA = 0.
\end{aligned} \tag{6.5.59}$$

By (6.5.55), (6.5.58) and the divergence theorem, we obtain

$$\int_L p_\alpha ds + \int_\Sigma P_\alpha dA = \int_\Sigma t^*_{3\alpha}(x_1, x_2, 0)dA,$$

$$\int_L (x_1 p_2 - x_2 p_1 + q)ds + \int_\Sigma (x_1 P_2 - x_2 P_1 + L_3)dA =$$

$$= \int_\Sigma [x_1 t^*_{32} - x_2 t^*_{31} + m^*_{33}]_{x_3=0} dA,$$

so that the conditions (6.5.59) are satisfied because the functions t^*_{ij} and m^*_{ij} satisfy the conditions (6.5.4), (6.5.7). In what follows we assume that $v_\alpha, w, \gamma_{\alpha\beta}, \sigma_{\alpha\beta}, \mu_{\alpha 3}$ are known.

The functions F, G_α satisfy the equations (6.5.54) and the boundary conditions (6.5.57). Let us study the existence of the solution of this problem. We denote by F', G'_α a particular solution of the equations (6.5.54) and write

$$F = F^0 + F', \quad G_\alpha = G^0_\alpha + G'_\alpha.$$

The functions F^0, G^0_α satisfy the homogeneous system (6.5.54) and the boundary conditions

$$\Lambda(F^0, G^0_1, G^0_2) = Q^0, \quad \Lambda_\rho(F^0, G^0_1, G^0_2) = Q^0_\rho \text{ on } L,$$

where

$$Q^0 = Q - \Lambda(F', G'_1, G'_2), \quad Q^0_\rho = Q_\rho - \Lambda_\rho(F', G'_1, G'_2). \tag{6.5.60}$$

The necessary and sufficient condition for the existence of the functions F^0 and G^0_α is [163], [181]

$$\int_L Q^0 ds = 0. \tag{6.5.61}$$

By (6.5.52), (6.5.55), (6.5.60) and the divergence theorem we obtain

$$\int_L Q^0 ds = \int_L Q ds - \int_\Sigma H dA = \int_\Sigma t^*_{33}(x_1, x_2, 0)dA,$$

so that the condition (6.5.61) is satisfied because t^*_{33} satisfies the condition (6.5.5).

Let us determine the constant τ. We introduce the functions χ and χ_α on Σ by

$$F = \tau\varphi - \chi, \quad G_\alpha = \tau\psi_\alpha - \frac{1}{2}\tau x_\alpha - \chi_\alpha, \tag{6.5.62}$$

where φ and ψ_α are the torsion functions [163]. The torsion functions satisfy the equations

$$\mathcal{L}(\varphi,\psi_1,\psi_2)=0,\quad \mathcal{L}_\rho(\varphi,\psi_1,\psi_2)=0 \text{ on } \Sigma \tag{6.5.63}$$

and the boundary conditions

$$\begin{aligned}&\Lambda(\varphi,\psi_1,\psi_2)=\tfrac{1}{2}(2\mu+\kappa)(x_2n_1-x_1n_2),\\&\Lambda_\rho(\varphi,\psi_1,\psi_2)=\tfrac{1}{2}(\beta+\gamma)n_\rho \text{ on } L.\end{aligned} \tag{6.5.64}$$

From (6.5.54), (6.5.57), (6.5.62)-(6.5.64) it follows that the functions χ and χ_α satisfy the following equations

$$\mathcal{L}(\chi,\chi_1,\chi_2)=-H,\quad \mathcal{L}_\rho(\chi,\chi_1,\chi_2)=-H_\rho \text{ on } \Sigma \tag{6.5.65}$$

and the boundary conditions

$$\begin{aligned}&\Lambda(\chi,\chi_1,\chi_2)=\mu n_\alpha u_\alpha^*(x_1,x_2,0),\\&\Lambda_\rho(\chi,\chi_1,\chi_2)=\alpha n_\rho\varphi_3^*(x_1,x_2,0),(x_1,x_2)\in L.\end{aligned} \tag{6.5.66}$$

Obviously, the necessary and sufficient condition for the existence of the functions χ and χ_α is satisfied. From (6.5.51) and (6.5.62) we obtain

$$\begin{aligned}t_{32}=&(n+1)\{\int_0^{x_3}t_{32}^*dx_3+\tau[\mu\varphi_{,2}+\kappa\psi_1+\frac{1}{2}(2\mu+\kappa)x_1]-\\&-\mu\chi_{,2}-\kappa\chi_1+(\mu+\kappa)u_2^*(x_1,x_2,0)\},\\t_{31}=&(n+1)\{\int_0^{x_3}t_{31}^*dx_3+\tau[\mu\varphi_{,1}-\kappa\psi_2-\frac{1}{2}(2\mu+\kappa)x_2]-\\&-\mu\chi_{,1}+\kappa\chi_2+(\mu+\kappa)u_1^*(x_1,x_2,0)\},\\m_{33}=&(n+1)\{\int_0^{x_3}m_{33}^*dx_3+\tau(\alpha\psi_{\rho,\rho}+\beta+\gamma)-\alpha\chi_{\rho,\rho}+\\&+(\alpha+\beta+\gamma)\varphi_3^*(x_1,x_2,0)\}.\end{aligned} \tag{6.5.67}$$

By (6.5.7) and (6.5.67) we get

$$\tau D=M, \tag{6.5.68}$$

where D is the micropolar torsional rigidity [163]

$$\begin{aligned}D=&\int_\Sigma[\mu(x_1\varphi_{,2}-x_2\varphi_{,1})+\kappa(x_1\psi_1+x_2\psi_2)+\\&+\frac{1}{2}(2\mu+\kappa)(x_1^2+x_2^2)+\alpha(\psi_{1,1}+\psi_{2,2})+\beta+\gamma]dA\end{aligned} \tag{6.5.69}$$

and we have used the notation

$$M=\int_\Sigma \{\mu(x_1\chi_{,2}-x_2\chi_{,1})+\kappa(x_2\chi_2+x_1\chi_1)+(\mu+\kappa)[x_2u_1^*(x_1,x_2,0)-$$
$$-x_1u_2^*(x_1,x_2,0)]+\alpha\chi_{\rho\rho}-(\alpha+\beta+\gamma)\varphi_3^*(x_1,x_2,0)\}dA.$$

It is known that $D>0$, so that from (6.5.68) we can determine the constant τ. Let us determine the constants a_i. From (6.5.47) and (6.5.5) we obtain

$$a_3=-a_1x_1^*-a_2x_2^*+\frac{1}{EA}R, \tag{6.5.70}$$

where

$$R=\int_\Sigma KdA,\quad K=-\lambda\gamma_{\alpha\alpha}-(\lambda+2\mu+\kappa)u_3^*(x_1,x_2,0)$$

and x_α^* are given by (6.5.26). It follows from (6.5.51) and (6.5.70) that

$$\begin{aligned} t_{33}=(n+1)\{&\int_0^{x_3}t_{33}^*dx_3+a_1[E(x_1-x_1^*)+\lambda e_{\alpha\alpha}^{(1)}]+\\ &+a_2[E(x_2-x_2^*)+\lambda e_{\alpha\alpha}^{(2)}]+\frac{1}{S}R-K\}. \end{aligned} \tag{6.5.71}$$

Taking into account (6.5.71), the conditions (6.5.6) become

$$L_{\alpha\beta}a_\beta=N_\alpha, \tag{6.5.72}$$

where $L_{\alpha\beta}$ are given by (6.5.29) and

$$N_\alpha=\int_\Sigma\{\varepsilon_{\alpha\beta3}[\mu_{3\beta}+\gamma\varphi_\beta^*(x_1,x_2,0)]+x_\alpha(K-\frac{1}{A}R)\}dA.$$

From (6.5.72) we can determine the constants a_α. The constant a_3 is given by (6.5.70).

The conditions (6.5.4) are identically satisfied on the basis of the equilibrium equations and the boundary conditions. For example, for the first of them we have

$$\int_\Sigma t_{31}dA=\int_\Sigma(t_{13}-m_{i2,i})dA=\int_\Sigma(t_{13}-m_{i2,i}+x_1t_{i3,i})dA=$$
$$=\int_\Sigma[x_1t_{33,3}-m_{32,3}+(x_1t_{\alpha3})_{,\alpha}-m_{\alpha2,\alpha}]dA=$$
$$=(n+1)\int_\Sigma(x_1t_{33}^*-m_{32}^*)dA+\int_L(x_1t_{\alpha3}n_\alpha-m_{\alpha2}n_\alpha)ds=0.$$

Thus, the considered problem is solved. The above results have been established in [167].

c) **Application**. Let us use the above results to study the thermoelastic deformation of a circular cylinder which is subjected to a prescribed temperature field. We assume that the generic cross-section of the cylinder is $\Sigma = \{x : x_1^2 + x_2^2 < a^2, x_3 = s\}, (a > 0, s = const.)$, and that the temperature field has the form $\theta = f(r)$, where $r = (x_1^2 + x_2^2)^{1/2}$. The solution of this problem can be expressed by (6.5.18) where $(u_\alpha^{(\rho)}, \varphi^{(\rho)})$ is the solution of the problem $\mathcal{D}^{(\rho)}, (\rho = 1, 2), v_\alpha$ and w satisfy the problem (6.5.20), (6.5.21), (6.5.23), (6.5.4) and the constants a_i are given by (6.5.25), (6.5.28).

The solutions of the problems $\mathcal{D}^{(\rho)}, (\rho = 1, 2)$, for a circular domain have been established by Krishna Reddy and Venkatasubramanian [226]. In what follows we shall use the polar coordinates and the relations of Section 6.3(a). To determine the solutions of the problems $\mathcal{D}^{(\rho)}$ we study the isothermal plane strain problem $\mathcal{A}$ characterized by the external data system

$$f_\alpha = 0, \quad g_3 = 0, \quad \theta = 0, \widetilde{t}_\alpha = 0, \quad \widetilde{m}_3 = s_1 n_1 + s_2 n_2, \tag{6.5.73}$$

where s_1 and s_2 are given constants. If we take

$$s_1 = 0, \quad s_2 = \beta + \gamma\nu, \tag{6.5.74}$$

then the problem $\mathcal{A}$ reduces to the problem $\mathcal{D}^{(1)}$. The problem $\mathcal{D}^{(2)}$ is obtained from $\mathcal{A}$ if we take

$$s_1 = -(\beta + \gamma\nu_1), s_2 = 0. \tag{6.5.75}$$

The problem $\mathcal{A}$ consists in the finding of the functions u_r, u_ϑ and φ on Σ which satisfy the equations (6.3.9), (6.3.10), (6.3.11), where $f_r = f_\vartheta = 0, g_3 = 0, \theta = 0$, and the boundary conditions

$$t_{rr} = 0, \quad t_{r\vartheta} = 0, \quad m_{rz} = s_1 \cos\vartheta + s_2 \sin\vartheta \;\text{ for } r = a. \tag{6.5.76}$$

We seek the solution of this problem in the form

$$u_r = u^{(1)}(r) \cos\vartheta + u^{(2)}(r) \sin\vartheta, \quad u_\vartheta = v^{(1)}(r) \cos\vartheta + v^{(2)}(r) \sin\vartheta,$$
$$\varphi = \psi^{(1)}(r) \cos\vartheta + \psi^{(2)}(r) \sin\vartheta, \tag{6.5.77}$$

where $u^{(\alpha)}, v^{(\alpha)}$ and $\varphi^{(\alpha)}$ are functions only on r. It follows from (6.3.11), (6.3.10) and (6.5.77) that

$$\begin{aligned}
t_{rr} &= t_{rr}^{(1)} \cos\vartheta + t_{rr}^{(2)} \sin\vartheta, \quad t_{\vartheta\vartheta} = t_{\vartheta\vartheta}^{(1)} \cos\vartheta + t_{\vartheta\vartheta}^{(2)} \sin\vartheta, \\
t_{r\vartheta} &= t_{r\vartheta}^{(1)} \cos\vartheta + t_{r\vartheta}^{(2)} \sin\vartheta, \quad t_{\vartheta r} = t_{\vartheta r}^{(1)} \cos\vartheta + t_{\vartheta r}^{(2)} \sin\vartheta, \\
m_{rz} &= m_{rz}^{(1)} \cos\vartheta + m_{rz}^{(2)} \sin\vartheta, \quad m_{\vartheta z} = m_{\vartheta z}^{(1)} \cos\vartheta + m_{\vartheta z}^{(2)} \sin\vartheta,
\end{aligned} \tag{6.5.78}$$

where

$$\begin{aligned}
t_{rr}^{(1)} &= (\lambda + 2\mu + \kappa)\frac{du^{(1)}}{dr} + \lambda(u^{(1)} + v^{(2)})r^{-1}, \\
t_{rr}^{(2)} &= (\lambda + 2\mu + \kappa)\frac{du^{(2)}}{dr} + \lambda(u^{(2)} - v^{(1)})r^{-1}, \\
t_{\vartheta\vartheta}^{(1)} &= (\lambda + 2\mu + \kappa)(u^{(1)} + v^{(2)})r^{-1} + \lambda\frac{du^{(1)}}{dr}, \\
t_{\vartheta\vartheta}^{(2)} &= (\lambda + 2\mu + \kappa)(u^{(2)} - v^{(1)})r^{-1} + \lambda\frac{du^{(2)}}{dr}, \\
t_{r\vartheta}^{(1)} &= (\mu + \kappa)\frac{dv^{(1)}}{dr} + \mu(u^{(2)} - v^{(1)})r^{-1} - \kappa\psi^{(1)}, \\
t_{r\vartheta}^{(2)} &= (\mu + \kappa)\frac{dv^{(2)}}{dr} - \mu(u^{(1)} + v^{(2)})r^{-1} - \kappa\psi^{(2)}, \\
t_{\vartheta r}^{(1)} &= \mu\frac{dv^{(1)}}{dr} + (\mu + \kappa)(u^{(2)} - v^{(1)})r^{-1} + \kappa\psi^{(1)}, \\
t_{\vartheta r}^{(2)} &= \mu\frac{dv^{(2)}}{dr} - (\mu + \kappa)(u^{(1)} + v^{(2)})r^{-1} + \kappa\psi^{(2)}, \\
m_{rz}^{(1)} &= \gamma\frac{d\psi^{(1)}}{dr}, \; m_{rz}^{(2)} = \gamma\frac{d\psi^{(2)}}{dr}, \; m_{\vartheta z}^{(1)} = \gamma r^{-1}\psi^{(2)}, \\
m_{\vartheta z}^{(2)} &= -\gamma r^{-1}\psi^{(1)}.
\end{aligned} \tag{6.5.79}$$

The equilibrium equations (6.3.9) reduce to

$$\begin{aligned}
&\frac{dt_{rr}^{(1)}}{dr} + (t_{\vartheta r}^{(2)} + t_{rr}^{(1)} - t_{\vartheta\vartheta}^{(1)})r^{-1} = 0, \\
&\frac{dt_{r\vartheta}^{(1)}}{dr} + (t_{\vartheta\vartheta}^{(2)} + t_{r\vartheta}^{(1)} + t_{\vartheta r}^{(1)})r^{-1} = 0, \\
&\frac{dm_{rz}^{(1)}}{dr} + (m_{\vartheta z}^{(2)} + m_{rz}^{(1)})r^{-1} + t_{r\vartheta}^{(1)} - t_{\vartheta r}^{(1)} = 0, \\
&\frac{dt_{rr}^{(2)}}{dr} + (t_{rr}^{(2)} - t_{\vartheta\vartheta}^{(2)} - t_{\vartheta r}^{(1)})r^{-1} = 0,
\end{aligned} \tag{6.5.80}$$

$$\frac{dt_{r\vartheta}^{(2)}}{dr} + (t_{r\vartheta}^{(2)} + t_{\vartheta r}^{(2)} - t_{\vartheta\vartheta}^{(1)})r^{-1} = 0,$$

$$\frac{dm_{rz}^{(2)}}{dr} + (m_{rz}^{(2)} - m_{\vartheta z}^{(1)})r^{-1} + t_{r\vartheta}^{(2)} - t_{\vartheta r}^{(2)} = 0.$$

Substituting (6.5.79) into (6.5.80) we obtain the equations

$$\begin{aligned}
&r^2\frac{d^2u^{(1)}}{dr^2} + r\frac{du^{(1)}}{dr} - (1+d_1)u^{(1)} + (1-d_1)r\frac{dv^{(2)}}{dr} - \\
&\quad -(1+d_1)v^{(2)} = -d_2 r\psi^{(2)},\\
&d_1(r^2\frac{d^2v^{(2)}}{dr^2} + r\frac{dv^{(2)}}{dr}) - (1+d_1)v^{(2)} - (1-d_1)r\frac{du^{(1)}}{dr} - \\
&\quad -(1+d_1)u^{(1)} = d_2 r^2\frac{d\psi^{(2)}}{dr},\\
&r^2\frac{d^2\psi^{(2)}}{dr^2} + r\frac{d\psi^{(2)}}{dr} - (1+2d_3r^2)\psi^{(2)} = -\, d_3 r(r\frac{dv^{(2)}}{dr} + v^{(2)}) - \\
&\quad -d_3 r u^{(1)}
\end{aligned} \tag{6.5.81}$$

and

$$\begin{aligned}
&d_1(r^2\frac{d^2v^{(1)}}{dr^2} + r\frac{dv^{(1)}}{dr}) - (1+d_1)v^{(1)} + r(1-d_1)\frac{du^{(2)}}{dr} + \\
&\quad +(1+d_1)u^{(2)} = d_2 r^2\frac{d\psi^{(1)}}{dr},\\
&r^2\frac{d^2\psi^{(1)}}{dr^2} + r\frac{d\psi^{(1)}}{dr} - (1+2d_3r^2)\psi^{(1)} = -\, d_3 r(r\frac{dv^{(1)}}{dr} + v^{(1)}) + d_3 r u^{(2)},\\
&r^2\frac{d^2u^{(2)}}{dr^2} + r\frac{du^{(2)}}{dr} - (1+d_1)u^{(2)} - (1-d_1)r\frac{dv^{(1)}}{dr} + \\
&\quad +(1+d_1)v^{(1)} = d_2 r\psi^{(1)},
\end{aligned} \tag{6.5.82}$$

where d_j are defined in (6.4.6).

Let us study the system (6.5.81). If we introduce the notations

$$r = e^t, \quad Y = \frac{d}{dt},$$

then the first two equations from (6.5.81) become

$$\begin{aligned}
&[Y^2 - (1+d_1)]u^{(1)} + [(1-d_1)Y - (1+d_1)]v^{(2)} = -\, d_2 e^t\psi^{(2)},\\
&\left[d_1Y^2 - (1+d_1)\right] v^{(2)} + [(d_1-1)Y - (1+d_1)]u^{(1)} = d_2 e^t Y\psi^{(2)}.
\end{aligned} \tag{6.5.83}$$

The general solution which correspond to a non-rigid displacement is

$$
\begin{aligned}
u^{(1)}&=A_1t+A_2e^{2t}+A_3e^{-2t}+\frac{1}{2d_1}d_2\int^t(e^{3s-2t}-e^s)\psi^{(2)}(s)ds,\\
v^{(2)}&=\frac{d_1-1}{d_1+1}A_1-A_1t-\frac{3-d_1}{1-3d_1}A_2e^{2t}+A_3e^{-2t}+\\
&+\frac{1}{2d_1}d_2\int^t(e^{3s-2t}+e^s)\psi^{(2)}(s)ds,
\end{aligned}
\tag{6.5.84}
$$

where A_i are arbitrary constants. The functions $u^{(1)}$ and $v^{(2)}$ must be finite for $r=0$ so that from (6.5.84) we obtain

$$
\begin{aligned}
u^{(1)}&=A_2r^2-\frac{1}{2d_1}d_2[\int_0^r\psi^{(2)}(x)dx-r^{-2}\int_0^r x^2\psi^{(2)}(x)dx],\\
v^{(2)}&=\frac{d_1-3}{1-3d_1}A_2r^2+\frac{1}{2d_1}d_2[\int_0^r\psi^{(2)}(x)dx+r^2\int_0^r x^2\psi^{(2)}(x)dx].
\end{aligned}
\tag{6.5.85}
$$

If we substitute $u^{(1)}$ and $v^{(2)}$ from (6.5.85) into $(6.5.81)_3$ then we find the equation

$$
r^2\frac{d^2\psi^{(2)}}{dr^2}+r\frac{d\psi^{(2)}}{dr}-(1+\delta^2r^2)\psi^{(2)}=\frac{8r^2}{1-3d_1}d_3A_2. \tag{6.5.86}
$$

The solution of (6.5.86), which is finite for $r=0$, has the form

$$
\psi^{(2)}=A_4I_1(\delta r)-\frac{8(\mu+\kappa)r}{(2\mu+\kappa)(1-3d_1)}A_2, \tag{6.5.87}
$$

where δ is given by (6.2.17) and A_4 is an arbitrary constant. In view of (6.5.87), from (6.5.85) we obtain

$$
\begin{aligned}
u^{(1)}&=Q_1A_2r^2+\frac{1}{2d_1\delta}d_2A_4[I_2(\delta r)-I_0(\delta d)],\\
v^{(2)}&=-Q_2A_2r^2+\frac{1}{2\delta_1\delta}d_2A_4[I_2(\delta r)+I_0(\delta r)],
\end{aligned}
\tag{6.5.88}
$$

where

$$
Q_1=1+\frac{\kappa}{(2\mu+\kappa)(1-3d_1)},\quad Q_2=\frac{1}{1-3d_1}(3-d_1+\frac{3\kappa}{2\mu+\kappa}). \tag{6.5.89}
$$

The solution of the system (6.5.82) can be determined in a similar way. Thus, we get

$$\begin{aligned} u^{(2)} &= Q_1 B_2 r^2 + \frac{1}{2d_1\delta} d_2 B_4 [I_0(\delta r) - I_2(\delta r)], \\ v^{(1)} &= Q_2 B_2 r^2 + \frac{1}{2d_1\delta} d_2 B_4 [I_0(\delta r) + I_2(\delta r)], \\ \psi^{(1)} &= B_4 I_1(\delta r) + \frac{8(\mu+\kappa)r}{(2\mu+\kappa)(1-3d_1)} B_2, \end{aligned} \qquad (6.5.90)$$

where B_2 and B_4 are arbitrary constants. It follows from (6.5.79), (6.5.87), (6.5.88) and (6.5.90) that

$$\begin{aligned} &t_{rr}^{(1)} = N_1 A_2 r - \delta\gamma r^{-1} A_4 I_2(\delta r), \quad t_{rr}^{(2)} = N_1 B_2 r + \delta\gamma r^{-1} B_4 I_2(\delta r), \\ &t_{r\vartheta}^{(1)} = N_2 B_2 r - \delta\gamma r^{-1} B_4 I_2(\delta r), \quad t_{r\vartheta}^{(2)} = -N_2 A_2 r - \delta\gamma r^{-1} A_4 I_2(\delta r), \\ &m_{rz}^{(1)} = \delta\gamma B_4 I_1'(\delta r) + \gamma Q_3 B_2, \quad m_{rz}^{(2)} = \delta\gamma A_4 I_1'(\delta r) - \gamma Q_3 A_2, \end{aligned} \qquad (6.5.91)$$

where

$$\begin{aligned} &N_1 = (3\lambda+4\mu+2\kappa)Q_1 - \lambda Q_2, \; N_2 = \mu Q_1 + (\mu+2\kappa)Q_2 - \kappa Q_3, \\ &Q_3 = 8(\mu+\kappa)[(2\mu+\kappa)(1-3d_1)]^{-1}. \end{aligned} \qquad (6.5.92)$$

We note that

$$t_{rr}^{(1)} = t_{r\vartheta}^{(2)}, \quad t_{rr}^{(2)} = -t_{r\vartheta}^{(1)}. \qquad (6.5.93)$$

The problem $\mathcal{D}^{(1)}$ is characterized by the boundary conditions

$$t_{rr} = 0, \quad t_{r\vartheta} = 0, \quad m_{rz} = (\beta + \gamma\nu)\sin\vartheta \quad \text{for } r = a. \qquad (6.5.94)$$

From (6.5.78), (6.5.91) and (6.5.93) we conclude that the conditions (6.5.94) are satisfied if and only if

$$A_2 = Z_2, \quad A_4 = Z_1, \quad B_2 = 0, \quad B_4 = 0, \qquad (6.5.95)$$

where

$$Z_1 = \frac{N_1 a^2 (\beta + \gamma\nu)}{\delta\gamma\, [N_1 a^2 I_1'(\delta a) - \gamma Q_3 I_2(\delta a)]}, \quad Z_2 = \frac{\delta\gamma}{N_1 a^2} Z_1 I_2(\delta a). \qquad (6.5.96)$$

Thus, the solution of the problem $\mathcal{D}^{(1)}$ is

$$u_r = u^{(1)} \cos\vartheta, \quad u_\vartheta = v^{(2)} \sin\vartheta, \quad \varphi^{(1)} = \psi^{(2)} \sin\vartheta, \qquad (6.5.97)$$

where $u^{(1)}, v^{(2)}$ and $\varphi^{(2)}$ are given by (6.5.87), (6.5.88) and the constants A_2 and A_4 are defined by (6.5.95), (6.5.96).

The problem $\mathcal{D}^{(2)}$ is characterized by the following boundary conditions

$$t_{rr} = 0, \quad t_{r\vartheta} = 0, \quad m_{rz} = -(\beta + \gamma\nu)\sin\vartheta \ \text{ for } r = a. \tag{6.5.98}$$

It follows from (6.5.78), (6.5.91), (6.5.93) and (6.5.98) that

$$A_2 = 0, \quad A_4 = 0, \quad B_2 = Z_2, \quad B_4 = -Z_1. \tag{6.5.99}$$

The solution of the problem $\mathcal{D}^{(2)}$ is

$$u_r = u^{(2)}\sin\vartheta, \quad u_\vartheta = v^{(1)}\cos\vartheta, \quad \varphi^{(2)} = \psi^{(1)}\cos\vartheta, \tag{6.5.100}$$

where $u^{(2)}, v^{(1)}$ and $\psi^{(1)}$ are given by (6.5.90) with the constants B_2 and B_4 defined by (6.5.99) and (6.5.96).

We note that the divergence of the displacement vector field and the components m_{zr} and $m_{z\vartheta}$ of the couple stress tensor for the problem $\mathcal{D}^{(1)}$ are given by

$$\begin{aligned}
&\operatorname{div}\mathbf{u} = (3Q_1 - Q_2)Z_2 r\cos\vartheta,\\
&m_{zr} = \beta[\delta Z_1 I_1'(\delta r) - Q_3 Z_2]\sin\vartheta,\\
&m_{z\vartheta} = \beta r^{-1}[Z_1 I_1(\delta r) - Q_3 Z_2 r]\cos\vartheta.
\end{aligned} \tag{6.5.101}$$

In the case of the problem $\mathcal{D}^{(2)}$ we have

$$\begin{aligned}
&\operatorname{div}\mathbf{u} = (3Q_1 - Q_2)Z_2 r\sin\vartheta,\\
&m_{zr} = -\beta[\delta Z_1 I_1'(\delta r) - Q_3 Z_2]\cos\vartheta,\\
&m_{z\vartheta} = \beta r^{-1}[Z_1 I_1(\delta r) - Q_3 Z_2 r]\sin\vartheta.
\end{aligned} \tag{6.5.102}$$

The functions $(u_\alpha^{(\rho)}, \varphi^{(\rho)})$ which satisfy the problems $\mathcal{D}^{(\rho)}, (\rho = 1, 2)$, are

$$\begin{aligned}
&u_1^{(1)} = u^{(1)}\cos^2\vartheta - v^{(2)}\sin^2\vartheta, \quad u_2^{(1)} = (u^{(1)} + v^{(2)})\sin\vartheta\cos\vartheta,\\
&u_1^{(2)} = (u^{(2)} - v^{(1)})\sin\vartheta\cos\vartheta, \quad u_2^{(2)} = u^{(2)}\sin^2\vartheta + v^{(1)}\cos^2\vartheta,\\
&\varphi^{(1)} = \psi^{(2)}\sin\vartheta, \quad \varphi^{(2)} = \psi^{(1)}\cos\vartheta,
\end{aligned} \tag{6.5.103}$$

where $u^{(\alpha)}, v^{(\alpha)}$ and $\psi^{(\alpha)}$ are defined in (6.5.87), (6.5.88) and (6.5.90) with $A_2 = B_2 = Z_2, A_4 = -B_4 = Z_1$. We now can determine x_α^* and $L_{\alpha\beta}$ from (6.5.26) and (6.5.29). Thus, we obtain

$$\begin{aligned}
&x_\alpha^* = 0, \quad L_{12} = 0,\\
&L_{11} = L_{22} = \tfrac{1}{4}\pi E a^4 + \pi(2\gamma + \beta\nu + \beta Q_3 Z_2)a^2 - \pi a Z_1 \beta I_1(\delta a),
\end{aligned} \tag{6.5.104}$$

where E is defined in (6.2.17). Here we have used the relations

$$2I_1(\delta r) + \delta r I_2(\delta r) = \delta r I_0(\delta r),$$

$$\int_\Sigma m_{32}^{(1)} dA = -\int_\Sigma m_{31}^{(2)} dA = -\beta \int_\Sigma [\nu + Q_3 Z_2 - \frac{1}{2}\delta Z_1 I_0(\delta r)] dA.$$

Let us study now the thermoelastic plane strain problem characterized by the equations (6.5.20), (6.5.21), (6.5.23) and the boundary conditions (6.5.24). We seek the solution of this problem in the form

$$v_r = V(r), \quad v_\vartheta = 0, \quad w = 0, \tag{6.5.105}$$

where

$$v_1 = v_r \cos\vartheta - v_\vartheta \sin\vartheta, \quad v_2 = v_r \sin\vartheta + v_\vartheta \cos\vartheta.$$

It follows from (6.3.12) that V satisfies the equation

$$(\lambda + 2\mu + \kappa)\frac{d}{dr}\left[\frac{1}{r}\frac{d}{dr}(rV)\right] = bf'.$$

The solution, which is finite for $r = 0$, is

$$V(r) = b_0 r^{-1} G(r) + C_1 r, \tag{6.5.106}$$

where b_0 is defined in (6.4.21), C_1 is an arbitrary constant and

$$G(r) = \int_0^r x f(x) dx. \tag{6.5.107}$$

The boundary conditions (6.5.24) are satisfied if and only if

$$C_1 = \frac{2\mu + \kappa}{(2\lambda + 2\mu + \kappa)a^2} b_0 G(a). \tag{6.5.108}$$

Thus the solution of the thermoelastic plane strain problem is

$$v_\alpha = x_\alpha r^{-1} V(r), \quad w = 0, \tag{6.5.109}$$

where V is given by (6.5.106)-(6.5.108). Clearly, we have $\mu_{3\alpha} = 0$ so that from (6.5.29) we find that $M_\alpha = 0$. We conclude from (6.5.28), (6.5.26) and (6.5.25) that

$$\begin{aligned} &a_1 = a_2 = 0, \quad F = bf - \lambda v_{\alpha,\alpha} = (2\mu + \kappa) b_0 [f - \frac{2\nu}{a^2} G(a)], \\ &a_3 = \frac{2}{a^2 E}(2\mu + \kappa) b_0 (1 - \nu) G(a). \end{aligned} \tag{6.5.110}$$

Thus, by (6.5.18), (6.5.103), (6.5.105) and (6.5.110) we obtain

$$u_\alpha = x_\alpha[\frac{1}{r}V(r) - a_3\nu], \quad u_3 = a_3x_3, \quad \varphi_i = 0,$$

where a_3 is given by (6.5.110) and V is defined by (6.5.106)-(6.5.108).

The results of Section 6.4 can be used to obtain the solution for a circular cylinder subject to a uniform heat flux vector.

6.6 Cylinders Composed of Different Materials

The problem of deformation of composite cylinders within the isothermal classical elasticity has been studied by Muskhelishvili [270], p. 546. The thermoelastic deformation of heterogeneous bodies has been studied in various papers (see, e.g., Blevins [20], Hsu, Reddy and Bert [155], Orthwein [296], Ma [248], Barber and Zhang [10], Ieşan [165] and the literature cited therein). In this section we study the problem of thermal stresses for composite micropolar cylinders.

We assume that the cross-section Σ of the right cylinder B is a C^1-smooth domain (Fichera [110], p. 369). Let L_1 and L_2 be complementary subsets of L and let Γ be a curve contained in Σ with the property that $\overline{\Gamma} \cup \overline{L}_\rho, (\rho = 1,2)$, is the boundary of a regular domain Σ_ρ contained in Σ such that $\Sigma_1 \cap \Sigma_2 = \emptyset$. We denote by B_ρ the cylinder defined by $B_\rho = \{\mathbf{x} : (x_1, x_2) \in \Sigma_\rho, 0 < x_3 < h\}, (\rho = 1,2)$. Let $\Pi_0 = \{\mathbf{x} : (x_1, x_2) \in \Gamma, 0 \leq x_3 \leq h\}$.

a) **Isotropic components.** We assume that B_ρ is occupied by an isotropic thermoelastic micropolar material with the constitutive coefficients $\lambda^{(\rho)}, \mu^{(\rho)}, \kappa^{(\rho)}, \alpha^{(\rho)}, \beta^{(\rho)}, \gamma^{(\rho)}$ and $b^{(\rho)}, (\rho = 1,2)$. We suppose that the constitutive coefficients are independent of x_3,

$$\lambda^{(\rho)}{=}\lambda^{(\rho)}(x_1,x_2), \mu^{(\rho)}{=}\mu^{(\rho)}(x_1,x_2), \ldots, b^{(\rho)}{=}b^{(\rho)}(x_1,x_2), \\ (x_1,x_2){\in}\Sigma. \tag{6.6.1}$$

The above functions are supposed to belong to C^∞.

We can consider that the cylinder B is composed of two materials which are welded together along the surface Π_0. We assume that the elastic potential corresponding to the material that occupies the domain B_ρ is positive definite.

The displacement vector, the microrotation vector, the stress vector and the couple-stress vector must be continuous in passing from one medium to another, so that we have the conditions

$$\begin{aligned}&[u_i]_1 = [u_i]_2\,, \quad [\varphi_i]_1 = [\varphi_i]_2\,,\\ &[t_{\alpha i}]_1\,\nu_\alpha = [t_{\alpha i}]_2\,\nu_\alpha, \quad [m_{\alpha i}]_1\,\nu_\alpha = [m_{\alpha i}]_2\,\nu_\alpha \text{ on } \Pi_0,\end{aligned} \tag{6.6.2}$$

where we have indicated that the expressions in brackets are calculated for the material corresponding to the regions B_1 and B_2 and ν_α are the direction cosines of the vector normal to Γ, outward to Σ_1. The cylinder is supposed to be free of lateral loading so that we have the conditions (6.5.3) on Π. We assume that the loading applied on each end is statically equivalent to zero. Thus, for $x_3 = 0$ we have the conditions (6.5.4)-(6.5.7). The problem consists in finding of the functions u_i and φ_i which satisfy the equations (6.5.2), (6.2.3) and (6.1.4) on B_ρ, the conditions (6.6.2) on Π_0, the conditions (6.5.3) on Π and the conditions (6.5.4)-(6.5.7) for $x_3 = 0$, where the temperature field has the form (6.5.1). We shall use the method presented in Section 6.5.

We first consider the equilibrium problem of isothermal plane strain for the domain Σ, with the body forces $f_\alpha^{(\rho)} \in C^\infty(\overline{\Sigma}_\rho)$ and the body couple $g_3^{(\rho)} \in C^\infty(\overline{\Sigma}_\rho)$. The basic equations are the equilibrium equations

$$t_{\beta\alpha,\beta} + \rho_0 f_\alpha^{(\rho)} = 0, \quad m_{\alpha 3,\alpha} + \varepsilon_{3\alpha\beta} t_{\alpha\beta} + \rho_0 g_3^{(\rho)} = 0 \text{ on } \Sigma_\rho, \tag{6.6.3}$$

the constitutive equations

$$\begin{aligned}&t_{\alpha\beta} = \lambda^{(\rho)} e_{\eta\eta}\delta_{\alpha\beta} + (\mu^{(\rho)} + \kappa^{(\rho)}) e_{\alpha\beta} + \mu^{(\rho)} e_{\beta\alpha},\\ &m_{\alpha 3} = \gamma^{(\rho)} \kappa_{\alpha 3} \text{ on } \Sigma_\rho\end{aligned} \tag{6.6.4}$$

and the geometrical equations (6.3.5) on Σ_ρ. If the displacement vector, the microrotation vector, the stress vector and the couple-stress vector are continuous on Σ, then we have the conditions

$$\begin{aligned}&[u_\alpha]_1 = [u_\alpha]_2\,, \quad [\varphi_3]_1 = [\varphi_3]_2\,,\\ &[t_{\alpha\beta}]_1\nu_\alpha = [t_{\alpha\beta}]_2\nu_\alpha, \quad [m_{\alpha 3}]_1\nu_\alpha = [m_{\alpha 3}]_2\nu_\alpha \text{ on } \Gamma.\end{aligned} \tag{6.6.5}$$

We consider the boundary conditions

$$[t_{\beta\alpha} n_\beta]_\rho = \widetilde{t}_\alpha^{(\rho)}, \quad [m_{\alpha 3} n_\alpha]_\rho = \widetilde{m}_3^{(\rho)} \text{ on } L_\rho, \tag{6.6.6}$$

where $\widetilde{t}_\alpha^{(\rho)}$ and $\widetilde{m}_3^{(\rho)}$ are C^∞ functions.

If the domains Σ and Σ_ρ satisfy some conditions of regularity (see Fichera [110], p. 369) then the boundary value problem (6.6.3)-(6.6.6) has a solution $u_\alpha \in C^\infty(\Sigma_1 \cup L_1) \cap C^\infty(\Sigma_2 \cup L_2) \cap C^0(\Sigma)$ if and only if

$$\begin{aligned} &\sum_{\rho=1}^{2}\Big(\int_{\Sigma_\rho} \rho_0 f_\alpha^{(\rho)} dA + \int_{L_\rho} \tilde{t}_\alpha^{(\rho)} ds\Big) = 0, \\ &\sum_{\rho=1}^{2}\Big[\int_{\Sigma_\rho} \rho_0(\varepsilon_{3\alpha\beta} x_\alpha f_\beta^{(\rho)} + g_3^{(\rho)}) dA + \int_{L_\rho} (\varepsilon_{3\alpha\beta} x_\alpha \tilde{t}_\beta^{(\rho)} + \tilde{m}_3^{(\rho)}) ds\Big] = 0. \end{aligned} \tag{6.6.7}$$

In what follows we assume that the domains Σ and Σ_ρ satisfy the conditions that ensure this existence theorem. If the conditions (6.6.5) are replaced by

$$\begin{aligned} &[u_\alpha]_1 = [u_\alpha]_2, \quad [\varphi_3]_1 = [\varphi_3]_2, \\ &[t_{\alpha\beta}]_1 \nu_\alpha = [t_{\alpha\beta}]_2 \nu_\alpha + p_\beta, \quad [m_{\alpha 3}]_1 \nu_\alpha = [m_{\alpha 3}]_2 \nu_\alpha + q \text{ on } \Gamma, \end{aligned} \tag{6.6.8}$$

where p_α and q are C^∞ functions, then the conditions (6.6.7) become

$$\begin{aligned} &\sum_{\rho=1}^{2}\Big(\int_{\Sigma_\rho} \rho_0 f_\alpha^{(\rho)} dA + \int_{L_\rho} \tilde{t}_\alpha^{(\rho)} ds\Big) + \int_\Gamma p_\alpha ds = 0, \\ &\sum_{\rho=1}^{2}\Big[\int_{\Sigma_\rho} \rho_0(\varepsilon_{3\alpha\beta} x_\alpha f_\beta^{(\rho)} + g_3^{(\rho)}) dA + \int_{L_\rho} (\varepsilon_{3\alpha\beta} x_\alpha \tilde{t}_\beta^{(\rho)} + \tilde{m}_3^{(\rho)}) ds\Big] + \\ &\quad + \int_\Gamma (\varepsilon_{3\alpha\beta} x_\alpha p_\beta + q) ds = 0. \end{aligned} \tag{6.6.9}$$

In the classical elasticity these conditions have been investigated by Sherman [341] and Muskhelishvili [270], p. 547.

We shall have occasion to use three special problems $A^{(s)} (s = 1, 2, 3)$ of isothermal plane strain. In what follows we denote by $u_\alpha^{(s)}, \varphi_3^{(s)}, e_{ij}^{(s)}$, $t_{ij}^{(s)}$ and $m_{ij}^{(s)}$ the functions $u_\alpha, \varphi_3, e_{ij}, t_{ij}$ and m_{ij} for the problems $A^{(s)}$. The problems $A^{(s)}$ are characterized by the equations

$$\begin{aligned} &t_{\beta\alpha,\beta}^{(\eta)} + (\lambda^{(\rho)} x_\eta)_{,\alpha} = 0, \quad t_{\beta\alpha,\beta}^{(3)} + \lambda_{,\alpha}^{(\rho)} = 0, (\eta = 1, 2), \\ &m_{\beta 3,\beta}^{(\eta)} + \varepsilon_{3\alpha\beta} t_{\alpha\beta}^{(\eta)} + \varepsilon_{3\alpha\eta} \beta_{,\alpha}^{(\rho)} = 0, \quad m_{\beta 3,\beta}^{(3)} + \varepsilon_{3\alpha\beta} t_{\alpha\beta}^{(3)} = 0, \\ &t_{\alpha\beta}^{(s)} = \lambda^{(\rho)} e_{\nu\nu}^{(s)} \delta_{\alpha\beta} + (\mu^{(\rho)} + \kappa^{(\rho)}) e_{\alpha\beta}^{(s)} + \mu^{(\rho)} e_{\beta\alpha}^{(s)}, \\ &m_{\alpha 3}^{(s)} = \gamma^{(\rho)} \varphi_{3,\alpha}^{(s)}, \quad e_{\alpha\beta}^{(s)} = u_{\beta,\alpha}^{(s)} + \varepsilon_{\beta\alpha 3} \varphi_3^{(s)} \end{aligned} \tag{6.6.10}$$

and the conditions

$$[u_\alpha^{(s)}]_1 = [u_\alpha^{(s)}]_2, \quad [\varphi_3^{(s)}]_1 = [\varphi_3^{(s)}]_2,$$
$$[t_{\beta\alpha}^{(s)}]_1\nu_\beta = [t_{\beta\alpha}^{(s)}]_2\nu_\beta + p_\alpha^{(s)}, [m_{\alpha 3}^{(s)}]_1\nu_\alpha = [m_{\alpha 3}^{(s)}]_2\nu_\alpha + q^{(s)} \text{ on } \Gamma, \tag{6.6.11}$$

$$[t_{\beta\alpha}^{(\eta)} n_\beta]_\rho = -\lambda^{(\rho)} x_\eta n_\alpha, \quad [t_{\beta\alpha}^{(3)} n_\beta]_\rho = -\lambda^{(\rho)} n_\alpha,$$
$$[m_{\alpha 3}^{(\eta)} n_\alpha]_\rho = \varepsilon_{3\eta\nu}\beta^{(\rho)} n_\nu, \quad [m_{\alpha 3}^{(3)} n_\alpha]_\rho = 0 \text{ on } L_\rho, (\rho = 1, 2), \tag{6.6.12}$$

where

$$p_\alpha^{(\eta)} = (\lambda^{(2)} - \lambda^{(1)}) x_\eta \nu_\alpha \quad p_\alpha^{(3)} = (\lambda^{(2)} - \lambda^{(1)})\nu_\alpha,$$
$$q^{(\eta)} = \varepsilon_{3\eta\alpha}(\beta^{(1)} - \beta^{(2)})\nu_\alpha, \quad q^{(3)} = 0.$$

It is easy to prove that the necessary and sufficient conditions (6.6.9) for the existence of the solution are satisfied for each boundary value problem $A^{(s)}$. In what follows we assume that the functions $u_\alpha^{(s)}$ and $\varphi_3^{(s)}$ are known.

We seek the solution of the problem $P^{(0)}$ in the form

$$u_\alpha = -\tfrac{1}{2} a_\alpha x_3^2 + \sum_{s=1}^{3} a_s u_\alpha^{(s)} + v_\alpha(x_1, x_2),$$
$$u_3 = (a_1 x_1 + a_2 x_2 + a_3) x_3, \tag{6.6.13}$$
$$\varphi_\alpha = \varepsilon_{3\alpha\beta} a_\beta x_3, \quad \varphi_3 = \sum_{s=1}^{3} a_s \varphi_3^{(s)} + w,$$

where $u_\alpha^{(s)}$ and $\varphi_3^{(s)}$ are the solutions of the problems $A^{(s)}$, v_α and w are unknown functions, and the a_s are unknown constants. It follows from (6.3.5), (6.6.4) and (6.6.13) that

$$t_{\alpha\beta} = \lambda^{(\rho)}(a_1 x_1 + a_2 x_2 + a_3)\delta_{\alpha\beta} + \sum_{s=1}^{3} a_s t_{\alpha\beta}^{(s)} + \sigma_{\alpha\beta},$$
$$t_{33} = (\lambda^{(\rho)} + 2\mu^{(\rho)} + \kappa^{(\rho)})(a_1 x_1 + a_2 x_2 + a_3) +$$
$$+ \lambda^{(\rho)} \sum_{s=1}^{3} a_s e_{\alpha\alpha}^{(s)} + \lambda^{(\rho)}\gamma_{\alpha\alpha} - b^{(\rho)}\theta,$$
$$t_{\alpha 3} = t_{3\alpha} = 0, \quad m_{\alpha\beta} = m_{33} = 0, \tag{6.6.14}$$
$$m_{\alpha 3} = \varepsilon_{3\alpha\nu}\beta^{(\rho)} a_\nu + \sum_{s=1}^{3} a_s m_{\alpha 3}^{(s)} + \mu_{\alpha 3},$$
$$m_{3\alpha} = \varepsilon_{3\alpha\nu}\gamma^{(\rho)} a_\nu + \sum_{s=1}^{3} a_s m_{3\alpha}^{(s)} + \mu_{3\alpha} \text{ on } B_\rho,$$

where

$$\sigma_{\alpha\beta} = \lambda^{(\rho)}\gamma_{\eta\eta}\delta_{\alpha\beta} + (\mu^{(\rho)} + \kappa^{(\rho)})\gamma_{\alpha\beta} + \mu^{(\rho)}\gamma_{\beta\alpha} - b^{(\rho)}\theta\delta_{\alpha\beta},$$
$$\mu_{\alpha 3} = \gamma^{(\rho)}w_{,\alpha}, \quad \mu_{3\alpha} = \beta^{(\rho)}w_{,\alpha}, \quad \gamma_{\alpha\beta} = v_{\beta,\alpha} + \varepsilon_{\beta\alpha 3}w. \tag{6.6.15}$$

With the help of (6.6.10), (6.6.14) and (6.6.15), the equilibrium equations (6.5.2) become

$$\sigma_{\alpha\beta,\alpha} = 0 \quad \mu_{\alpha 3,\alpha} + \varepsilon_{\alpha\beta 3}\sigma_{\alpha\beta} = 0 \quad \text{on } \Sigma. \tag{6.6.16}$$

In view of (6.6.10)-(6.6.15), we see that the conditions (6.6.2) and (6.5.3) reduce to

$$[v_\alpha]_1 = [v_\alpha]_2, \quad [w]_1 = [w]_2,$$
$$[\sigma_{\alpha\beta}]_1\nu_\alpha = [\sigma_{\alpha\beta}]_2\nu_\alpha, [\mu_{\alpha 3}]_1\nu_\alpha = [\mu_{\alpha 3}]_2\nu_\alpha \text{ on } \Gamma, \tag{6.6.17}$$
$$[\sigma_{\beta\alpha}n_\beta]_\rho = 0, \quad [\mu_{\alpha 3}n_\alpha]_\rho = 0 \text{ on } L_\rho, (\rho = 1, 2).$$

From (6.6.15)-(6.6.17) it follows that the functions v_α and w satisfy the equations and boundary conditions of the thermoelastic plane-strain problem corresponding to the temperature distribution f. In what follows we assume that this problem is solved.

The conditions (6.5.4) and (6.5.7) are satisfied on the basis of relations (6.6.14). From (6.5.5) and (6.5.6) we obtain for the constants a_s the system

$$A_{rs}a_s = d_s, \tag{6.6.18}$$

where

$$A_{\alpha\beta} = \sum_{\rho=1}^{2}\int_{\Sigma_\rho}\{x_\alpha[(\lambda^{(\rho)} + 2\mu^{(\rho)} + \kappa^{(\rho)})x_\beta + \lambda^{(\rho)}e_{\eta\eta}^{(\beta)}] -$$
$$-\varepsilon_{3\alpha\lambda}(\varepsilon_{3\lambda\beta}\gamma^{(\rho)} + m_{3\lambda}^{(\beta)})\}dA,$$
$$A_{\alpha 3} = \sum_{\rho=1}^{2}\int_{\Sigma_\rho}[x_\alpha(\lambda^{(\rho)} + 2\mu^{(\rho)} + \kappa^{(\rho)} + \lambda^{(\rho)}e_{\eta\eta}^{(3)}) - \varepsilon_{3\alpha\lambda}m_{3\lambda}^{(3)}]dA, \tag{6.6.19}$$
$$A_{3\alpha} = \sum_{\rho=1}^{2}\int_{\Sigma_\rho}[(\lambda^{(\rho)} + 2\mu^{(\rho)} + \kappa^{(\rho)})x_\alpha + \lambda^{(\rho)}e_{\eta\eta}^{(\alpha)}]dA,$$
$$A_{33} = \sum_{\rho=1}^{2}\int_{\Sigma_\rho}(\lambda^{(\rho)} + 2\mu^{(\rho)} + \kappa^{(\rho)} + \lambda^{(\rho)}e_{\eta\eta}^{(3)})dA,$$

$$d_\alpha = -\sum_{\rho=1}^{2} \int_{\Sigma_\rho} x_\alpha (\lambda^{(\rho)} \gamma_{\nu\nu} - b^{(\rho)} T) dA,$$

$$d_3 = -\sum_{\rho=1}^{2} \int_{\Sigma_\rho} (\lambda^{(\rho)} \gamma_{\alpha\alpha} - b^{(\rho)} T) dA.$$

As in Section 6.5 we can prove that $\det(A_{rs}) \neq 0$ so that the system (6.6.18) determines the constants a_s. Thus the problem $P^{(0)}$ is solved.

We now study the recurrence process. The problem consists in finding of the functions $u_i, \varphi_i, e_{ij}, \kappa_{ij}, t_{ij}$ and m_{ij} which satisfy the equations (6.5.2), (6.2.3) and (6.1.4) on B_ρ, the conditions (6.6.2) on Π_0, the conditions (6.5.3) on Π and the conditions (6.5.4)-(6.5.7) where the temperature has the form (6.5.45), assuming that we know the functions $u_i^*, \varphi_i^*, e_{ij}^*, \kappa_{ij}^*, t_{ij}^*$ and m_{ij}^* which satisfy the equations (6.5.2), (6.2.3) and (6.1.4) on B_ρ, the conditions (6.6.2) on Π_0, the conditions (6.5.3) on Π, and the conditions (6.5.4)-(6.5.7) for the temperature field (6.5.46).

We seek the solution of the problem in the form

$$\begin{aligned}
u_\alpha &= (n+1)\Big[\int_0^{x_3} u_\alpha^* dx_3 - \frac{1}{2} b_\alpha x_3^2 + \sum_{s=1}^{3} b_s u_\alpha^{(s)} + \varepsilon_{3\beta\alpha} \tau x_\beta x_3 + v_\alpha(x_1, x_2)\Big],\\
u_3 &= (n+1)\Big[\int_0^{x_3} u_3^* dx_3 + (b_1 x_1 + b_2 x_2 + b_3) x_3 + G(x_1, x_2)\Big],\\
\varphi_\alpha &= (n+1)\Big[\int_0^{x_3} \varphi_\alpha^* dx_3 + \varepsilon_{3\alpha\beta} b_\beta x_3 + G_\alpha(x_1, x_2)\Big],\\
\varphi_3 &= (n+1)\Big[\int_0^{x_3} \varphi_3^* dx_3 + \tau x_3 + \sum_{s=1}^{3} b_s \varphi_3^{(s)} + w(x_1, x_2)\Big],
\end{aligned} \tag{6.6.20}$$

where $u_\alpha^{(s)}$ and $\varphi_3^{(s)}$ are the solutions of the problems $A^{(s)}, v_\alpha, w, G$ and G_α are unknown functions, and b_i and τ are unknown constants. If we introduce the notation

$$\begin{aligned}
\sigma_{\alpha\beta} &= \lambda^{(\rho)} \gamma_{\nu\nu} \delta_{\alpha\beta} + (\mu^{(\rho)} + \kappa^{(\rho)}) \gamma_{\alpha\beta} + \mu^{(\rho)} \gamma_{\beta\alpha},\\
\mu_{\alpha 3} &= \gamma^{(\rho)} w_{,\alpha}, \quad \mu_{3\alpha} = \beta^{(\rho)} w_{,\alpha}, \quad \gamma_{\alpha\beta} = v_{\beta,\alpha} + \varepsilon_{\beta\alpha 3} w,
\end{aligned} \tag{6.6.21}$$

then we obtain

$$t_{\alpha\beta} = (n+1)\Big\{\int_0^{x_3} t_{\alpha\beta}^* dx_3 + \lambda^{(\rho)} [b_1 x_1 + b_2 x_2 + b_3 + u_3^*(x_1, x_2, 0)] \delta_{\alpha\beta} + \sum_{s=1}^{3} b_s t_{\alpha\beta}^{(s)} + \sigma_{\alpha\beta}\Big\},$$

$$t_{33} = (n+1)\{\int_0^{x_3} t_{33}^* dx_3 + (\lambda^{(\rho)} + 2\mu^{(\rho)} + \kappa^{(\rho)})[b_1x_1 + b_2x_2 + b_3 +$$

$$+ u_3^*(x_1, x_2, 0)] + \lambda^{(\rho)}\gamma_{\alpha\alpha} + \lambda^{(\rho)} \sum_{s=1}^{3} b_s e_{\alpha\alpha}^{(s)}\},$$

$$t_{\alpha 3} = (n+1)[\int_0^{x_3} t_{\alpha 3}^* dx_3 + (\mu^{(\rho)} + \kappa^{(\rho)})G_{,\alpha} + \varepsilon_{\alpha\beta 3}(\kappa^{(\rho)} G_\beta -$$

$$-\mu^{(\rho)}\tau x_\beta) + \mu^{(\rho)} u_\alpha^*(x_1, x_2, 0)],$$

$$t_{3\alpha} = (n+1)\{\int_0^{x_3} t_{3\alpha}^* dx_3 + \mu^{(\rho)} G_{,\alpha} + \varepsilon_{\beta\alpha 3}[\kappa^{(\rho)} G_\beta + (\mu^{(\rho)} + \kappa^{(\rho)})\tau x_\beta] +$$

$$+ (\mu^{(\rho)} + \kappa^{(\rho)}) u_\alpha^*(x_1, x_2, 0)\},$$

$$m_{\lambda\nu} = (n+1)\{\int_0^{x_3} m_{\lambda\nu}^* dx_3 + \alpha^{(\rho)} G_{\kappa,\kappa}\delta_{\lambda\nu} + \beta^{(\rho)} G_{\lambda,\nu} + \gamma^{(\rho)} G_{\nu,\lambda} +$$

$$+ \alpha^{(\rho)}[\tau + \varphi_3^*(x_1, x_2, 0)]\delta_{\lambda\nu}\},$$

$$m_{33} = (n+1)\{\int_0^{x_3} m_{33}^* dx_3 + \alpha^{(\rho)} G_{\lambda,\lambda} + (\alpha^{(\rho)} + \beta^{(\rho)} +$$

$$+ \gamma^{(\rho)})[\tau + \varphi_3^*(x_1, x_2, 0)]\},$$

$$m_{\alpha 3} = (n+1)[\int_0^{x_3} m_{\alpha 3}^* dx_3 + \varepsilon_{3\alpha\nu}\beta^{(\rho)} b_\nu + \sum_{s=1}^{3} b_s m_{\alpha 3}^{(s)} +$$

$$+ \mu_{\alpha 3} + \beta^{(\rho)} \varphi_\alpha^*(x_1, x_2, 0)],$$

$$m_{3\alpha} = (n+1)[\int_0^{x_3} m_{3\alpha}^* dx_3 + \varepsilon_{3\alpha\nu}\gamma^{(\rho)} b_\nu + \sum_{s=1}^{3} b_s m_{3\alpha}^{(s)} + \mu_{3\alpha} +$$

$$+ \gamma^{(\rho)} \varphi_\alpha^*(x_1, x_2, 0)] \text{ on } B_\rho.$$

(6.6.22)

Let $\Omega = (G, G_1, G_2)$ be the ordered triplet of functions G, G_1, G_2. We introduce the operators

$$\mathcal{L}^{(\rho)}\Omega = [(\mu^{(\rho)} + \kappa^{(\rho)})G_{,\alpha}]_{,\alpha} + \varepsilon_{3\alpha\beta}[\kappa^{(\rho)} G_\beta]_{,\alpha},$$

$$\mathcal{L}_\nu^{(\rho)}\Omega = [\alpha^{(\rho)} G_{\lambda,\lambda}]_{,\nu} + [\beta^{(\rho)} G_{\lambda,\nu}]_{,\lambda} + [\gamma^{(\rho)} G_{\nu,\lambda}]_{,\lambda} + \qquad (6.6.23)$$

$$+ \varepsilon_{\nu\beta 3}\kappa^{(\rho)} G_{,\beta} - 2\kappa^{(\rho)} G_\nu,$$

$$\Lambda^{(\rho)}\Omega = (\mu^{(\rho)} + \kappa^{(\rho)})G_{,\alpha}n_\alpha + \kappa^{(\rho)}\varepsilon_{3\alpha\beta}G_\beta n_\alpha,$$
$$\Lambda_\nu^{(\rho)}\Omega = \alpha^{(\rho)}G_{\lambda,\lambda}n_\nu + \beta^{(\rho)}G_{\lambda,\nu}n_\lambda + \gamma^{(\rho)}G_{\nu,\lambda}n_\lambda.$$

The equilibrium equations (6.5.2) reduce to

$$\sigma_{\alpha\beta,\alpha} + P_\beta^{(\rho)} = 0, \quad \mu_{\alpha 3,\alpha} + \varepsilon_{\alpha\beta 3}\sigma_{\alpha\beta} + L_3^{(\rho)} = 0, \tag{6.6.24}$$

$$\mathcal{L}^{(\rho)}\Omega = H^{(\rho)}, \quad \mathcal{L}_\alpha^{(\rho)}\Omega = H_\alpha^{(\rho)} \text{ on } \Sigma_\rho, \tag{6.6.25}$$

where

$$\begin{aligned}
P_\beta^{(\rho)} &= \lambda^{(\rho)}u^*_{3,\beta}(x_1,x_2,0) + t^*_{3\beta}(x_1,x_2,0),\\
L_3^{(\rho)} &= \beta^{(\rho)}\varphi^*_{\lambda,\lambda}(x_1,x_2,0) + m^*_{33}(x_1,x_2,0),\\
H^{(\rho)} &= -[\mu^{(\rho)}u^*_\alpha(x_1,x_2,0)]_{,\alpha} - t^*_{33}(x_1,x_2,0) + \varepsilon_{\alpha\beta 3}\tau[\mu^{(\rho)}x_\beta]_{,\alpha},\\
H_\nu^{(\rho)} &= -[\alpha^{(\rho)}\varphi^*_3(x_1,x_2,0)]_{,\nu} - m^*_{3\nu}(x_1,x_2,0) +\\
&\quad + \kappa^{(\rho)}\varepsilon_{\nu\lambda 3}u^*_\lambda(x_1,x_2,0) + \tau(x_\nu\kappa^{(\rho)} - \alpha^{(\rho)}_{,\nu}).
\end{aligned} \tag{6.6.26}$$

The boundary conditions (6.5.3) reduce to

$$[\sigma_{\alpha\beta}n_\alpha]_\rho = p_\beta^{(\rho)}, \quad [\mu_{\alpha 3}n_\alpha]_\rho = q^{(\rho)}, \tag{6.6.27}$$

$$\Lambda^{(\rho)}\Omega = Q^{(\rho)}, \quad \Lambda_\nu^{(\rho)}\Omega = Q_\nu^{(\rho)} \text{ on } L_\rho, \tag{6.6.28}$$

where

$$\begin{aligned}
p_\beta^{(\rho)} &= -\lambda^{(\rho)}n_\beta u^*_3(x_1,x_2,0), \quad q^{(\rho)} = -\beta^{(\rho)}\varphi^*_\alpha(x_1,x_2,0)n_\alpha,\\
Q^{(\rho)} &= \varepsilon_{\alpha\beta 3}\mu^{(\rho)}\tau n_\alpha x_\beta - \mu^{(\rho)}n_\alpha u^*_\alpha(x_1,x_2,0),\\
Q_\nu^{(\rho)} &= -\alpha^{(\rho)}[\tau + \varphi^*_3(x_1,x_2,0)]n_\nu.
\end{aligned} \tag{6.6.29}$$

From (6.6.20), (6.6.26), (6.6.10)-(6.6.12) it follows that the conditions (6.6.2) become

$$\begin{aligned}
&[v_\alpha]_1 = [v_\alpha]_2, \quad [w]_1 = [w]_2,\\
&[\sigma_{\alpha\beta}]_1\nu_\alpha - [\sigma_{\alpha\beta}]_2\nu_\alpha = p_\beta, \quad [\mu_{\alpha 3}]_1\nu_\alpha - [\mu_{\alpha 3}]_2\nu_\alpha = q \text{ on } \Gamma,
\end{aligned} \tag{6.6.30}$$

$$\begin{aligned}
&[G]_1 = [G]_2, \quad [G_\alpha]_1 = [G_\alpha]_2,\\
&\left[\Lambda^{(1)}\Omega\right](\nu) - \left[\Lambda^{(2)}\Omega\right](\nu) = N,\\
&\left[\Lambda_\lambda^{(1)}\Omega\right](\nu) - \left[\Lambda_\lambda^{(2)}\Omega\right](\nu) = (\alpha^{(2)} - \alpha^{(1)})\,[\tau +\\
&\qquad + \varphi^*_3(x_1,x_2,0)]\,\nu_\lambda \text{ on } \Gamma,
\end{aligned} \tag{6.6.31}$$

where $[\Lambda^{(\rho)}\Omega](\nu)$ denotes $\Lambda^{(\rho)}\Omega$ for $n_\alpha = \nu_\alpha$ and

$$p_\beta = (\lambda^{(2)} - \lambda^{(1)})\nu_\beta u_3^*(x_1, x_2, 0), \quad q = (\beta^{(2)} - \beta^{(1)})\nu_\alpha \varphi_\alpha^*(x_1, x_2, 0),$$
$$N = (\mu^{(1)} - \mu^{(2)})[\varepsilon_{\alpha\beta 3}\tau x_\beta - u_\alpha^*(x_1, x_2, 0)]\nu_\alpha. \tag{6.6.32}$$

It follows from (6.6.21), (6.6.24), (6.6.27), (6.6.30) that the functions v_α and w satisfy the equations of a plane-strain problem. The necessary and sufficient conditions to solve this problem are

$$\sum_{\rho=1}^{2}\Big(\int_{\Sigma} P_\alpha^{(\rho)} dA + \int_{L_\rho} p_\alpha^{(\rho)} ds\Big) + \int_\Gamma p_\alpha ds = 0,$$
$$\sum_{\rho=1}^{2}\Big[\int_{\Sigma_\rho} (\varepsilon_{3\alpha\beta} x_\alpha P_\beta^{(\rho)} + L_3^{(\rho)}) dA + \int_{L_\rho} (\varepsilon_{3\alpha\beta} x_\alpha p_\beta^{(\rho)} + q^{(\rho)}) ds\Big] + \tag{6.6.33}$$
$$+ \int_L (\varepsilon_{3\alpha\beta} x_\alpha p_\beta + q) ds = 0.$$

By (6.6.26), (6.6.29), (6.6.32) and the divergence theorem the above conditions become

$$\sum_{\rho=1}^{2}\int_{\Sigma_\rho} t_{3\alpha}^*(x_1, x_2, 0) dA = 0,$$
$$\sum_{\rho=1}^{2}\int_{\Sigma_\rho} [\varepsilon_{3\alpha\beta} x_\alpha t_{3\beta}^*(x_1, x_2, 0) + m_{33}^*(x_1, x_2, 0)] dA = 0,$$

so that the conditions (6.6.33) are satisfied because the functions t_{ij}^* and m_{ij}^* satisfy the conditions (6.5.4) and (6.5.7). In what follows we assume that v_α and w are known functions.

The functions G, G_1 and G_2 satisfy (6.6.25) and the conditions (6.6.28) and (6.6.31). The necessary and sufficient condition for the existence of the solution of this problem is (see Fichera [110], p. 369; [170])

$$\sum_{\rho=1}^{2}\Big(\int_{\Sigma_\rho} H^{(\rho)} dA - \int_{L_\rho} Q^{(\rho)} ds\Big) - \int_\Gamma N ds = 0. \tag{6.6.34}$$

Taking into account (6.6.26), (6.6.29) and (6.6.32) we can write (6.6.34) in the form

$$\sum_{\rho=1}^{2}\int_{\Sigma_\rho} t_{33}^*(x_1, x_2, 0) dA = 0.$$

Thus the condition (6.6.34) is satisfied since t_{33}^* satisfies the condition (6.5.5).

Let us determine the constant τ. We consider the ordered triplet of functions $(\Phi, \Phi_1, \Phi_2) = V$, such that

$$\mathcal{L}^{(\rho)}V = \varepsilon_{\alpha\beta 3}[\mu^{(\rho)}x_\beta]_{,\alpha}, \quad \mathcal{L}_\nu^{(\rho)}V = x_\nu\kappa^{(\rho)} - \alpha_{,\nu}^{(\rho)} \text{ on } \Sigma_\rho, \tag{6.6.35}$$

$$\begin{aligned}
&[\Phi]_1 = [\Phi]_2, \quad [\Phi_\alpha]_1 = [\Phi_\alpha]_2,\\
&[\Lambda^{(1)}V](\nu) - [\Lambda^{(2)}V](\nu) = (\mu^{(1)} - \mu^{(2)})\varepsilon_{\alpha\beta 3}x_\beta\nu_\alpha,\\
&[\Lambda_\lambda^{(1)}V](\nu) - [\Lambda_\lambda^{(2)}V](\nu) = (\alpha^{(2)} - \alpha^{(1)})\nu_\lambda \text{ on } \Gamma
\end{aligned} \tag{6.6.36}$$

and

$$\Lambda^{(\rho)}V = \varepsilon_{\alpha\beta 3}n_\alpha\mu^{(\rho)}x_\beta, \quad \Lambda_\lambda^{(\rho)}V = -\alpha^{(\rho)}n_\lambda \text{ on } L_\rho. \tag{6.6.37}$$

Obviously, the necessary and sufficient condition for the existence of the functions Φ and Φ_α is satisfied. We introduce the functions Ψ and Ψ_α by the relations

$$G = \tau\Phi - \Psi \quad G_\alpha = \tau\Phi_\alpha - \Psi_\alpha. \tag{6.6.38}$$

It is easy to see that Ψ and Ψ_α are independent of τ. From (6.6.22) and (6.6.38) we obtain

$$\begin{aligned}
t_{3\alpha} =& (n+1)\{\int_0^{x_3} t_{3\alpha}^* dx_3 + \tau[\mu^{(\rho)}\Phi_{,\alpha} + \varepsilon_{\beta\alpha 3}(\mu^{(\rho)} + \kappa^{(\rho)})x_\beta + \varepsilon_{\beta\alpha 3}\kappa^{(\rho)}\Phi_\beta] +\\
&+ (\mu^{(\rho)} + \kappa^{(\rho)})u_\alpha^*(x_1, x_2, 0) - \mu^{(\rho)}\Psi_{,\alpha} - \varepsilon_{\beta\alpha 3}\kappa^{(\rho)}\Psi_\beta\},\\
m_{33} =& (n+1)[\int_0^{x_3} m_{33}^* dx_3 + \tau(\alpha^{(\rho)}\Phi_{\lambda,\lambda} + \alpha^{(\rho)} + \beta^{(\rho)} + \gamma^{(\rho)}) +\\
&+ (\alpha^{(\rho)} + \beta^{(\rho)} + \gamma^{(\rho)})\varphi_3^*(x_1, x_2, 0) - \alpha^{(\rho)}\Psi_{\lambda,\lambda}],
\end{aligned}$$

so that from (6.5.7) we get

$$\tau D = M, \tag{6.6.39}$$

where

$$\begin{aligned}
D =& \sum_{\rho=1}^{2}\int_{\Sigma_\rho}\{\varepsilon_{3\alpha\beta}x_\alpha\mu^{(\rho)}\Phi_{,\beta} + \varepsilon_{3\alpha\beta}\varepsilon_{3\nu\beta}[\kappa^{(\rho)}\Phi_\nu + (\mu^{(\rho)} + \kappa^{(\rho)})x_\nu]x_\alpha +\\
&+ \alpha^{(\rho)}\Phi_{\lambda,\lambda} + \alpha^{(\rho)} + \beta^{(\rho)} + \gamma^{(\rho)}\}dA,\\
M =& \sum_{\rho=1}^{2}\int_{\Sigma_\rho}\{\varepsilon_{3\alpha\beta}x_\alpha[\mu^{(\rho)}\Psi_{,\beta} + \varepsilon_{\eta\beta 3}\kappa^{(\rho)}\Psi_\eta - (\mu^{(\rho)} + \kappa^{(\rho)})u_\beta^*(x_1, x_2, 0)] +\\
&+ \alpha^{(\rho)}\Psi_{\lambda,\lambda} - (\alpha^{(\rho)} + \beta^{(\rho)} + \gamma^{(\rho)})\varphi_3^*(x_1, x_2, 0)\}dA.
\end{aligned}$$

It is known [171] that $D \neq 0$ so we can determine the constant τ from (6.6.39). Let us determine the constants b_i. From (6.5.5), (6.5.6) and (6.6.22) we find

$$A_{rs} b_s = h_r, \tag{6.6.40}$$

where the A_{rs} are given by (6.6.19) and we have used the notations

$$h_\alpha = -\sum_{\rho=1}^{2} \int_{\Sigma_\rho} \{x_\alpha [\lambda^{(\rho)} \gamma_{\lambda\lambda} + (\lambda^{(\rho)} + 2\mu^{(\rho)} + \kappa^{(\rho)}) u_3^*(x_1, x_2, 0)]$$
$$-\varepsilon_{3\alpha\beta} [\mu_{3\beta} + \gamma^{(\rho)} \varphi_\beta^*(x_1, x_2, 0)]\} dA,$$
$$h_3 = -\sum_{\rho=1}^{2} \int_{\Sigma_\rho} [\lambda^{(\rho)} \gamma_{\alpha\alpha} + (\lambda^{(\rho)} + 2\mu^{(\rho)} + \kappa^{(\rho)}) u_3^*(x_1, x_2, 0)] dA.$$

From (6.6.40) we can determine the constants b_i. As in Section 6.5 we can prove that the conditions (6.5.4) are identically satisfied. The problem is therefore solved.

It is easy to extend the solution to the case in which B consists of n bodies with different elasticities.

b) **Anisotropic cylinders.** We now assume that B_ρ is occupied by an anisotropic material with the constitutive coefficients $A_{ijrs}^{(\rho)}, B_{ijrs}^{(\rho)}, C_{ijrs}^{(\rho)}$, $\beta_{ij}^{(\rho)}$ and $\xi_{ij}^{(\rho)}$, $(\rho = 1, 2)$. The constitutive coefficients are supposed to be independent of the axial coordinate and to belong to C^∞. We assume that the elastic potential corresponding to the material that occupies B_ρ is positive definite. In this case the problem consists in finding of the functions u_i and φ_i which satisfy the equations (6.5.2), $(6.1.3)_{1,2}$, (6.1.4) on B_ρ, the conditions (6.6.2) on Π_0, the conditions (6.5.3) on Π and the conditions (6.5.4)-(6.5.7) for $x_3 = 0$, where the temperature field θ has the form (6.5.1).

Let us consider the problem of the isothermal generalized plane strain for the domain Σ, with the body loads $f_i^{(\rho)}, g_i^{(\rho)} \in C^\infty(\overline{\Sigma}_\rho)$ (see Section 6.3 (e)). The basic equations are the equilibrium equations

$$t_{\alpha i,\alpha} + \rho_0 f_i^{(\beta)} = 0, m_{\alpha i,\alpha} + \varepsilon_{ijk} t_{jk} + \rho_0 g_i^{(\beta)} = 0 \text{ on } \Sigma_\beta, \tag{6.6.41}$$

the constitutive equations

$$t_{\alpha i} = A_{\alpha ijk}^{(\rho)} e_{jk} + B_{\alpha i\beta j}^{(\rho)} \kappa_{\beta j}, \quad t_{3\alpha} = A_{3\alpha jk}^{(\rho)} e_{jk} + B_{3\alpha\beta j}^{(\rho)} \kappa_{\beta j},$$
$$m_{\alpha i} = B_{jk\alpha i}^{(\rho)} e_{jk} + C_{\alpha i\beta j}^{(\rho)} \kappa_{\beta j} \text{ on } \Sigma_\rho, \tag{6.6.42}$$

and the geometrical equations (6.3.39) on Σ_ρ. If the vector fields $\mathbf{u}, \varphi, \mathbf{t}$ and $\mathbf{m}$ are continuous on Σ, then we have the conditions

$$\begin{aligned} &[u_i]_1 = [u_i]_2\,, \quad [\varphi_i]_1 = [\varphi_i]_2\,, \\ &[t_{\alpha i}]_1\,\nu_\alpha = [t_{\alpha i}]_2\,\nu_\alpha, \quad [m_{\alpha i}]_1\,\nu_\alpha = [m_{\alpha i}]_2\,\nu_\alpha \text{ on } \Gamma. \end{aligned} \tag{6.6.43}$$

The boundary conditions (6.3.44) become

$$[t_{\beta i}n_\beta]_\rho = \tilde{t}_i^{(\rho)}, \quad [m_{\beta i}n_\beta]_\rho = \tilde{m}_i^{(\rho)} \text{ on } L_\rho, \tag{6.6.44}$$

where $\tilde{t}_i^{(\rho)}$ and $\tilde{m}_i^{(\rho)}$ are C^∞ functions. It is known (cf. Fichera [110], p. 369) that the boundary value problem (6.6.41)-(6.6.44) has a solution $u_i, \varphi_i \in C^\infty(\Sigma_1 \cup L_1) \cap C^\infty(\Sigma_2 \cup L_2) \cap C^\infty(\Sigma)$ if and only if

$$\begin{aligned} &\sum_{\alpha=1}^{2}\Big[\int_{\Sigma_\alpha} \rho_0 f_i^{(\alpha)} dA + \int_{L_\alpha} \tilde{t}_i^{(\alpha)} ds\Big] = 0, \\ &\sum_{\nu=1}^{2}\Big[\int_{\Sigma_\nu} \rho_0(\varepsilon_{3\alpha\beta}x_\alpha f_\beta^{(\nu)} + g_3^{(\nu)})dA + \int_{L_\nu}(\varepsilon_{3\alpha\beta}x_\alpha \tilde{t}_\beta^{(\nu)} + \tilde{m}_3^{(\nu)})ds\Big] = 0. \end{aligned} \tag{6.6.45}$$

If the conditions (6.6.43) are replaced by

$$\begin{aligned} &[u_i]_1 = [u_i]_2\,, \quad [\varphi_i]_1 = [\varphi_i]_2\,, \\ &[t_{\alpha i}]_1\,\nu_\alpha = [t_{\alpha i}]_2\,\nu_\alpha + F_i, \quad [m_{\alpha i}]_1\,\nu_\alpha = [m_{\alpha i}]_2\,\nu_\alpha + G_i \text{ on } \Gamma, \end{aligned} \tag{6.6.46}$$

where F_i and G_i are C^∞ functions, then the conditions (6.6.45) are replaced by

$$\begin{aligned} &\sum_{\alpha=1}^{2}\Big[\int_{\Sigma_\alpha} \rho_0 f_i^{(\alpha)} dA + \int_{L_\alpha} \tilde{t}_i^{(\alpha)} ds\Big] + \int_\Gamma F_i ds = 0, \\ &\sum_{\nu=1}^{2}\Big[\int_{\Sigma_\nu} \rho_0(\varepsilon_{3\alpha\beta}x_\alpha f_\beta^{(\nu)} + g_3^{(\nu)})dA + \int_{L_\nu}(\varepsilon_{3\alpha\beta}x_\alpha \tilde{t}_\beta^{(\nu)} + \tilde{m}_3^{(\nu)})ds\Big] + \\ &\qquad + \int_\Sigma (\varepsilon_{3\alpha\beta}x_\alpha F_\beta + G_3)ds = 0. \end{aligned} \tag{6.6.47}$$

In what follows we shall use four special problems $B^{(s)}, (s = 1, 2, 3, 4)$, of isothermal plane strain. The problems $B^{(s)}$ are characterized by the geometrical equations

$$e_{\alpha i}^{(s)} = u_{i,\alpha}^{(s)} + \varepsilon_{i\alpha k}\varphi_k^{(s)}, \quad e_{3i}^{(s)} = \varepsilon_{i3k}\varphi_k^{(s)}, \quad \kappa_{\alpha i}^{(s)} = \varphi_{i,\alpha}^{(s)}, \tag{6.6.48}$$

the constitutive equations

$$t_{\alpha i}^{(s)} = A_{\alpha ijk}^{(\rho)} e_{jk}^{(s)} + B_{\alpha i\beta j}^{(\rho)} \kappa_{\beta j}^{(s)}, \quad t_{3\alpha}^{(s)} = A_{3\alpha jk}^{(\rho)} e_{jk}^{(s)} + B_{3\alpha\beta j}^{(\rho)} \kappa_{\beta j}^{(s)},$$
$$m_{\alpha i}^{(s)} = B_{jk\alpha i}^{(\rho)} e_{jk}^{(s)} + C_{\alpha i\beta j}^{(\rho)} \kappa_{\beta j}^{(s)} \text{ on } \Sigma_\rho, \tag{6.6.49}$$

the equilibrium equations

$$t_{\alpha i,\alpha}^{(s)} + \rho_0 f_{si}^{(\rho)} = 0, \quad m_{\alpha i,\alpha}^{(s)} + \varepsilon_{ijk} t_{jk}^{(s)} + \rho_0 g_{si}^{(\rho)} = 0 \text{ on } \Sigma_\rho, \tag{6.6.50}$$

the conditions on Γ

$$[u_i^{(s)}]_1 = [u_i^{(s)}]_2, \ [\varphi_i^{(s)}]_1 = [\varphi_i^{(s)}]_2, \ [t_{\alpha i}^{(s)}]_1 \nu_\alpha = [t_{\alpha i}^{(s)}]_2 \nu_\alpha + F_{si},$$
$$[m_{\alpha i}^{(s)}]_1 \nu_\alpha = [m_{\alpha i}^{(s)}]_2 \nu_\alpha + G_{si} \text{ on } \Gamma \tag{6.6.51}$$

and the boundary conditions

$$[t_{\beta i}^{(s)} n_\beta]_\rho = \tilde{t}_{si}^{(\rho)}, \quad [m_{\beta i}^{(s)} n_\beta]_\rho = \tilde{m}_{si}^{(\rho)} \text{ on } L_\rho, \tag{6.6.52}$$

where

$$\rho_0 f_{\beta i}^{(\rho)} = (A_{\alpha i33}^{(\rho)} x_\beta + \varepsilon_{3\nu\beta} B_{\alpha i3\nu}^{(\rho)})_{,\alpha}, \quad \rho_0 f_{3i}^{(\rho)} = A_{\alpha i33,\alpha}^{(\rho)},$$
$$\rho_0 f_{4i}^{(\rho)} = (A_{\alpha i3\nu}^{(\rho)} \varepsilon_{3\beta\nu} x_\beta + B_{\alpha i33}^{(\rho)})_{,\alpha},$$
$$\rho_0 g_{\beta i}^{(\rho)} = (B_{33\alpha i}^{(\rho)} x_\beta + \varepsilon_{3\nu\beta} C_{\alpha i3\nu}^{(\rho)})_{,\alpha} + \varepsilon_{ijk}(A_{jk33}^{(\rho)} x_\beta + \varepsilon_{3\nu\beta} B_{jk3\nu}^{(\rho)}),$$
$$\rho_0 g_{3i}^{(\rho)} = B_{33\alpha i,\alpha}^{(\rho)} + \varepsilon_{ijk} A_{jk33}^{(\rho)},$$
$$\rho_0 g_{4i}^{(\rho)} = (B_{3\nu\alpha i}^{(\rho)} \varepsilon_{3\beta\nu} x_\beta + C_{\alpha i33}^{(\rho)})_{,\alpha} + \varepsilon_{ijk}(A_{jk3\nu}^{(\rho)} \varepsilon_{3\beta\nu} x_\beta + B_{jk33}^{(\rho)}),$$
$$\tilde{t}_{\beta i}^{(\rho)} = -(A_{\alpha i33}^{(\rho)} x_\beta + \varepsilon_{3\nu\beta} B_{\alpha i33}^{(\rho)}) n_\alpha, \quad \tilde{t}_{3i}^{(\rho)} = -A_{\alpha i33}^{(\rho)} n_\alpha,$$
$$\tilde{t}_{4i}^{(\rho)} = -(A_{\alpha i3\nu}^{(\rho)} \varepsilon_{3\beta\nu} x_\beta + B_{\alpha i33}^{(\rho)}) n_\alpha,$$
$$\tilde{m}_{\beta i}^{(\rho)} = -(B_{33\alpha i}^{(\rho)} x_\beta + \varepsilon_{3\nu\beta} C_{\alpha i3\nu}^{(\rho)}) n_\alpha, \quad \tilde{m}_{3i}^{(\rho)} = -B_{33\alpha i}^{(\rho)} n_\alpha,$$
$$\tilde{m}_{4i}^{(\rho)} = -(B_{3\nu\alpha i}^{(\rho)} \varepsilon_{3\beta\nu} x_\beta + C_{\alpha i33}^{(\rho)}) n_\alpha,$$
$$F_{\beta i} = -[(A_{\alpha i33}^{(1)} - A_{\alpha i33}^{(2)}) x_\beta + \varepsilon_{3\nu\beta}(B_{\alpha i3\nu}^{(1)} - B_{\alpha i3\nu}^{(2)})] \nu_\alpha,$$
$$F_{3i} = -(A_{\alpha i33}^{(1)} - A_{\alpha i33}^{(2)}) \nu_\alpha,$$
$$F_{4i} = -[(A_{\alpha i3\nu}^{(1)} - A_{\alpha i3\nu}^{(2)}) \varepsilon_{3\beta\nu} x_\beta + (B_{\alpha i33}^{(1)} - B_{\alpha i33}^{(2)})] \nu_\alpha,$$
$$G_{\beta i} = -[(B_{33\alpha i}^{(1)} - B_{33\alpha i}^{(2)}) x_\beta + \varepsilon_{3\nu\beta}(C_{\alpha i3\nu}^{(1)} - C_{\alpha i3\nu}^{(2)})] \nu_\alpha,$$
$$G_{3i} = -(B_{33\alpha i}^{(1)} - B_{33\alpha i}^{(2)}) \nu_\alpha,$$
$$G_{4i} = -[(B_{3\nu\alpha i}^{(1)} - B_{3\nu\alpha i}^{(2)}) \varepsilon_{3\beta\nu} x_\beta + (C_{\alpha i33}^{(1)} - C_{\alpha i33}^{(2)})] \nu_\alpha. \tag{6.6.53}$$

The necessary and sufficient conditions (6.6.47) for the existence of the solution are satisfied for each boundary value problem $B^{(s)}$, $(s = 1,2,3,4)$. In what follows we assume that the functions $u_i^{(s)}$ and $\varphi_i^{(s)}$ are determined.

We seek the solution of the problem $P^{(0)}$ in the form

$$
\begin{aligned}
u_\alpha &= -\frac{1}{2}b_\alpha x_3^2 + \varepsilon_{3\beta\alpha} b_4 x_\beta x_3 + \sum_{s=1}^{4} b_s u_\alpha^{(s)} + v_\alpha(x_1,x_2),\\
u_3 &= (b_1 x_1 + b_2 x_2 + b_3)x_3 + \sum_{s=1}^{4} b_s u_3^{(s)} + v_3(x_1,x_2),\\
\varphi_\alpha &= \varepsilon_{3\alpha\beta} b_\beta x_3 + \sum_{s=1}^{4} b_s \varphi_\alpha^{(s)} + \psi_\alpha(x_1,x_2),\\
\varphi_3 &= b_4 x_3 + \sum_{s=1}^{4} b_s \varphi_3^{(s)} + \psi_3(x_1,x_2),
\end{aligned}
\tag{6.6.54}
$$

where $u_i^{(s)}$ and $\varphi_i^{(s)}$ are the solutions of the problems $B^{(s)}$, v_i and ψ_i are unknown functions, and b_s, $(s = 1,2,3,4)$, are unknown constants. It follows from (6.6.54), (6.1.4), $(6.1.3)_{1,2}$, (6.6.48) and (6.6.49) that

$$
t_{ij} = \sum_{s=1}^{4} b_s \pi_{ij}^{(s)} + s_{ij},\quad m_{ij} = \sum_{s=1}^{4} b_s \mu_{ij}^{(s)} + \tau_{ij},
\tag{6.6.55}
$$

where

$$
\begin{aligned}
&\pi_{ij}^{(\alpha)} = t_{ij}^{(\alpha)} + A_{ij33}^{(\rho)} x_\alpha + \varepsilon_{3\nu\alpha} B_{ij3\nu}^{(\rho)},\quad \pi_{ij}^{(3)} = t_{ij}^{(3)} + A_{ij33}^{(\rho)},\\
&\pi_{ij}^{(4)} = t_{ij}^{(4)} + A_{ij3\nu}^{(\rho)} \varepsilon_{3\beta\nu} x_\beta + B_{ij33}^{(\rho)},\\
&\mu_{ij}^{(\alpha)} = m_{ij}^{(\alpha)} + B_{33ij}^{(\rho)} x_\alpha + \varepsilon_{3\nu\alpha} C_{ij3\nu}^{(\rho)},\quad \mu_{ij}^{(3)} = m_{ij}^{(3)} + B_{33ij}^{(\rho)},\\
&\mu_{ij}^{(4)} = m_{ij}^{(4)} + B_{3\nu ij}^{(\rho)} \varepsilon_{3\beta\nu} x_\beta + C_{ij33}^{(\rho)} \text{ on } \Sigma_\rho
\end{aligned}
\tag{6.6.56}
$$

and

$$
\begin{aligned}
&s_{ij} = A_{ijmn}^{(\rho)} \gamma_{mn} + B_{ijmn}^{(\rho)} \nu_{mn} - \beta_{ij}^{(\rho)} f,\\
&\tau_{ij} = B_{mnij}^{(\rho)} \gamma_{mn} + C_{ijmn}^{(\rho)} \nu_{mn} - \xi_{ij}^{(\rho)} f,\\
&\gamma_{\alpha i} = v_{i,\alpha} + \varepsilon_{i\alpha k}\psi_k,\quad \gamma_{3i} = \varepsilon_{i3k}\psi_k,\quad \nu_{\alpha i} = \psi_{i,\alpha},\quad \nu_{3i} = 0.
\end{aligned}
\tag{6.6.57}
$$

With the help of (6.6.50), (6.6.53) and (6.6.55), the equilibrium equations (6.5.2) reduce to

$$
s_{\alpha i,\alpha} = 0,\quad \tau_{\alpha i,\alpha} + \varepsilon_{ijk} s_{jk} = 0 \text{ on } \Sigma_\rho.
\tag{6.6.58}
$$

In view of (6.6.51), (6.6.53)-(6.6.56), the conditions (6.6.2) reduce to

$$[v_i]_1 = [v_i]_2, [\psi_i]_1 = [\psi_i]_2, \quad [s_{\alpha i}]_1 \nu_\alpha = [s_{\beta i}]_2 \nu_\beta, \\ [\tau_{\alpha i}]_1 \nu_\alpha = [\tau_{\beta i}]_2 \nu_\beta \text{ on } \Gamma. \tag{6.6.59}$$

With the help of (6.6.52), (6.6.53), (6.6.55) and (6.6.56), the boundary conditions (6.5.3) become

$$[s_{\alpha i} n_\alpha]_\rho = 0, \quad [\tau_{\alpha i} n_\alpha]_\rho = 0 \text{ on } L_\rho. \tag{6.6.60}$$

Thus, the functions v_i and ψ_i are the components of the displacement and microrotation vector fields in the thermoelastic generalized plane strain problem (6.6.57)-(6.6.60), corresponding to the temperature field f. In what follows we assume that this problem is solved.

The conditions (6.5.4) are satisfied on the basis of the equilibrium equations and boundary conditions,

$$\int_\Sigma t_{31} dA = \int_\Sigma (t_{13} - m_{j2,j}) dA = \int_\Sigma (t_{13} + x_1 t_{j3,j} - m_{j2,j}) dA = \\ = \int_\Sigma [(x_1 t_{\alpha 3})_{,\alpha} - m_{\alpha 2,\alpha}] dA = \int_L (x_1 t_{\alpha 3} n_\alpha - m_{\alpha 2} n_\alpha) ds = 0,$$

$$\int_\Sigma t_{32} dA = \int_\Sigma (t_{23} + m_{j1,j}) dA = \int_\Sigma (t_{23} + x_2 t_{j3,j} + m_{j1,j}) dA = \\ = \int_\Sigma [(x_2 t_{\alpha 3})_{,\alpha} + m_{\alpha 1,\alpha}] dA = \int_L (x_2 t_{\alpha 3} n_\alpha + m_{\alpha 1} n_\alpha) ds = 0.$$

In view of (6.6.55), the conditions (6.5.5)-(6.5.7) lead to the following system for the constants b_s,

$$\sum_{s=1}^{4} D_{rs} b_s = M_r, \quad (r = 1, 2, 3, 4), \tag{6.6.61}$$

where

$$D_{\alpha s} = \sum_{\rho=1}^{2} \int_{\Sigma_\rho} (x_\alpha \pi_{33}^{(s)} + \varepsilon_{3\beta\alpha} \mu_{3\beta}^{(s)}) dA, \quad D_{3s} = \sum_{\rho=1}^{2} \int_{\Sigma_\rho} \pi_{33}^{(s)} dA,$$

$$D_{4s} = \sum_{\rho=1}^{2} \int_{\Sigma_\rho} (\varepsilon_{3\alpha\beta} x_\alpha \pi_{3\beta}^{(s)} + \mu_{33}^{(s)}) dA,$$

$$M_\alpha = -\sum_{\rho=1}^{2} \int_{\Sigma_\rho} (x_\alpha s_{33} - \varepsilon_{3\alpha\beta} \tau_{3\beta}) dA, \quad M_3 = -\sum_{\rho=1}^{2} \int_{\Sigma_\rho} s_{33} dA,$$

$$M_4 = -\sum_{\rho=1}^{2} \int_{\Sigma_\rho} (\varepsilon_{3\alpha\beta} x_\alpha s_{3\beta} + \tau_{33}) dA. \qquad (6.6.62)$$

We can prove (see Section 6.5) that $\det(D_{rs}) \neq 0$ so that the system (6.6.61) determines the constants $b_k, (k = 1, 2, 3, 4)$. Thus, the problem $P^{(0)}$ is solved.

Consider now the recurrence process. We assume that we know the functions $u_i^*, \varphi_i^*, e_{ij}^*, \kappa_{ij}^*, t_{ij}^*$ and m_{ij}^* which satisfy the equations (6.5.2), $(6.1.3)_{1,2}$ (6.1.4) on B_ρ, the conditions (6.6.2) on Π_0, the conditions (6.5.3) on Π and the conditions (6.5.4)-(6.5.7), where the temperature field θ has the form (6.5.46). The problem consists in the finding of the functions $u_i, \varphi_i, e_{ij}, \kappa_{ij}, t_{ij}$ and m_{ij} which satisfy the equations (6.5.2), $(6.1.3)_{1,2}$, (6.1.4) on B_ρ, the conditions (6.6.2) on Π_0, the conditions (6.5.3) on Π and the conditions (6.5.4)-(6.5.7) assuming that the temperature distribution is (6.5.45). We seek the solution of this problem in the form

$$\begin{aligned}
u_\alpha &= (n+1)[\int_0^{x_3} u_\alpha^* dx_3 - \frac{1}{2} c_\alpha x_3^2 + \varepsilon_{3\beta\alpha} c_4 x_3 x_\beta + \sum_{s=1}^{4} c_s u_\alpha^{(s)} + v_\alpha(x_1, x_2)],\\
u_3 &= (n+1)[\int_0^{x_3} u_3^* dx_3 + (c_1 x_1 + c_2 x_2 + c_3) x_3 + \sum_{s=1}^{4} c_s u_3^{(s)} + v_3(x_1, x_2)],\\
\varphi_\alpha &= (n+1)[\int_0^{x_3} \varphi_\alpha^* dx_3 + \varepsilon_{3\alpha\beta} c_\beta x_3 + \sum_{s=1}^{4} c_s \varphi_\alpha^{(s)} + \psi_\alpha(x_1, x_2)],\\
\varphi_3 &= (n+1)[\int_0^{x_3} \varphi_3^* dx_3 + c_4 x_3 + \sum_{s=1}^{4} c_s \varphi_3^{(s)} + \psi_3(x_1, x_2)],
\end{aligned} \qquad (6.6.63)$$

where $u_i^{(s)}$ and $\varphi_i^{(s)}$ are the displacements and microrotations in the problems $B^{(s)}, v_i$ and ψ_i are unknown functions, and $c_s, (s = 1, 2, 3, 4)$, are unknown constants. It follows from (6.6.63) and the constitutive equations that

$$\begin{aligned}
t_{ij} &= (n+1)[\int_0^{x_3} t_{ij}^* dx_3 + \sum_{s=1}^{4} c_s \pi_{ij}^{(s)} + s_{ij} + T_{ij}^{(\rho)}],\\
m_{ij} &= (n+1)[\int_0^{x_3} m_{ij}^* dx_3 + \sum_{s=1}^{4} c_s \mu_{ij}^{(s)} + \mu_{ij} + M_{ij}^{(\rho)}] \text{ on } B_\rho,
\end{aligned} \qquad (6.6.64)$$

where $\pi_{ij}^{(s)}$ and $\mu_{ij}^{(s)}$ are given by (6.6.56), s_{ij} and μ_{ij} are defined by

$$s_{ij} = A_{ijmn}^{(\rho)}\gamma_{mn} + B_{ijmn}^{(\rho)}\nu_{mn}, \quad \mu_{ij} = B_{mnij}^{(\rho)}\gamma_{mn} + C_{ijmn}^{(\rho)}\nu_{mn},$$
$$\gamma_{\alpha i} = v_{i,\alpha} + \varepsilon_{i\alpha k}\psi_k, \quad \gamma_{3i} = \varepsilon_{i3k}\psi_k, \nu_{\alpha i} = \psi_{i,\alpha}, \quad \nu_{3i} = 0 \tag{6.6.65}$$

and we have used the notations

$$T_{ij}^{(\rho)} = A_{ij3k}^{(\rho)}u_k^*(x_1, x_2, 0) + B_{ij3k}^{(\rho)}\psi_k^*(x_1, x_2, 0),$$
$$M_{ij}^{(\rho)} = B_{3kij}^{(\rho)}u_k^*(x_1, x_2, 0) + C_{ij3k}^{(\rho)}\psi_k^*(x_1, x_2, 0). \tag{6.6.66}$$

With the help of (6.6.50), (6.6.53) and (6.6.64), the equilibrium equations (6.5.2) reduce to

$$s_{\alpha i,\alpha} + \Phi_i^{(\rho)} = 0, \quad \mu_{\alpha i,\alpha} + \varepsilon_{ijk}s_{jk} + \Lambda_i^{(\rho)} = 0 \text{ on} \Sigma_\rho, \tag{6.6.67}$$

where

$$\Phi_i^{(\rho)} = T_{\alpha i,\alpha}^{(\rho)} + t_{3i}^*(x_1, x_2, 0), \quad \Lambda_i^{(\rho)} = M_{\alpha i,\alpha}^{(\rho)} + m_{3i}^*(x_1, x_2, 0). \tag{6.6.68}$$

In view of (6.6.51), (6.6.53), (6.6.63) and (6.6.64), the conditions (6.6.2) become

$$[v_i]_1 = [v_i]_2, \quad [\psi_i]_1 = [\psi_i]_2,$$
$$[s_{\alpha i}]_1\nu_\alpha = [s_{\beta i}]_2\nu_\beta + p_i, \quad [\mu_{\alpha i}]_1\nu_\alpha = [\mu_{\beta i}]_2\nu_\beta + q_i \text{ on } \Gamma, \tag{6.6.69}$$

where

$$p_i = (T_{\alpha i}^{(2)} - T_{\alpha i}^{(1)})\nu_\alpha, \quad q_i = (M_{\alpha i}^{(2)} - M_{\alpha i}^{(1)})\nu_\alpha. \tag{6.6.70}$$

The boundary conditions (6.5.3) reduce to

$$[s_{\alpha i}n_\alpha]_\rho = P_i^{(\rho)}, \quad [\mu_{\alpha i}n_\alpha]_\rho = Q_i^{(\rho)} \text{ on } L_\rho, \tag{6.6.71}$$

where

$$P_i^{(\rho)} = -T_{\alpha i}^{(\rho)}n_\alpha, \quad Q_i^{(\rho)} = -M_{\alpha i}^{(\rho)}n_\alpha.$$

Thus, the functions v_i are the components of the displacement vector and ψ_i are the components of the microrotation vector in the generalized isothermal plane strain problem (6.6.65), (6.6.67), (6.6.69) and (6.6.71).

It is easy to see that

$$\sum_{\rho=1}^{2}[\int_{\Sigma_\rho}\Phi_i^{(\rho)}dA+\int_{L_\rho}P_i^{(\rho)}ds]+\int_\Gamma p_i ds=\int_\Sigma t_{3i}^*(x_1,x_2,0)dA,$$

$$\sum_{\rho=1}^{2}[\int_{\Sigma_\rho}(\varepsilon_{3\alpha\beta}x_\alpha\Phi_\beta^{(\rho)}+\Lambda_3^{(\rho)})dA+\int_{L_\rho}(\varepsilon_{3\alpha\beta}x_\alpha P_\beta^{(\rho)}+Q_3^{(\rho)})ds]+$$

$$+\int_\Gamma(\varepsilon_{3\alpha\beta}x_\alpha p_\beta+q_3)ds=\int_\Sigma[\varepsilon_{3\alpha\beta}x_\alpha t_{3\beta}^*(x_1,x_2,0)+m_{33}^*(x_1,x_2,0)]dA. \tag{6.6.72}$$

The functions t_{ij}^* and m_{ij}^* satisfy the conditions (6.5.4)-(6.5.7), so that the relations (6.6.72) show that the necessary and sufficient conditions for the existence of solution of the boundary value problem (6.6.65), (6.6.67), (6.6.69), (6.6.71) are satisfied. In what follows we shall treat the functions v_i and ψ_i as known functions.

We note that on the basis of the equilibrium equations (6.5.2) and the boundary conditions (6.5.3) we obtain

$$\int_\Sigma t_{31}dA=\int_\Sigma(t_{13}+x_1t_{j3,j}-m_{j2,j})dA=\int_\Sigma[(x_1t_{\alpha3})_{,\alpha}-$$

$$-m_{\alpha2,\alpha}+x_1t_{33,3}-m_{32,3}]dA=(n+1)\int_\Sigma(x_1t_{33}^*-m_{32}^*)dA, \tag{6.6.73}$$

$$\int_\Sigma t_{32}dA=(n+1)\int_\Sigma(x_2t_{33}^*+m_{31}^*)dA.$$

The functions t_{33}^* and $m_{3\alpha}^*$ satisfy the conditions (6.5.6). Thus from (6.6.73) we conclude that $t_{3\alpha}$ satisfy the conditions (6.5.4).

From (6.5.5)-(6.5.7) and (6.6.64) we obtain the following equations for the constants c_s,

$$\sum_{s=1}^{4}D_{rs}c_s=-N_r,(r=1,2,3,4), \tag{6.6.74}$$

where D_{rs} are given by (6.6.62) and

$$N_\alpha=\sum_{\rho=1}^{2}\int_{\Sigma_\rho}[x_\alpha(s_{33}+T_{33}^{(\rho)})-\varepsilon_{3\alpha\beta}(\mu_{3\beta}+M_{3\beta}^{(\rho)})]dA,$$

$$N_3=\sum_{\rho=1}^{2}\int_{\Sigma_\rho}(s_{33}+T_{33}^{(\rho)})dA,N_4=\sum_{\rho=1}^{2}\int_{\Sigma_\rho}[\varepsilon_{3\alpha\beta}x_\alpha(s_{3\beta}+T_{3\beta}^{(\rho)})+\mu_{33}+M_{33}^{(\rho)}]dA.$$

The system (6.6.74) determines the constants $c_s, (s = 1,2,3,4)$. Thus, the problem is solved. The results presented in this section have been established in [176].

c) **Application.** We now use the method given in Section 6.6(a) to study the deformation of a heterogeneous circular cylinder which is subjected to a constant temperature variation $\theta = T^* =$ constant. We assume that the domain Σ_1 is bounded by two concentric circles L and Γ of radius R_1 and R_2, respectively, where $R_2 < R_1$. The domain Σ_2 is bounded by the circle Γ. We suppose that the domains Σ_1 and Σ_2 are occupied by two different homogeneous and isotropic elastic materials.

In this case the solution of the problem has the form (6.6.13). The functions v_α and w satisfy the equations (6.6.15), (6.6.16) on Σ and the conditions (6.6.17). We can see that in this case we have

$$\begin{aligned} &v_\alpha = K_1 x_\alpha \text{ on } \Sigma_2, \quad v_\alpha = (K_2 + K_3 r^{-2}) x_\alpha \text{ on } \Sigma_1, \\ &w = 0 \text{ on } \Sigma, \end{aligned} \tag{6.6.75}$$

where

$$\begin{aligned} K_1 &= \frac{1}{2} T^* m C^{(2)} [C^{(1)} b^{(1)} (R_1^2 - R_2^2) + C^{(1)} b^{(2)} R_2^2 + 2 b^{(2)} \omega^{(1)} R_1^2], \\ K_2 &= \frac{1}{2} T^* m C^{(1)} [C^{(2)} b^{(1)} (R_1^2 - R_2^2) + C^{(2)} b^{(2)} R_2^2 + 2 b^{(1)} \omega^{(1)} R_1^2], \\ K_3 &= T^* m \omega^{(1)} [b^{(2)} C^{(2)} - b^{(1)} C^{(1)}] R_1^2 R_2^2, \quad r = (x_1^2 + x_2^2)^{1/2}, \\ m^{-1} &= C^{(2)} (R_1^2 - R_2^2) + C^{(1)} R_1^2 + 2\omega^{(1)} R_1^2, \\ C^{(\alpha)} &= 2(2\lambda^{(\alpha)} + 2\mu^{(\alpha)} + \kappa^{(\alpha)})^{-1}, \quad \omega^{(\alpha)} = (2\mu^{(\alpha)} + \kappa^{(\alpha)})^{-1}. \end{aligned} \tag{6.6.76}$$

The solutions of the problems $A^{(s)}, (s = 1,2,3)$, and the constants A_{rs} introduced by (6.6.19) have been established in [174], p. 384. We note that

$$\begin{aligned} &A_{12} = A_{21} = A_{3\alpha} = A_{\alpha 3} = 0, \\ &A_{33} = (\lambda^{(1)} + 2\mu^{(1)} + \kappa^{(1)} + 2\lambda^{(1)} G_1) S_1 + \\ &\quad + (\lambda^{(2)} + 2\mu^{(2)} + \kappa^{(2)} + 2\lambda^{(2)} G_2) S_2, \end{aligned} \tag{6.6.77}$$

where

$$S_1 = \pi (R_1^2 - R_2^2), \quad S_2 = \pi R_2^2, G_1 = (1 - 2\nu^{(1)}) G_3 R_1^{-2} - \nu^{(1)},$$
$$G_2 = [R_2^{-2} + R_1^{-2} - 2\nu^{(1)} R_1^{-2}] G_3, \nu^{(\alpha)} = \lambda^{(\alpha)} (2\lambda^{(\alpha)} + 2\mu^{(\alpha)} + \kappa^{(\alpha)})^{-1},$$

$$\begin{aligned} G_3 = &(\nu^{(1)} - \nu^{(2)}) [(2\mu^{(1)} + \kappa^{(1)}) \nu^{(2)} (R_2^{-2} - R_1^{-2}) \frac{1}{\lambda^{(2)}} + \\ &+ R_1^{-2} + R_2^{-2} - 2\nu^{(1)} R_1^{-2}]^{-1}. \end{aligned}$$

It follows from (6.6.19) and (6.6.75) that

$$d_\alpha = 0, d_3 = b^{(2)}S_2 + b^{(1)}S_1 - 2\lambda^{(2)}K_1S_2 - 2\lambda^{(1)}K_2S_1. \tag{6.6.78}$$

In view of (6.6.77) and (6.6.78), the system (6.6.18) implies that

$$a_\alpha = 0, \quad a_3 = \frac{1}{A_{33}}d_3. \tag{6.6.79}$$

The functions $u_\alpha^{(3)}$ and $\varphi_3^{(3)}$ are given by

$$\begin{aligned} u_\alpha^{(3)} &= G_2x_\alpha \text{ on } \Sigma_2, u_\alpha^{(3)} = (G_1 + G_3r^{-2})x_\alpha \text{ on } \Sigma_1, \\ \varphi_3^{(3)} &= 0 \text{ on } \Sigma. \end{aligned} \tag{6.6.80}$$

Thus, from (6.6.13), (6.6.75) and (6.6.80) we obtain

$$u_\alpha = a_3u_\alpha^{(3)} + v_\alpha, \quad u_3 = a_3x_3, \quad \varphi_i = 0,$$

where the constant a_3 is given by (6.6.79).

Chapter 7

NONSIMPLE MATERIALS

7.1 The Nonlinear Thermoelasticity

In this section we study the theory of thermoelastic media where the second-order displacement gradient is added to the classical set of independent constitutive variables. The theory of nonsimple elastic materials has been a subject of intensive study. The equations of motion, constitutive equations, and the boundary conditions of the theory of nonsimple elastic bodies were first given in a general, nonlinear form by Toupin [364], [365]. The theory of nonsimple thermoelastic materials has been discussed in various papers (see, for example, Ahmadi and Firoozbakhsh [2], [3], Batra [13], Ieşan [175], Martinez and Quintanilla [254]).

In what follows we consider the motion of a body described by the mapping (1.1.1). We postulate an energy balance in the form

$$\int_P \rho_0(\dot{e} + v_i\dot{v}_i)dV = \int_P \rho_0(f_i v_i + S)dV + \int_{\partial P}(T_i v_i + M_{ji}v_{i,j} + Q)dA, \tag{7.1.1}$$

for every part P of B and every time. Here, ρ_0 is the density in the reference configuration B, e is the internal energy per unit mass, f_i is the body force per unit mass, T_i is the stress vector associated with the surface $\partial\mathcal{P}$ but measured per unit area of $\partial P, M_{ij}$ is the hypertraction associated with the surface $\partial\mathcal{P}$ measured per unit undeformed area, Q is the heat flux across $\partial\mathcal{P}$, measured per unit area of ∂P and $v_i = \dot{x}_i$.

We use the method given by Green and Rivlin [125] to obtain the basic equations from the balance of energy and the invariance requirements under superposed rigid body motions.

We suppose that the continuum has arrived at the given state at time t through some prescribed motion. We consider a second motion which differs from the given motion only by a constant superposed rigid translational velocity. We assume that $\rho_0, \dot{e}, f_i, T_i, M_{ij}, S$ and Q are unaltered by such superposed rigid velocity. If we use the equation

(7.1.1) with v_i replaced by $v_i + a_i$ where a_i are arbitrary constants, we obtain (1.1.6). If the components of the first Piola-Kirchhoff stress tensor are T_{Ki}, it follows from (1.1.6) that

$$T_i = T_{Ai}N_A. \tag{7.1.2}$$

The local form of the equation (1.1.6) is

$$T_{Ai,A} + \rho_0 f_i = \rho_0 \dot{v}_i \text{ on } B \times (t_0, t_1). \tag{7.1.3}$$

Use of the relations (7.1.2) and (7.1.3) in (7.1.1) leads to the following form of the energy balance

$$\int_P \rho_0 \dot{e} dV = \int_P (T_{ki}v_{i,B} + \rho_0 S)dV + \int_{\partial P} (M_{ji}v_{i,j} + Q)dA. \tag{7.1.4}$$

We apply this equation to a region which in the reference state was a tetrahedron bounded by coordinate planes through the point X and by a plane whose unit normal is N_K, to obtain the result

$$(M_{ji} - M_{Kji}N_K)v_{i,j} + Q - Q_A N_A = 0, \tag{7.1.5}$$

where Q_K is the heat flux and M_{Kji} is the hyperstress tensor associated with surfaces which were originally coordinate planes perpendicular to the X_K-axes through the point X, measured per unit undeformed area. Then, using (7.1.5) in (7.1.4) and applying the resulting equation to an arbitrary region, we obtain

$$\rho_0 \dot{e} = (T_{Ai} + M_{Kji,K}X_{A,j})v_{i,A} + M_{Kji}X_{L,j}v_{i,LK} + Q_{A,A} + \rho_0 S. \tag{7.1.6}$$

We introduce the notation

$$H_{KLi} = M_{Kji}X_{L,j}, \quad S_{Ai} = T_{Ai} + H_{KAi,K}. \tag{7.1.7}$$

Then, the equation (7.1.6) becomes

$$\rho_0 \dot{e} = S_{Ai}v_{i,A} + H_{KLi}v_{i,KL} + Q_{A,A} + \rho_0 S. \tag{7.1.8}$$

This equation may be written in the form

$$\rho_0 \dot{e} = p_{ij}v_{i,j} + H_{KLi}v_{i,KL} + Q_{A,A} + \rho_0 S, \tag{7.1.9}$$

where

$$p_{ij} = S_{Ai}x_{j,A}. \tag{7.1.10}$$

Let us now consider a motion of the body which differs from the given motion only by a superposed uniform rigid body angular velocity, and let us assume that $\rho_0, \dot{e}, T_{Ai}, H_{KLi}, Q_A$ and S are not affected by such motion. Then, the equation (7.1.9) leads to

$$p_{ij} + H_{KLi}x_{j,KL} = p_{ji} + H_{KLj}x_{i,KL}. \qquad (7.1.11)$$

Let $H_{(MN)i}$ be the part of H_{MNi}, symmetric with respect to M and N. The skew symmetric part $H_{[MN]i}$ makes no contribution to the rate of work over any closed surface in the body, or over the boundary. Thus,

$$\begin{aligned}
&\int_{\partial P} (T_i v_i + M_{ji} v_{i,j}) dA = \int_{\partial P} (T_{Ki} v_i + H_{KLi} v_{i,L}) N_K dA = \\
&= \int_{\partial P} [(S_{Ki} - H_{MKi,M}) v_i + H_{KLi} v_{i,L}] N_K dA = \\
&= \int_{\partial P} [(S_{Ki} - H_{(MK)i,M}) v_i + H_{(KL)i} v_{i,L}] N_K dA + \\
&+ \int_{\partial P} (H_{[KL]i} v_i)_{,L} N_K dA = \int_{\partial P} [(S_{Ki} - H_{(MK)i,M}) v_i + \\
&+ H_{(KL)i} v_{i,L}] N_K dA.
\end{aligned}$$

In what follows we shall assume that $H_{[MN]i} = 0$.

We postulate the entropy production inequality in the form (1.1.18) for every part P of B and any time. If we introduce the Helmholtz free energy $\psi = e - T\eta$, then the equation (7.1.9) becomes

$$\rho_0(\dot{\psi} + \dot{T}\eta + T\dot{\eta}) = (T_{Ai} + H_{BAi,B})\dot{x}_{i,A} + H_{BAi}\dot{x}_{i,AB} + Q_{L,L} + \rho_0 S. \quad (7.1.12)$$

A thermoelastic material is defined as one for which the following constitutive equations hold

$$\begin{aligned}
&\psi = \widehat{\psi}(x_{i,K}, x_{i,KL}, T, T_{,M}, X_P), \\
&T_{Aj} = \widehat{T}_{Aj}(x_{i,K}, x_{i,KL}, T, T_{,M}, X_P), \\
&H_{ABj} = \widehat{H}_{ABj}(x_{i,K}, x_{i,KL}, T, T_{,M}, X_P), \\
&\eta = \widehat{\eta}(x_{i,K}, x_{i,KL}, T, T_{,M}, X_P), \qquad (7.1.13)\\
&Q_A = \widehat{Q}_A(x_{i,K}, x_{i,KL}, T, T_{,M}, X_P), \\
&Q = \widehat{Q}(x_{i,k}, x_{i,KL}, T, T_{,M}, X_P, N_S), \\
&M_{rs} = \widehat{M}_{rs}(x_{i,K}, x_{i,KL}, T, T_{,M}, X_P, N_S),
\end{aligned}$$

where $\widehat{\psi}, \widehat{T}_{Aj}, \widehat{H}_{ABj}, \widehat{\eta}, \widehat{Q}_A, \widehat{Q}$ and $\widehat{M}_{rs}$ are smooth functions. In particular, we assume that the partial derivatives of $\widehat{\psi}, \widehat{T}_{Ai}, \widehat{H}_{ABj}, \widehat{\eta}$ and $\widehat{Q}_A$, at any fixed state $(x_{i,K}, x_{i,KL}, T, T_{,M}, X_N)$ are bounded functions on $\overline{B} \times [t_0, t_1]$ where t_1 is some time instant.

For a given deformation, $v_{i,j}$ in (7.1.5) may be chosen arbitrary so that, on the basis of the constitutive equations, we get

$$Q = Q_K N_K, \quad M_{ji} = M_{Kji} N_K. \tag{7.1.14}$$

If we use (7.1.13) in (1.1.18) then we obtain the following local form of the second law of thermodynamics

$$\rho_0 T \dot{\eta} - \rho_0 S - Q_{L,L} + \frac{1}{T} Q_K T_{,K} \geq 0. \tag{7.1.15}$$

From (7.1.12) and (7.1.15) we find

$$\begin{gathered}(T_{Ai} + H_{BAi,B} - \rho_0 \frac{\partial \psi}{\partial x_{i,A}}) \dot{x}_{i,A} - \rho_0 (\eta + \frac{\partial \psi}{\partial T}) \dot{T} + \\ +(H_{BAi} - \rho_0 \frac{\partial \psi}{\partial x_{i,AB}}) \dot{x}_{i,AB} - \rho_0 \frac{\partial \psi}{\partial T_{,A}} \dot{T}_{,A} + \frac{1}{T} Q_A T_{,A} \geq 0.\end{gathered} \tag{7.1.16}$$

If we use the notation $\sigma = \rho_0 \psi$ and standard arguments, then the inequality (7.1.16) implies that

$$\begin{gathered}\sigma = \widehat{\sigma}(x_{i,K}, x_{i,KL}, T, X_N), \quad \rho_0 \eta = -\frac{\partial \widehat{\sigma}}{\partial T}, \\ T_{Ai} = \frac{\partial \widehat{\sigma}}{\partial x_{i,A}} - \left(\frac{\partial \widehat{\sigma}}{\partial x_{i,AK}} \right)_{,K}, \quad H_{ABi} = \frac{\partial \widehat{\sigma}}{\partial x_{i,AB}}\end{gathered} \tag{7.1.17}$$

and

$$\widehat{Q}_A T_{,A} \geq 0. \tag{7.1.18}$$

Clearly, from (7.1.7) and (7.1.17) we get

$$S_{Ai} = \frac{\partial \widehat{\sigma}}{\partial x_{i,A}}. \tag{7.1.19}$$

With the aid of (7.1.17) the energy equation (7.1.12) reduces to

$$\rho_0 T \dot{\eta} = Q_{L,L} + \rho_0 S. \tag{7.1.20}$$

We note that the inequality (7.1.18) implies

$$\widehat{Q}_K (x_{i,K}, x_{i,KL}, T, 0, X_N) = 0. \tag{7.1.21}$$

The functions $\widehat{\sigma}$ and $\widehat{Q}_A$ must be invariant under Euclidean displacements. It can be shown that every invariant functions σ and Q_A are expressible in the form (cf. Toupin [365])

$$\begin{aligned} \sigma &= \widetilde{\sigma}(E_{KL}, G_{KLM}, T, X_M), \\ Q_A &= \widetilde{Q}_A(E_{KL}, G_{KLM}, T, T_{,K}, X_M), \end{aligned} \tag{7.1.22}$$

where

$$2E_{KL} = x_{i,K}x_{i,L} - \delta_{KL}, \quad G_{KLM} = x_{i,K}x_{i,LM}. \tag{7.1.23}$$

From (7.1.17), (7.1.19), (7.1.22) and (7.1.23) we obtain

$$S_{Ki} = \frac{\partial \widetilde{\sigma}}{\partial E_{KL}} x_{i,L} + \frac{\partial \widetilde{\sigma}}{\partial G_{KLM}} x_{i,LM}, \quad H_{KLi} = \frac{\partial \widetilde{\sigma}}{\partial G_{MKL}} x_{i,M}. \tag{7.1.24}$$

To avoid ambiguity we assume that $\widetilde{\sigma}$ in (7.1.24) is arranged as a symmetric function of E_{AB}, and E_{AB} is understood to mean $(E_{AB}+E_{BA})/2$ in $\partial\widetilde{\sigma}/\partial E_{AB}$.

We note that if S_{Ki} and H_{KLi} have the form (7.1.24) then the equation (7.1.11) is identically satisfied. From (7.1.7) and (7.1.24) we obtain

$$T_{Ai} = \frac{\partial \widetilde{\sigma}}{\partial E_{AL}} x_{i,L} + \frac{\partial \widetilde{\sigma}}{\partial G_{ALM}} x_{i,LM} - \left(\frac{\partial \widetilde{\sigma}}{\partial G_{MKA}} x_{i,M} \right)_{,K}. \tag{7.1.25}$$

Clearly, from (7.1.17) we get

$$\rho_0 \eta = -\frac{\partial \widetilde{\sigma}}{\partial T}. \tag{7.1.26}$$

Thus, the basic equations of the theory consist of the equations of motion (7.1.3), the equation of energy (7.1.20), the constitutive equations (7.1.22), (7.1.24)-(7.1.26), and the geometrical equations (7.1.23). To the system of basic equations we must adjoin boundary conditions and initial conditions. Following Toupin [365] and Mindlin [263], to derive the form of mechanical boundary conditions we study the total rate of work of the surface forces over the surface ∂B. By (7.1.2), (7.1.7) and (7.1.14) we find

$$\int_{\partial B} (T_i v_i + M_{ji} v_{i,j}) dA = \int_{\partial B} (T_{Ki} v_i + H_{KLi} v_{i,L}) N_K dA. \tag{7.1.27}$$

In the last integral of (7.1.27), $v_{i,L}$ is not independent of v_i on ∂B; only its normal component $Dv_i = v_{i,L} N_L$ is independent. If we introduce the surface gradient D_K defined by

$$D_K = (\delta_{KL} - N_K N_L) \frac{\partial}{\partial X_L},$$

then

$$H_{KLi}v_{i,L} = H_{KLi}N_K N_L D v_i - v_i D_L(H_{KLi}N_K) + D_L(H_{KLi}N_K v_i) =$$
$$= H_{KLi}N_K N_L D v_i - v_i D_L(H_{KLi}N_K) + (D_M N_M) N_K N_L H_{KLi} v_i +$$
$$+ \varepsilon_{QPM} N_Q (\varepsilon_{MAL} N_A N_K H_{KLi} v_i)_{,P}, \tag{7.1.28}$$

where ε_{KLM} is the alternating symbol.

We assume that B is a bounded region with Lipschitz boundary ∂B which consists of a finite number of smooth surfaces. Let C_α be intersections of two adjoined smooth surfaces and $C = \cup\, C_\alpha$. By Stokes' theorem we obtain

$$\int_{\partial B} \varepsilon_{QPM} N_Q (\varepsilon_{MAL} N_A N_K H_{KLi} v_i)_{,P} dA = \int_C \Pi_i v_i dS, \tag{7.1.29}$$

where

$$\Pi_i = < H_{KLi} N_K M_L >, \quad M_L = \varepsilon_{LKA} s_K N_A. \tag{7.1.30}$$

Here, $< f >$ denotes the difference in values of f as a given point on an edge is approached from either side, and s_K are the components of the unit vector tangent to C.

It follows from (7.1.27)-(7.1.29) that

$$\int_{\partial B} (T_i v_i + M_{ji} v_{i,j}) dA = \int_{\partial B} (P_i v_i + R_i D v_i) dA + \int_C \Pi_i v_i dS, \tag{7.1.31}$$

where

$$\begin{aligned} P_i &= T_{Ki} N_K - D_L(H_{KLi} N_K) + (D_P N_P) H_{KLi} N_K N_L, \\ R_i &= H_{KLi} N_K N_L. \end{aligned} \tag{7.1.32}$$

If ∂B is smooth, then the line integral in (7.1.31) vanishes. Let B_{KL} be the second fundamental form of the surface ∂B. Then (see, for example, McConnel [259]) we have

$$D_K N_L = -B_{KL}$$

and P_i can be written in the following alternative form

$$P_i = T_{Ki} N_K - N_K D_L H_{KLi} + H_{KLi}(B_{KL} - B_{MM} N_K N_L). \tag{7.1.33}$$

It follows from (7.1.7) and (7.1.33) that

$$\begin{aligned} P_i = {} & S_{Ki} N_K - 2 N_K D_L H_{KLi} - N_K N_L D H_{KLi} + \\ & + (B_{KL} - B_{MM} N_K N_L) H_{KLi}. \end{aligned} \tag{7.1.34}$$

Let S_1 and S_2 be complementary subsurfaces of ∂B. We consider the following boundary conditions

$$\begin{aligned} &x_i = \widetilde{x}_i, \quad Dx_i = \widetilde{d}_i, \quad T = \widetilde{T} \text{ on } S_1 \times (t_0, t_1), \\ &P_i = \widetilde{P}_i, \quad R_i = \widetilde{R}_i, \quad Q = \widetilde{Q} \text{ on } S_2 \times (t_0, t_1), \\ &\Pi_i = \widetilde{\Pi}_i \text{ on } C \times (t_0, t_1), \end{aligned} \tag{7.1.35}$$

where $\widetilde{x}_i, \widetilde{d}_i, \widetilde{T}, \widetilde{P}_i, \widetilde{R}_i, \widetilde{Q}$ and $\widetilde{\Pi}_i$ are prescribed functions.

The initial conditions are

$$\mathbf{x}(\mathbf{X},0)=\mathbf{x}^0(\mathbf{X}), \dot{\mathbf{x}}(\mathbf{X},0)=\mathbf{v}^0(\mathbf{X}), \eta(\mathbf{X},0)=\eta^0(\mathbf{X}), \mathbf{X} \in \overline{B}, \tag{7.1.36}$$

where $\mathbf{x}^0, \mathbf{v}^0$ and η^0 are given. We assume that: (i) ρ_0 is continuous and strictly positive on $\overline{B}$; (ii) $\mathbf{f}$ and S are continuous on $\overline{B} \times [t_0, t_1)$; (iii) $\mathbf{x}^0, \mathbf{v}^0$ and η^0 are continuous on $\overline{B}$; (iv) $\widetilde{x}_i$ and $\widetilde{T}$ are continuous on $S_1 \times [t_0, t_1)$, and $\widetilde{d}_i$ are continuous in time and piecewise regular on $S_1 \times [t_0, t_1)$; (v) $\widetilde{P}_i, \widetilde{R}_i$ and $\widetilde{Q}$ are continuous in time and piecewise regular on $S_2 \times [t_0, t_1)$; (vi) $\widetilde{\Pi}_i$ are continuous in time and piecewise regular on $C \times [t_0, t_1)$.

7.2 Uniqueness and Continuous Dependence Results

Within the context of classical nonlinear thermoelasticity, Dafermos [80] has established continuous dependence results for materials which are non-conductor of heat. In this section we extend some of results derived by Dafermos [80] to the nonsimple thermoelastic materials. We assume that the material is non-conductor of heat and that the body is bounded by a smooth surface. Throughout this section we assume that the constitutive equations (7.1.13) hold.

We shall say that $U = (\mathbf{x}, T)$ is a smooth thermodynamic process if $\mathbf{x}$ and T are functions of class C^2 which satisfy the equations of motion (7.1.3) and the balance law of energy

$$\begin{aligned} \frac{\partial}{\partial t}\left(\frac{1}{2}\rho_0 \mathbf{v}^2 + \sigma + \rho_0 T\eta\right) = \rho_0 f_i v_i + (T_{Ai} v_i + \\ + H_{ABi} v_{i,B})_{,A} + \rho_0 S. \end{aligned} \tag{7.2.1}$$

In what follows we assume that

$$\rho_0(\mathbf{X}) \geq \rho_1 > 0, \quad \mathbf{X} \in \overline{B}. \tag{7.2.2}$$

A smooth thermodynamic process will be called admissible if it satisfies the Clausius-Duhem inequality (7.1.15). Since $Q_A = 0$, the energy equation (7.1.20) becomes

$$T\dot{\eta} = S. \tag{7.2.3}$$

Clearly, the inequality of Clausius-Duhem is satisfied by a smooth admissible process, as an equality. In what follows it will be convenient to denote the components of the deformation gradient $\mathbf{F} = (x_{i,A})$ by F_{iA} and the components of the second order gradient $\mathbf{K} = (x_{i,AB})$ by K_{iAB}. Moreover, the symbol $|\cdot|$ denotes a norm either in an Euclidean vector space or in a tensor space, while $\|\cdot\|$ denotes an L_2-norm.

Let $U = (\mathbf{x}, T)$ and $\overline{U} = (\overline{\mathbf{x}}, \overline{T})$ be two smooth admissible thermodynamic processes defined on $\overline{B} \times [0, t_1]$ corresponding to the supply terms $(\mathbf{f}, S)$ and $(\overline{\mathbf{f}}, \overline{S})$, respectively.

We introduce the function Φ on $[0, t_1]$ by

$$\begin{aligned}\Phi = \int_B [&\frac{1}{2}\rho_0(v_i - \overline{v}_i)(v_i - \overline{v}_i) + \sigma - \overline{\sigma} - \overline{S}_{Ai}(F_{iA} - \overline{F}_{iA}) - \\ &-\overline{H}_{ABi}(K_{iAB} - \overline{K}_{iAB}) + \rho_0\eta(T - \overline{T})]dV,\end{aligned} \tag{7.2.4}$$

where

$$\begin{aligned}&\overline{\sigma} = \widehat{\sigma}(\overline{x}_{i,A}, \overline{x}_{i,AB}, \overline{T}), \quad \overline{S}_{Ai} = \overline{T}_{Ai} + \overline{H}_{BAi,B}, \\ &\overline{T}_{Ai} = \widehat{T}_{Ai}(\overline{x}_{i,A}, \overline{x}_{i,AB}, \overline{T}), \quad \overline{H}_{ABi} = \widehat{H}_{ABi}(\overline{x}_{i,A}, \overline{x}_{i,AB}, \overline{T}).\end{aligned} \tag{7.2.5}$$

By using (7.1.17), (7.1.19) we can show that W is of quadratic order in $\|(\mathbf{v} - \overline{\mathbf{v}}, \mathbf{F} - \overline{\mathbf{F}}, \mathbf{K} - \overline{\mathbf{K}}, T - \overline{T})(\cdot, t)\|_{L^2(B)}$.

We now establish an evolutionary relation for Φ. Let E be defined by

$$E = [(T_{Ai} - \overline{T}_{Ai})(v_i - \overline{v}_i) + (H_{ABi} - \overline{H}_{ABi})(v_{i,B} - \overline{v}_{i,B})]N_A. \tag{7.2.6}$$

Theorem 7.2.1. *If U and $\overline{U}$ are two smooth admissible processes corresponding to the supply terms $(\mathbf{f}, S)$ and $(\overline{\mathbf{f}}, \overline{S})$ in $L^\infty(B \times [0, t_1])$ then*

$$\begin{aligned}\dot{\Phi} = &\int_B \rho_0[(f_i - \overline{f}_i)(v_i - \overline{v}_i) + \frac{1}{T}(S - \overline{S})(T - \overline{T})]dV + \\ &+ \int_{\partial B} E dA + \int_B \Big\{\dot{\overline{F}}_{iA}[S_{Ai} - \overline{S}_{Ai} - \frac{\partial \overline{S}_{Ai}}{\partial \overline{F}_{jB}}(F_{jB} - \overline{F}_{jB}) - \\ &- \frac{\partial \overline{S}_{Ai}}{\partial \overline{K}_{jMN}}(K_{jMN} - \overline{K}_{jMN}) - \frac{\partial \overline{S}_{Ai}}{\partial \overline{T}}(T - \overline{T})] + \dot{\overline{K}}_{iAB}[H_{ABi} - \\ &- \overline{H}_{ABi} - \frac{\partial \overline{H}_{ABi}}{\partial \overline{F}_{jM}}(F_{jM} - \overline{F}_{jM}) - \frac{\partial \overline{H}_{ABi}}{\partial \overline{K}_{jMN}}(K_{jMN} - \overline{K}_{jMN}) -\end{aligned}$$

$$-\frac{\partial\overline{H}_{ABi}}{\partial\overline{T}}(T-\overline{T})]-\rho_0\dot{\overline{T}}[\eta-\overline{\eta}-\frac{\partial\overline{\eta}}{\partial\overline{F}_{iA}}(F_{iA}-\overline{F}_{iA})-$$

$$-\frac{\partial\overline{\eta}}{\partial\overline{K}_{jMN}}(K_{jMN}-\overline{K}_{jMN})-\frac{\partial\overline{\eta}}{\partial\overline{T}}(T-\overline{T})]-\frac{1}{T\overline{T}}\rho_0\overline{S}(T-\overline{T})^2\Big\}\,dV. \tag{7.2.7}$$

Proof. By (7.2.4) we obtain

$$\begin{aligned}\dot{\Phi}=\int_B\{&\frac{\partial}{\partial t}[\frac{1}{2}\rho_0\mathbf{v}^2+\sigma+\rho_0T\eta]-\frac{\partial}{\partial t}[\frac{1}{2}\rho_0\overline{\mathbf{v}}^2+\overline{\sigma}+\\&+\rho_0\overline{T}\overline{\eta}]-\rho_0\dot{v}_i\overline{v}_i-\rho_0v_i\dot{\overline{v}}_i+2\rho_0\overline{v}_i\dot{\overline{v}}_i-\\&-\dot{\overline{S}}_{Ai}(F_{iA}-\overline{F}_{iA})-\dot{\overline{H}}_{ABi}(K_{iAB}-\overline{K}_{iAB})-\\&-\overline{S}_{Ai}(\dot{F}_{iA}-\dot{\overline{F}}_{iA})-\overline{H}_{ABi}(\dot{K}_{iAB}-\dot{\overline{K}}_{iAB})-\\&-\rho_0\dot{\overline{T}}(\eta-\overline{\eta})-\rho_0\overline{T}(\dot{\eta}-\dot{\overline{\eta}})\}dV.\end{aligned}$$

Using the equations (7.1.3), (7.2.1) and the divergence theorem we arrive at

$$\begin{aligned}\dot{\Phi}=&\int_B\rho_0[(f_i-\overline{f}_i)(v_i-\overline{v}_i)+S-\overline{S}]dV+\int_{\partial B}EdA+\\&+\int_B[\dot{\overline{F}}_{iA}(S_{Ai}-\overline{S}_{Ai})-\dot{\overline{S}}_{Ai}(F_{iA}-\overline{F}_{iA})+\dot{\overline{K}}_{iAB}(H_{ABi}-\overline{H}_{ABi})-\\&-\dot{\overline{H}}_{ABi}(K_{iAB}-\overline{K}_{iAB})-\rho_0\dot{\overline{T}}(\eta-\overline{\eta})-\rho_0\overline{T}(\dot{\eta}-\dot{\overline{\eta}})]dV.\end{aligned}\tag{7.2.8}$$

By (7.1.17), (7.1.19),

$$\begin{aligned}\dot{\overline{S}}_{Ai}&=\frac{\partial\overline{S}_{Bj}}{\partial\overline{F}_{iA}}\dot{\overline{F}}_{jB}+\frac{\partial\overline{H}_{MNj}}{\partial\overline{F}_{iA}}\dot{\overline{K}}_{jMN}-\rho_0\frac{\partial\overline{\eta}}{\partial\overline{F}_{iA}}\dot{\overline{T}},\\\dot{\overline{H}}_{ABi}&=\frac{\partial\overline{S}_{Lm}}{\partial\overline{K}_{iAB}}\dot{\overline{F}}_{mL}+\frac{\partial\overline{H}_{MNr}}{\partial\overline{K}_{iAB}}\dot{\overline{K}}_{rMN}-\rho_0\frac{\partial\overline{\eta}}{\partial\overline{K}_{iAB}}\dot{\overline{T}},\end{aligned}$$

so that

$$\begin{aligned}&\dot{\overline{F}}_{iA}(S_{Ai}-\overline{S}_{Ai})+\dot{\overline{K}}_{iAB}(H_{ABi}-\overline{H}_{ABi})-\dot{\overline{S}}_{Ai}(F_{iA}-\overline{F}_{iA})-\\&-\dot{\overline{H}}_{ABi}(K_{iAB}-\overline{K}_{iAB})-\rho_0\dot{T}(\eta-\overline{\eta})-\rho_0\overline{T}(\dot{\eta}-\dot{\overline{\eta}})=\\&=\dot{\overline{F}}_{iA}[S_{Ai}-\overline{S}_{Ai}-\frac{\partial\overline{S}_{Ai}}{\partial\overline{F}_{jB}}(F_{jB}-\overline{F}_{jB})-\frac{\partial\overline{S}_{Ai}}{\partial\overline{K}_{jMN}}(K_{jMN}-\overline{K}_{jMN})-\\&-\frac{\partial\overline{S}_{Ai}}{\partial\overline{T}}(T-\overline{T})]+\dot{\overline{K}}_{iAB}[H_{ABi}-\overline{H}_{ABi}-\end{aligned}$$

$$
\begin{aligned}
&-\frac{\partial \overline{H}_{ABi}}{\partial \overline{F}_{jM}}(F_{jM}-\overline{F}_{jM})-\frac{\partial \overline{H}_{ABi}}{\partial \overline{K}_{jMN}}(K_{jMN}-\overline{K}_{jMN})-\\
&-\frac{\partial \overline{H}_{ABi}}{\partial \overline{T}}(T-\overline{T})]-\rho_0\dot{T}[\eta-\overline{\eta}-\frac{\partial \overline{\eta}}{\partial \overline{F}_{iA}}(F_{iA}-\overline{F}_{iA})-\\
&-\frac{\partial \overline{\eta}}{\partial \overline{K}_{jMN}}(K_{jMN}-\overline{K}_{jMN})-\frac{\partial \overline{\eta}}{\partial \overline{T}}(T-\overline{T})]-\\
&-\rho_0\dot{\overline{\eta}}(T-\overline{T})-\rho_0\overline{T}(\dot{\eta}-\dot{\overline{\eta}}).
\end{aligned}
\tag{7.2.9}
$$

With the help of (7.2.3) we obtain

$$
\begin{aligned}
&\rho_0 S-\rho_0\overline{S}-\rho_0\dot{\overline{\eta}}(T-\overline{T})-\rho_0 T(\dot{\eta}-\dot{\overline{\eta}})=\\
&=\frac{1}{T}\rho_0(S-\overline{S})(T-\overline{T})+\frac{1}{T\overline{T}}\rho_0\overline{S}(T-\overline{T})^2.
\end{aligned}
\tag{7.2.10}
$$

By (7.2.8), (7.2.9) and (7.2.10) we obtain the desired result.□

We note that the right-hand side of (7.2.7) is of quadratic order in

$$
\|(\mathbf{v}-\overline{\mathbf{v}},\mathbf{F}-\overline{\mathbf{F}},\mathbf{K}-\overline{\mathbf{K}},T-\overline{T},\mathbf{f}-\overline{\mathbf{f}},S-\overline{S})(\cdot,t)\|_{L^2(B)}.
$$

We say that a smooth admissible process $(\overline{\mathbf{x}},\overline{T})(\mathbf{X},t)$ resides in the convexity region of internal energy if: (α) for each $(\mathbf{X},t)\in\overline{B}\times[0,t_1]$ there exists a positive constant λ such that

$$
\begin{aligned}
&\frac{\partial^2\overline{\sigma}}{\partial \overline{F}_{iA}\partial \overline{F}_{jB}}\Lambda_{iA}\Lambda_{jB}+2\frac{\partial^2\overline{\sigma}}{\partial \overline{F}_{iA}\partial \overline{K}_{jMN}}\Lambda_{iA}\Gamma_{jMN}+\\
&+\frac{\partial^2\overline{\sigma}}{\partial \overline{K}_{iAB}\partial \overline{K}_{jMN}}\Gamma_{iAB}\Gamma_{jMN}\geq\lambda(|\mathbf{\Lambda}|^2+|\mathbf{\Gamma}|^2),
\end{aligned}
\tag{7.2.11}
$$

for all Λ_{iA} and Γ_{jMN};

$$
(\beta)\quad \frac{\partial\overline{\eta}}{\partial\overline{T}}>0 \text{ on } \overline{B}\times[0,t_1]. \tag{7.2.12}
$$

Theorem 7.2.2. *Let $(\overline{\mathbf{x}},\overline{T})$ be a smooth admissible process on $\overline{B}\times[0,t_1]$, residing in the convexity region of internal energy, with supply terms $(\overline{\mathbf{f}},\overline{S})(\mathbf{X},t)\in L^\infty(B\times[0,t_1])$. Then there are positive constants δ,α_1,M_1, and N_1 with the following property:*

If $(\mathbf{x},T)$ is any smooth admissible process on $B\times[0,t_1]$ corresponding to the supply terms $(\mathbf{f},S)\in L^\infty(B\times[0,t_1])$, such that

$$
\begin{aligned}
&|\mathbf{F}(\mathbf{X},t)-\overline{\mathbf{F}}(\mathbf{X},t)|+|\mathbf{K}(\mathbf{X},t)-\overline{\mathbf{K}}(\mathbf{X},t)|+\\
&+|T(\mathbf{X},t)-\overline{T}(\mathbf{X},t)|<\delta,
\end{aligned}
\tag{7.2.13}
$$

$(\mathbf{X},t) \in B \times [0,t_1]$, *and*

$$(P_i - \overline{P}_i)(v_i - \overline{v}_i) + (R_i - \overline{R}_i)(Dv_i - D\overline{v}_i) = 0 \text{ on } \partial B \times [0,t_1], \quad (7.2.14)$$

then we have, for any $s \in [0,t_1]$,

$$\begin{aligned}&\|(\mathbf{v}-\overline{\mathbf{v}}, \mathbf{F}-\overline{\mathbf{F}}, \mathbf{K}-\overline{\mathbf{K}}, T-\overline{T})(\cdot,s)\|_{L^2(B)} \leq \\ &\leq \{M_1\|(\mathbf{v}-\overline{\mathbf{v}}, \mathbf{F}-\overline{\mathbf{F}}, \mathbf{K}-\overline{\mathbf{K}}, T-\overline{T})(\cdot,0)\|_{L^2(B)}+ \\ &+N_1\int_0^s \|(\mathbf{f}-\overline{\mathbf{f}}, S-\overline{S})(\cdot,t)\|_{L^2(B)}dt\}\exp(\alpha_1 s).\end{aligned} \quad (7.2.15)$$

Proof. Clearly,

$$\int_{\partial B} E dA = \int_{\partial B} [(P_i - \overline{P}_i)(v_i - \overline{v}_i) + (R_i - \overline{R}_i)(Dv_i - D\overline{v}_i)]dA. \quad (7.2.16)$$

By (7.1.17), (7.1.19), (7.2.14), (7.2.15), Schwarz inequality and (7.2.7) it follows that one can determine the positive constants c_1 and c_2 such that for all $t \in [0,t_1]$, one has

$$\begin{aligned}&\dot{\Phi} \leq c_1\|(\mathbf{F}-\overline{\mathbf{F}}, \mathbf{K}-\overline{\mathbf{K}}, T-\overline{T})(\cdot,t)\|^2_{L^2(B)}+ \\ &+c_2\|(\mathbf{f}-\overline{\mathbf{f}}, S-\overline{S})(\cdot,t)\|_{L^2(B)}\|(\mathbf{v}-\overline{\mathbf{v}}, T-\overline{T})(\cdot,t)\|_{L^2(B)}.\end{aligned} \quad (7.2.17)$$

Let

$$Z(t) = \|(\mathbf{v}-\overline{\mathbf{v}}, \mathbf{F}-\overline{\mathbf{F}}, \mathbf{K}-\overline{\mathbf{K}}, T-\overline{T})(\cdot,t)\|_{L^2(B)},$$

$t \in [0,t_1]$. If we fix $s \in [0,t_1]$ and integrate the inequality (7.2.15) over $[0,\tau], \tau \in [0,s]$, we obtain

$$\Phi(\tau) \leq \Phi(0) + c_1\int_0^\tau Z^2(t)dt + c_2\int_0^\tau \|(\mathbf{f}-\overline{\mathbf{f}}, S-\overline{S})(\cdot,t)\|_{L^2(B)}Z(t)dt. \quad (7.2.18)$$

By (7.1.17), (7.1.19), we have

$$\begin{aligned}&\sigma - \overline{\sigma} - \overline{S}_{Ai}(F_{iA} - \overline{F}_{iA}) - \overline{H}_{ABi}(K_{iAB} - \overline{K}_{iAB}) + \rho_0\eta(T-\overline{T}) = \\ &= \frac{1}{2}\left[\frac{\partial^2\overline{\sigma}}{\partial\overline{F}_{iA}\partial\overline{F}_{jB}}(F_{iA} - \overline{F}_{iA})(F_{jB} - \overline{F}_{jB}) + \right. \\ &+2\frac{\partial^2\overline{\sigma}}{\partial\overline{F}_{iA}\partial\overline{K}_{jMN}}(F_{iA} - \overline{F}_{iA})(K_{jMN} - \overline{K}_{jMN})+ \\ &\left.+\frac{\partial^2\overline{\sigma}}{\partial\overline{K}_{iAB}\overline{K}_{jMN}}(K_{iAB} - \overline{K}_{iAB})(K_{jMN} - \overline{K}_{jMN})\right] + \\ &+\rho_0(T-\overline{T})^2\frac{\partial\overline{\eta}}{\partial\overline{T}} + \mathrm{o}(|\mathbf{F}-\overline{\mathbf{F}}|^2 + |\mathbf{K}-\overline{\mathbf{K}}|^2 + |T-\overline{T}|^2).\end{aligned} \quad (7.2.19)$$

From (7.2.2), (7.2.4), (7.2.11), (7.2.18) and (7.2.19), it follows that there exists a positive constant δ such that whenever (7.2.13) holds, we have

$$\begin{aligned}&\rho_1\|(\mathbf{v}-\overline{\mathbf{v}})(\cdot,\tau)\|^2_{L^2(B)}+\\&+\lambda\|(\mathbf{F}-\overline{\mathbf{F}},\mathbf{K}-\overline{\mathbf{K}},T-\overline{T})(\cdot,\tau)\|^2_{L^2(B)}\leq 2\Phi(0)+\\&+2c_1\int_0^\tau Z^2(t)dt+2c_2\int_0^\tau\|(\mathbf{f}-\overline{\mathbf{f}},S-\overline{S})(\cdot,t)\|_{L^2(B)}Z(t)dt.\end{aligned}\tag{7.2.20}$$

If we use the estimate

$$\Phi(0)\leq c_3\|(\mathbf{v}-\overline{\mathbf{v}},\mathbf{F}-\overline{\mathbf{F}},\mathbf{K}-\overline{\mathbf{K}},T-\overline{T})(\cdot,0)\|^2_{L^2(B)},c_3>0$$

and the notations

$$m=\min(\rho_1,\lambda),M_1^2=2c_3/m,\alpha_1=c_1/m,N_1=c_2/m,$$

then (7.2.20) implies that

$$Z^2(t)\leq M_1^2Z^2(0)+2\int_0^\tau[\alpha_1 Z^2(t)+N_1\|(\mathbf{f}-\overline{\mathbf{f}},S-\overline{S})(\cdot,t)\|_{L^2(B)}Z(t)]dt.$$

By Lemma 1.4.1 we arrive at the desired result.□

The next uniqueness theorem is a direct consequence of Theorem 7.2.2.

Theorem 7.2.3. *Let $(\overline{\mathbf{x}},\overline{T})$ and $(\mathbf{x},T)$ be as in Theorem 7.2.2. Assume that the corresponding supply terms $(\overline{\mathbf{f}},\overline{S})$ and $(\mathbf{f},S)$ coincide on $B\times[0,t_1]$ and that both processes originate from the same state. Then $(\overline{\mathbf{x}},\overline{T})$ and $(\mathbf{x},T)$ coincide on $B\times[0,t_1]$.*

The results presented in this section have been established in [64]. Following [80] we can extend the results for smooth processes in the strong ellipticity region.

7.3 Linear Theories

In the first part of this section we use the equations established in Section 7.1 to derive the linear theory of nonsimple thermoelastic bodies. Then we present a linear theory of thermoelasticity which is capable of predicting a finite speed of heat propagation.

a) **Basic equations.** In what follows we use the notation $X_i=\delta_{iK}X_K$ and $f_{,i}=\partial f/\partial X_i$, where δ_{iK} is the Kronecker delta. We denote

$$u_i=x_i-X_i,\quad \theta=T-T_0,$$

where T_0 is the constant temperature in the reference configuration. In the linear theory we assume that $u_i = \varepsilon u_i'$ and $\theta = \varepsilon\theta'$, where ε is a constant small enough for squares and higher powers to be neglected. The tensors E_{KL} and G_{KLM} reduce to

$$e_{ij} = \frac{1}{2}(u_{i,j} + u_{j,i}), \quad \kappa_{ijk} = u_{k,ij}, \tag{7.3.1}$$

respectively. In the linear theory, and assuming that the initial body is free from stress and hyperstress, we have

$$\begin{aligned} \widetilde{\sigma} = {} & \frac{1}{2}A_{ijrs}e_{ij}e_{rs} + B_{ijpqr}e_{ij}\kappa_{pqr} + \frac{1}{2}C_{ijkpqr}\kappa_{ijk}\kappa_{pqr} - \\ & -\frac{1}{2}d\theta^2 + a_{ij}e_{ij}\theta + c_{ijk}\kappa_{ijk}\theta, \end{aligned} \tag{7.3.2}$$

where the constitutive coefficients are prescribed functions on X_s. We note that

$$\begin{aligned} & A_{ijrs} = A_{rsij} = A_{jirs}, B_{ijpqr} = B_{jipqr} = B_{ijqpr}, \\ & C_{ijkpqr} = C_{pqrijk} = C_{jikpqr}, a_{ij} = a_{ji}, c_{ijk} = c_{jik}. \end{aligned} \tag{7.3.3}$$

In the framework of linear theory we shall employ the notations

$$t_{ji} = \delta_{jK}T_{Ki}, \mu_{rsi} = \delta_{rK}\delta_{sL}H_{KLi}, \tau_{ij} = \delta_{iA}S_{Aj}, q_i = \delta_{iK}Q_K. \tag{7.3.4}$$

It follows from (7.1.21), (7.1.22) and (7.3.4) that

$$q_i = k_{ij}\theta_{,j}. \tag{7.3.5}$$

In view of (7.1.24), (7.1.26) and (7.3.2) we obtain

$$\tau_{ij} = \frac{\partial\widetilde{\sigma}}{\partial e_{ij}}, \quad \mu_{ijk} = \frac{\partial\widetilde{\sigma}}{\partial\kappa_{ijk}}, \quad \rho_0\eta = -\frac{\partial\widetilde{\sigma}}{\partial\theta}. \tag{7.3.6}$$

Within the context of linear theory the relation (7.1.25) becomes

$$t_{ij} = \tau_{ij} - \mu_{kij,k}. \tag{7.3.7}$$

We note that (7.1.11) reduces to $\tau_{ij} = \tau_{ji}$.

The equations of motion (7.1.3) can be written in the form

$$\tau_{ji,j} - \mu_{kji,kj} + \rho_0 f_i = \rho_0\ddot{u}_i \quad \text{on } B \times (t_0, t_1). \tag{7.3.8}$$

In the linear theory the energy equation (7.1.20) reduces to

$$\rho_0 T_0\dot{\eta} = q_{i,i} + \rho_0 S \quad \text{on } B \times (t_0, t_1). \tag{7.3.9}$$

From (7.3.2), (7.3.6) and (7.3.5) we obtain the constitutive equations

$$\begin{aligned}
\tau_{ij} &= A_{ijrs}e_{rs} + B_{ijpqr}\kappa_{pqr} + a_{ij}\theta,\\
\mu_{ijk} &= B_{rsijk}e_{rs} + C_{ijkpqr}\kappa_{pqr} + c_{ijk}\theta,\\
\rho_0\eta &= d\theta - a_{ij}e_{ij} - c_{ijk}\kappa_{ijk},\\
q_i &= k_{ij}\theta_{,j}.
\end{aligned} \tag{7.3.10}$$

The basic equations of the linear theory are: the equations of motion (7.3.8), the equation of energy (7.3.9), the constitutive equations (7.3.10) and the geometrical equations (7.3.1).

By (7.1.30) and (7.1.32) we obtain

$$\begin{aligned}
P_i &= t_{ji}n_j - D_j(\mu_{kji}n_k) + (D_jn_j)\mu_{pqi}n_pn_q,\\
R_i &= \mu_{rsi}n_rn_s, \quad \Pi_i =< \mu_{pji}n_p\varepsilon_{jrq}s_rn_q >,
\end{aligned} \tag{7.3.11}$$

where ε_{ijk} is the alternating symbol, $n_i = \delta_{iK}N_K$ and D_i are the components of the surface gradient. The relations (7.1.33) and (7.1.34) reduce to

$$\begin{aligned}
P_i &= t_{ji}n_j - n_jD_k\mu_{jki} + (b_{rs} - b_{mm}n_rn_s)\mu_{rsi},\\
P_i &= \tau_{ji}n_j - 2n_jD_r\mu_{jri} - n_rn_sD\mu_{rsi} + (b_{rs} - b_{mm}n_rn_s)\mu_{rsi},
\end{aligned} \tag{7.3.12}$$

respectively, where $Df = f_{,i}n_i$ and $b_{ij} = \delta_{iK}\delta_{jL}B_{KL}$. We note that (7.1.27) and (7.1.31) imply that

$$\int_{\partial B}(t_{ji}v_i + \mu_{jri}v_{i,r})n_jdA = \int_{\partial B}(P_iv_i + R_iDv_i)dA + \int_C \Pi_iv_ids. \tag{7.3.13}$$

To the system of field equations we adjoin the boundary conditions

$$\begin{aligned}
&u_i = \widetilde{u}_i, Du_i = \widetilde{\nu}_i, \theta = \widetilde{\theta} \ \text{ on } S_1 \times (t_0, t_1)\\
&P_i = \widetilde{P}_i, R_i = \widetilde{R}_i, q_in_i = \widetilde{q} \ \text{ on } S_2 \times (t_0, t_1),\\
&\Pi_i = \widetilde{\Pi}_i \ \text{ on } C \times (t_0, t_1),
\end{aligned} \tag{7.3.14}$$

where $\widetilde{u}_i, \widetilde{\nu}_i, \widetilde{\theta}, \widetilde{P}_i, \widetilde{R}_i, \widetilde{q}$ and $\widetilde{\Pi}_i$ are given. The initial conditions can be written in the form

$$u_i(\mathbf{X}, 0) = u_i^0(\mathbf{X}), \dot{u}_i(\mathbf{X}, 0) = v_i^0(\mathbf{X}), \theta(\mathbf{X}, 0) = \theta^0(\mathbf{X}), \mathbf{X} \in \overline{B}, \tag{7.3.15}$$

where $\mathbf{u}^0, \mathbf{v}^0$ and θ^0 are prescribed functions.

The linear theory of nonsimple thermoelastic materials based on the Clausius-Duhem inequality has been established by Ahmadi and Firoozbakhsh [2].

b) **Thermoelasticity with second sound.** We now present a linear theory of thermoelasticity for nonsimple materials by discussing restrictions on constitutive equations with the help of an entropy production inequality proposed by Green and Laws [129]. For simple materials, the linear theory of thermoelasticity based on the entropy production inequality of Green and Laws [129] was derived by Green and Lindsay [130]. This theory is capable of predicting a finite speed of heat propagation. The results we give here have been presented in [175].

We postulate the energy balance in the form (7.1.1). With the help of notations from Section 7.3(a), the local form of the energy balance becomes

$$\rho_0 \dot{e} = \tau_{ij}\dot{e}_{ij} + \mu_{ijk}\dot{\kappa}_{ijk} + q_{i,i} + \rho_0 S. \tag{7.3.16}$$

We note that from (7.1.1) we have obtained the equations of motions and the relations $\tau_{ij} = \tau_{ji}$.

We restrict our attention to the linear theory of homogeneous bodies. In what follows we adopt the entropy production inequality proposed by Green and Laws [129]

$$\int_P \rho_0 \dot{\eta} dV - \int_P \rho_0 \frac{1}{\Phi} S dV - \int_{\partial P} \frac{1}{\Phi} q_i n_i dA \geq 0, \tag{7.3.17}$$

for every part P of B and every time, where η is the specific entropy and Φ is a strictly positive function. With the help of (7.3.16), the inequality (7.3.17) yields

$$\tau_{ij}\dot{e}_{ij} + \mu_{ijk}\dot{\kappa}_{ijk} - \rho_0(\dot{A} + \eta\dot{\Phi}) + \frac{1}{\Phi} q_i \Phi_{,i} \geq 0, \tag{7.3.18}$$

where $A = e - \eta\Phi$.

We assume that $A, \Phi, \tau_{ij}, \mu_{ijk}, \eta$ and q_i are functions of $e_{ij}, \kappa_{ijk}, \theta, \dot{\theta}$, $\theta_{,i}$, consistent with the assumption of the linear theory. It is easy to see that e_{ij} and κ_{ijk} are invariant under superposed rigid-body motions.

From the constitutive assumptions and (7.3.18) we obtain the following restrictions

$$\begin{gathered}
\tau_{ij} = \rho_0\left(\frac{\partial A}{\partial e_{ij}} + \eta\frac{\partial \Phi}{\partial e_{ij}}\right), \quad \mu_{ijk} = \rho_0\left(\frac{\partial A}{\partial \kappa_{ijk}} + \eta\frac{\partial \Phi}{\partial \kappa_{ijk}}\right), \\
\frac{\partial A}{\partial \dot{\theta}} + \eta\frac{\partial \Phi}{\partial \dot{\theta}} = 0, \quad -\rho_0\left(\frac{\partial A}{\partial \theta_{,i}} + \eta\frac{\partial \Phi}{\partial \theta_{,i}}\right) + \frac{1}{\Phi} q_i \frac{\partial \Phi}{\partial \dot{\theta}} = 0,
\end{gathered} \tag{7.3.19}$$

$$\frac{q_i}{\Phi}\left(\frac{\partial \Phi}{\partial \theta}\theta_{,i}+\frac{\partial \Phi}{\partial e_{rs}}\kappa_{irs}\right)-\rho_0\left(\frac{\partial A}{\partial \theta}+\eta\frac{\partial \Phi}{\partial \theta}\right)\dot{\theta}\geq 0, \tag{7.3.20}$$

$$q_i\frac{\partial \Phi}{\partial \theta_{,j}}+q_j\frac{\partial \Phi}{\partial \theta_{,i}}=0,\quad q_s\frac{\partial \Phi}{\partial \kappa_{ijk}}=0. \tag{7.3.21}$$

Provided $q_i \neq 0$ it can be shown from (7.3.21) that

$$\Phi=\Phi(\theta,\dot{\theta},e_{ij}). \tag{7.3.22}$$

Following [130], we add the condition

$$\Phi(\theta,0,e_{ij})=T_0+\theta. \tag{7.3.23}$$

In view of (7.3.22), from (7.3.19) we obtain

$$\begin{aligned}\tau_{ij}&=\frac{\partial \sigma^*}{\partial e_{ij}}+\rho_0\eta\frac{\partial \Phi}{\partial e_{ij}},\quad \mu_{ijk}=\frac{\partial \sigma^*}{\partial \kappa_{ijk}},\\ \rho_0\eta&=-\frac{\partial \sigma^*}{\partial \dot{\theta}}\Big/\frac{\partial \Phi}{\partial \dot{\theta}},\quad q_i=\Phi\frac{\partial \sigma^*}{\partial \theta_{,i}}\Big/\frac{\partial \Phi}{\partial \dot{\theta}},\end{aligned} \tag{7.3.24}$$

where $\sigma^*=\rho_0 A$ and we assumed that $\partial\Phi/\partial\dot{\theta}\neq 0$.

With the help of (7.3.19) and (7.3.22), the energy equation (7.3.16) reduces to

$$\rho_0\Phi\dot{\eta}-q_{i,i}-\rho_0 S+\left(\frac{\partial \sigma^*}{\partial \theta}+\rho_0\eta\frac{\partial \Phi}{\partial \theta}\right)\dot{\theta}+\frac{\partial \sigma^*}{\partial \theta_{,i}}\dot{\theta}_{,i}=0. \tag{7.3.25}$$

In the linear theory, and assuming that the initial body is free from stress and hyperstress and has zero heat flux, we have

$$\begin{aligned}\sigma^*&=\sigma_0-a\theta-b\dot{\theta}-\frac{1}{2}d\theta^2-e\theta\dot{\theta}-\frac{1}{2}f\dot{\theta}^2+a_i\theta\theta_{,i}+\\ &+\alpha b_i\dot{\theta}\theta_{,i}+\frac{1}{2}\alpha K_{ij}\theta_{,i}\theta_{,j}+a_{ijk}e_{ij}\theta_{,k}+a_{ij}e_{ij}\theta+\\ &+b_{ij}e_{ij}\dot{\theta}+b_{ijkl}\kappa_{ijk}\theta_{,l}+c_{ijk}\kappa_{ijk}\theta+d_{ijk}\kappa_{ijk}\dot{\theta}+\\ &+\frac{1}{2}A_{ijkl}e_{ij}e_{kl}+B_{ijklm}e_{ij}\kappa_{klm}+\frac{1}{2}C_{ijkrsm}\kappa_{ijk}\kappa_{rsm},\\ \Phi&=T_0+\theta+\alpha\dot{\theta}+\beta\theta\dot{\theta}+\frac{1}{2}\gamma\dot{\theta}^2+g_{ij}e_{ij}\dot{\theta},\end{aligned} \tag{7.3.26}$$

where the constitutive coefficients are prescribed constants. From (7.3.24) and (7.3.26) we obtain

$$\begin{aligned}
\tau_{ij} &= A_{ijkl}e_{kl} + B_{ijrsm}\kappa_{rsm} + a_{ij}\theta + a_{ijk}\theta_{,k} + (b_{ij} + \frac{b}{\alpha}g_{ij})\dot{\theta},\\
\mu_{ijk} &= B_{rsijk}e_{rs} + C_{ijkmnr}\kappa_{mnr} + c_{ijk}\theta + b_{ijkr}\theta_{,r} + d_{ijk}\dot{\theta},\\
\rho_0\eta &= \frac{1}{\alpha}[b + (e - \frac{b\beta}{\alpha})\theta + (f - \frac{b\gamma}{\alpha})\dot{\theta} - \alpha b_i\theta_{,i} - e_{ij}(b_{ij} + \frac{b}{\alpha}g_{ij}) - d_{ijk}\kappa_{ijk}],\\
q_i &= \frac{T_0}{\alpha}(a_i\theta + \alpha b_i\dot{\theta} + \alpha k_{ij}\theta_{,j} + a_{mni}e_{mn} + b_{mnri}\kappa_{mnr}).
\end{aligned} \tag{7.3.27}$$

The entropy inequality (7.3.20) then yields the following restrictions on the coefficients

$$\begin{aligned}
&a_i = 0, a_{ijk} = 0, b_{ijkr} = 0, b = a\alpha, e - a\beta = d\alpha,\\
&\alpha a_{ij} = b_{ij} + a g_{ij},\quad d_{ijk} = \alpha c_{ijk},
\end{aligned} \tag{7.3.28}$$

$$(d\alpha - h)\dot{\theta}^2 + 2b_i\dot{\theta}\theta_{,i} + K_{ij}\theta_{,i}\theta_{,j} \geq 0, \tag{7.3.29}$$

where $h\alpha = f - a\gamma$. The relations (7.3.27) reduce to

$$\begin{aligned}
\tau_{ij} &= A_{ijrs}e_{rs} + B_{ijpqr}\kappa_{pqr} + a_{ij}(\theta + \alpha\dot{\theta}),\\
\mu_{ijk} &= B_{rsijk}e_{rs} + C_{ijkmnr}\kappa_{mnr} + c_{ijk}(\theta + \alpha\dot{\theta}),\\
\rho_0\eta &= a + d\theta + h\dot{\theta} - b_i\theta_{,i} - a_{ij}e_{ij} - c_{ijk}\kappa_{ijk},\\
q_i &= T_0(b_i\dot{\theta} + K_{ij}\theta_{,j}).
\end{aligned} \tag{7.3.30}$$

The coefficients in (7.3.30) have the following symmetries

$$\begin{aligned}
&A_{ijrs} = A_{rsij} = A_{jirs},\quad B_{ijpqr} = B_{jipqr} = B_{ijqpr},\\
&C_{ijkpqr} = C_{pqrijk} = C_{jikpqr}, a_{ij} = a_{ji}, c_{ijk} = c_{jik}, K_{ij} = K_{ji}.
\end{aligned} \tag{7.3.31}$$

In the case of a centrosymmetric isotropic material the number of independent coefficients is greatly reduced and the relations (7.3.31) become

$$\begin{aligned}
\tau_{ij} &= \lambda e_{rr}\delta_{ij} + 2\mu e_{ij} - \beta(\theta + \alpha\dot{\theta})\delta_{ij},\\
\mu_{ijk} &= \frac{1}{2}\alpha_1(\kappa_{rri}\delta_{jk} + 2\kappa_{krr}\delta_{ij} + \kappa_{rrj}\delta_{ik}) +\\
&\quad + \alpha_2(\kappa_{irr}\delta_{jk} + \kappa_{jrr}\delta_{ik}) + 2\alpha_3\kappa_{rrk}\delta_{ij} +\\
&\quad + 2\alpha_4\kappa_{ijk} + \alpha_5(\kappa_{kji} + \kappa_{kij}),\\
\rho_0\eta &= a + d\theta + h\dot{\theta} + \beta e_{rr},\quad q_i = k\theta_{,i},
\end{aligned} \tag{7.3.32}$$

where δ_{ij} is Kronecker's delta. In the linear theory, the equation (7.3.25) reduces to

$$\rho_0 T_0 \dot{\eta} - q_{i,i} = \rho_0 S. \tag{7.3.33}$$

To the system of field equations (7.3.1), (7.3.8), (7.3.30), (7.3.33) we adjoin the boundary conditions (7.3.14) and the initial conditions

$$\begin{aligned} &u_i(\mathbf{X}, 0) = u_i^0(\mathbf{X}), \dot{u}_i(\mathbf{X}, 0) = v_i^0(\mathbf{X}), \eta(\mathbf{X}, 0) = \eta^0(\mathbf{X}), \\ &\theta(\mathbf{X}, 0) = \theta^0(\mathbf{X}), \mathbf{X} \in \overline{B}, \end{aligned} \tag{7.3.34}$$

where $u_i^0, v_i^0, \eta^0, \theta^0$ are prescribed functions.

c) **A uniqueness result.** We introduce the notation

$$2V = A_{ijrs} e_{ij} e_{rs} + 2B_{ijkrs} e_{ij} \kappa_{krs} + C_{ijkpqr} \kappa_{ijk} \kappa_{pqr}. \tag{7.3.35}$$

Theorem 7.3.1. *Assume that*

(i) ρ_0, h, α and T_0 are strictly positive;

(ii) V is a positive semi-definite quadratic form.

Then the boundary-initial-value problem of thermoelasticity has at most one solution.

Proof. With the help of (7.3.30) we have

$$\tau_{ij}\dot{e}_{ij} + \mu_{ijk}\dot{\kappa}_{ijk} + \rho_0\dot{\eta}(\theta + \alpha\dot{\theta}) = \dot{V} + (d\dot{\theta} + h\ddot{\theta} - b_i\dot{\theta}_{,i})(\theta + \alpha\dot{\theta}). \tag{7.3.36}$$

On the other hand, from (7.3.1), (7.3.8), (7.3.13) and (7.3.33), it follows that

$$\begin{aligned} \int_B [\tau_{ij}\dot{e}_{ij} + \mu_{ijk}\dot{\kappa}_{ijk} + \rho_0\dot{\eta}(\theta + \alpha\dot{\theta})]dV = &\int_B \rho_0[f_i v_i + \frac{1}{T_0} S(\theta + \alpha\dot{\theta})]dV + \\ &+ \int_{\partial B} [P_i v_i + R_i D v_i + \frac{1}{T_0} q(\theta + \alpha\dot{\theta})]dA + \int_C \Pi_i v_i ds - \\ &- \int_B [(b_i\dot{\theta} + K_{ij}\theta_{,j})(\theta_{,i} + \alpha\dot{\theta}_{,i}) + \rho_0 v_i \dot{v}_i]dV. \end{aligned} \tag{7.3.37}$$

From (7.3.36) and (7.3.37) we obtain the identity

$$\begin{aligned} &\frac{d}{dt}\int_B \{\frac{1}{2}\rho_0 v_i v_i + V + \frac{1}{2}\alpha[(\alpha d - h)(\theta/\alpha)^2 - 2b_i(\theta/\alpha)\theta_{,i} + \\ &\quad + K_{ij}\theta_{,i}\theta_{,j} + (h/2\alpha)(\theta + \alpha\dot{\theta})^2\}dV + \\ &\quad + \int_B [(d\alpha - h)\dot{\theta}^2 + 2b_i\dot{\theta}\theta_{,i} + K_{ij}\theta_{,i}\theta_{,j}]dV = \\ &= \int_B \rho_0[f_i v_i + \frac{1}{T_0} S(\theta + \alpha\dot{\theta})]dV + \int_{\partial B} [P_i v_i + R_i D v_i + \\ &\quad + \frac{1}{T_0} q(\theta + \alpha\dot{\theta})]dA + \int_C \Pi_i v_i ds. \end{aligned} \tag{7.3.38}$$

Let us note that from the residual entropy inequality (7.3.29) we have the condition

$$(d\alpha - h)x^2 + 2b_i x y_i + K_{ij} y_i y_j \geq 0, \tag{7.3.39}$$

for all arbitrary x and y_i. Suppose that there are two solutions. Then their difference $\{u_i^*, \theta^*\}$ corresponds to null data. From (7.3.38), by using (7.3.39) and standard arguments, it follows that $u_i^* = 0, \theta^* = 0$.□

Existence and uniqueness of solution in theory of nonsimple thermoelasticity have been investigated by various authors (see, e.g., Chiriţă and Rusu [56], Martinez and Quintanilla [254]).

7.4 Isotropic Bodies

From (7.3.1), (7.3.8), (7.3.32) and (7.3.33) we obtain the field equations in terms of the displacement and temperature, for centrosymmetric homogeneous and isotropic bodies

$$\begin{aligned} &c_2^2(1 - \ell_2^2\Delta)\Delta\mathbf{u} + [c_1^2 - c_2^2 - (c_1^2\ell_1^2 - c_2^2\ell_2^2)\Delta] \text{ grad div } \mathbf{u} - \\ &\quad -\frac{\beta}{\rho_0} \text{ grad } (\theta + \alpha\dot{\theta}) + \mathbf{f} = \ddot{\mathbf{u}}, \\ &k\Delta\theta - c(\dot{\theta} + \alpha^*\ddot{\theta}) - T_0\beta \text{ div } \dot{\mathbf{u}} = -\rho_0 S, \end{aligned} \tag{7.4.1}$$

where Δ is the Laplacian and

$$\begin{aligned} &c_1^2 = (\lambda + 2\mu)/\rho_0, \quad c_2^2 = \mu/\rho_0, \ell_1^2 = 2(\lambda + 2\mu)^{-1}\sum_{i=1}^{5}\alpha_i, \\ &\ell_2^2 = 2(\alpha_3 + \alpha_4)/\mu, \quad c = T_0 d, \alpha^* = h/d. \end{aligned} \tag{7.4.2}$$

We now define the following dimensionless quantities

$$x_i = (k/cc_1)x_i', t = (k/cc_1^2)t', u_i = (k/cc_1)u_i', \theta = \theta' T_0. \tag{7.4.3}$$

Introducing the above quantities in (7.4.1) and suppressing the primes we get

$$\begin{aligned} &c_2^2 D_2\mathbf{u} + [c_1^2(1 - \lambda_1^2\Delta) - c_2^2(1 - \lambda_2^2\Delta)] \text{ grad div } \mathbf{u} - \\ &\quad -\frac{1}{\rho_0}\beta T_0(1 + \tau_1\frac{\partial}{\partial t}) \text{ grad } \theta = -N\mathbf{f}, \\ &D_3\theta - \frac{\beta}{c} \text{ div } \dot{\mathbf{u}} = -KS, \end{aligned} \tag{7.4.4}$$

where $N = k/cc_1, K = k\rho_0/T_0c^2c_1^2$. Throughout this section we use the notations

$$D_\alpha = (1-\lambda_\alpha^2\Delta)\Delta - (c_1^2/c_\alpha^2)\frac{\partial^2}{\partial t^2},\ \lambda_\alpha^2 = \ell_\alpha^2(cc_1/k)^2,$$
$$D_3 = \Delta - \frac{\partial}{\partial t} - \tau_0\frac{\partial^2}{\partial t^2},$$
$$D_4 = [c_1^2(1-\lambda_1^2\Delta) - c_2^2(1-\lambda_2^2\Delta)]D_3 - c_1^2\varepsilon(1+\tau_1\frac{\partial}{\partial t})\frac{\partial}{\partial t}, \tag{7.4.5}$$
$$\tau_0 = \alpha^* cc_1^2/k,\ \tau_1 = \alpha cc_1^2/k,\ \varepsilon = \beta^2T_0/\rho_0cc_1^2.$$

We assume that $N\mathbf{f}$ has the Helmholtz decomposition $N\mathbf{f} = \text{grad}\,\vartheta + \text{curl}\,\gamma$.

Theorem 7.4.1. *Let* $\mathbf{u} = grad\ \varphi + curl\ \boldsymbol{\psi}$. *If* $\boldsymbol{\psi}, \varphi$ *and* θ *satisfy the equations*

$$D_2\boldsymbol{\psi} = -\frac{1}{c_2^2}\gamma,$$
$$D_1\varphi - \frac{\beta T_0}{\rho_0c_1^2}(1+\tau_1\frac{\partial}{\partial t})\theta = -\frac{1}{c_1^2}\vartheta, \tag{7.4.6}$$
$$D_3\theta - \frac{\beta}{c}\Delta\dot{\varphi} = -KS,$$

then $\mathbf{u}$ *and* θ *satisfy the field equations.*

This result follows by substituting the representation of $\mathbf{u}$ in the equations (7.4.4). If we introduce the notation

$$\varphi = \frac{\beta T_0}{\rho_0c_1^2}\varphi^*,$$

then from the equations (7.4.6) we obtain

$$D_1\varphi^* - (1+\tau_1\frac{\partial}{\partial t})\theta = -\frac{\rho_0}{\beta T_0}\vartheta,\ \ D_3\theta - \varepsilon\Delta\dot{\varphi}^* = -KS.$$

If $\vartheta = S = 0$, then the functions θ and φ^* satisfy the equation

$$[D_1D_3 - \varepsilon(1+\tau_1\frac{\partial}{\partial t})\Delta]\chi = 0.$$

a) **A formal solution.** We now present a representation of solutions of the field equations (7.4.1).

Theorem 7.4.2. *Let*

$$\mathbf{u} = c_1^2[D_1D_3 - \varepsilon(1+\tau_1\frac{\partial}{\partial t})\frac{\partial}{\partial t}\Delta]\mathbf{G} - D_4 \text{ grad div } \mathbf{G}+ \\ +\frac{\varepsilon c}{\beta}(1+\tau_1\frac{\partial}{\partial t}) \text{ grad } H, \qquad (7.4.7)$$

$$\theta = \frac{\beta c_2^2}{c} D_2 \text{ div } \dot{\mathbf{G}} + D_1 H,$$

where $\mathbf{G}$ *and* H *satisfy*

$$D_2\left[D_1D_3 - \varepsilon\left(1+\tau_1\frac{\partial}{\partial t}\right)\frac{\partial}{\partial t}\Delta\right]\mathbf{G} = -c_1^{-2}c_2^{-2}N\mathbf{f}, \\ \left[D_1D_3 - \varepsilon\left(1+\tau_1\frac{\partial}{\partial t}\right)\frac{\partial}{\partial t}\Delta\right]H = -KS. \qquad (7.4.8)$$

Then $\mathbf{u}$ *and* θ *meet the field equations.*

Proof. By (7.4.7) and $(7.4.4)_1$ we have

$$c_2^2 D_2\mathbf{u} + [c_1^2(1-\lambda_1^2\Delta) - c_2^2(1-\lambda_2^2\Delta)] \text{ grad div } \mathbf{u}- \\ -\frac{\beta T_0}{\rho_0}(1+\tau_1\frac{\partial}{\partial t}) \text{ grad } \theta = c_1^2c_2^2[D_1D_3 - \varepsilon(1+\tau_1\frac{\partial}{\partial t})\frac{\partial}{\partial t}\Delta]D_2\mathbf{G}- \\ -\{c_2^2D_2D_4 - [c_1^2(1-\lambda_1^2\Delta) - c_2^2(1-\lambda_2^2\Delta)][c_1^2D_1D_3- \\ -c_1^2\varepsilon(1+\tau_1\frac{\partial}{\partial t})\frac{\partial}{\partial t}\Delta - D_4\Delta] + \varepsilon c_1^2c_2^2(1+\tau_1\frac{\partial}{\partial t})\frac{\partial}{\partial t}D_2\} \text{ grad div } \mathbf{G}+ \\ +(1+\tau_1\frac{\partial}{\partial t})\{\frac{\varepsilon c c_2^2}{\beta}D_2 - \frac{\beta T_0}{\rho_0}D_1 + \frac{\varepsilon c}{\beta}\Delta[c_1^2(1-\lambda_1^2\Delta)- \\ -c_2^2(1-\lambda_2^2\Delta)]\} \text{ grad } H.$$

In view of $(7.4.8)_1$ and the identities

$$c_2^2D_2D_4 + \varepsilon c_1^2c_2^2(1+\tau_1\frac{\partial}{\partial t})\frac{\partial}{\partial t}D_2 = [c_1^2(1-\lambda_1^2\Delta)- \\ -c_2^2(1-\lambda_2^2\Delta)][c_1^2D_1D_3 - \varepsilon c_1^2(1+\tau_1\frac{\partial}{\partial t})\frac{\partial}{\partial t}\Delta - D_4\Delta], \\ \frac{\varepsilon c c_2^2}{\beta}D_2 - \frac{\beta T_0}{\rho_0}D_1 = -\frac{\varepsilon c}{\beta}[c_1^2(1-\lambda_1^2\Delta) - c_2^2(1-\lambda_2^2\Delta)]\Delta,$$

the right-hand side of the above relation reduce to $-N\mathbf{f}$ and we conclude that the equation $(7.4.4)_1$ is satisfied. By using the identity

$$c_2^2D_2D_3 - c_1^2[D_1D_3 - \varepsilon(1+\tau_1\frac{\partial}{\partial t})\frac{\partial}{\partial t}\Delta] + D_4\Delta = 0,$$

from $(7.4.4)_2$, $(7.4.7)$ and $(7.4.8)_2$ we obtain

$$D_3\theta - \frac{\beta}{c}\text{div }\dot{\mathbf{u}} = [D_1 D_3 - \varepsilon(1+\tau_1\frac{\partial}{\partial t})\frac{\partial}{\partial t}\Delta]H = -KS.$$

Thus the proof is complete.□

The above solutions can be used in order to obtain fundamental solutions.

b) **Plane waves.** In this section we briefly consider harmonic plane wave solutions for homogeneous and isotropic bodies with zero body force and heat supply. It is easy to see that transverse waves are independent of thermal effects, as in classical theory. We may, without loss of generality, take the x_1-axis along the direction of propagation and choose $u_1 = u(x_1,t), u_2 = u_3 = 0, \theta = \theta(x_1,t)$. If we assume that

$$u = \text{Re}\{u^* \exp[i(\xi x_1 - \chi t)]\}, \theta = \text{Re}\{\theta^* \exp[i(\xi x_1 - \chi t)]\},$$

then from (7.4.4), with $\mathbf{f} = \mathbf{0}, S = 0$, we obtain the characteristic equation

$$\begin{aligned}&\lambda_1^2\xi^6 + \xi^4[1 - \lambda_1^2(i\chi + \tau_0\chi^2)] - \xi^2[\chi^2(1+\tau_0+\varepsilon\tau_1)+\\&\quad +i\chi(1+\varepsilon)] + i\chi^3 + \tau_0\chi^4 = 0.\end{aligned} \tag{7.4.9}$$

We denote by $\pm\xi_j$ the roots of the equation (7.4.9). Let us examine the roots of (7.4.9) when $\chi \ll 1$. A straightforward calculation shows that

$$\begin{aligned}\xi_1 &= \frac{\chi}{\sqrt{1+\varepsilon}}(1 + \frac{1}{2}iA_1\chi + \frac{1}{8}A_2\chi^2) + O(\chi^4),\\ \xi_2 &= [\frac{1}{2}\chi(1+\varepsilon)]^{1/2}(1+i) + O(\chi^{3/2}),\\ \xi_3 &= -\frac{i}{\lambda_1} + \frac{1}{2}\varepsilon\lambda_1\chi - \frac{1}{8}\lambda_1 i(4B_1 + \lambda_1^2\varepsilon^2)\chi^2 + B_2\chi^3 + O(\chi^4),\end{aligned} \tag{7.4.10}$$

where

$$\begin{aligned}&A_1 = \varepsilon(1+\varepsilon)^{-2}[1 + (1+\varepsilon)(\tau_1 - \tau_0)], \quad A_2 = A_1^2 + 4(1+\varepsilon)A_3,\\ &A_3 = -\lambda_1^2(1+\varepsilon)^{-3} + \varepsilon(1+\varepsilon)^{-5}[1 + (\tau_1-\tau_0)(1+\varepsilon)][1 - \varepsilon-\\ &\qquad -(1+\varepsilon)(\tau_0 + \varepsilon\tau_1)],\\ &B_1 = \varepsilon\lambda_1^2(1+\varepsilon) + 1 + \varepsilon\tau_1,\\ &2B_2 = \lambda_1^3 - \frac{1}{8}\varepsilon\lambda_1^3(4Q + \lambda_1^2\varepsilon^2) - \varepsilon\lambda_1^3[1 + \tau_0 + \varepsilon\tau_1+\\ &\qquad +2\lambda_1^2\varepsilon(1+\varepsilon)] - \lambda_1^5(1+\varepsilon)[\varepsilon + \lambda_1^{-2}(1+\varepsilon\tau_1)].\end{aligned}$$

The phase velocities v_j and the attenuation coefficients ϑ_j are given by

$$v_j = c_1 \chi/\mathrm{Re}\xi_j, \quad \vartheta_j = (\omega^*/c_1)\mathrm{Im}\,\xi_j, \quad \omega^* = cc_1^2/k. \tag{7.4.11}$$

From (7.4.10) we get

$$\begin{aligned}
&v_1 \approx c_1(1+\varepsilon)^{1/2}(1-\frac{1}{8}A_2\chi^2),\\
&\vartheta_1 \approx (\omega^*/2c_1)\chi^2(1+\varepsilon)^{-5/2}\varepsilon[1+(\tau_1-\tau_0)(1+\varepsilon)],\\
&v_2 \approx c_1(2\chi)^{1/2}(1+\varepsilon)^{-1/2}[1-\frac{1}{2}M(1+\varepsilon)^{-1}\chi],\\
&\vartheta_2 \approx (\omega^*/c_1)[\frac{1}{2}\chi(1+\varepsilon)]^{1/2}[1-\frac{1}{2}M(1+\varepsilon)^{-1}\chi],\\
&M = \varepsilon(1+\varepsilon)\lambda_1^2 + 1 + \tau_0 + \varepsilon\tau_1 - \frac{1}{1+\varepsilon},\\
&v_3 \approx \frac{2c_1}{\varepsilon\lambda_1}(1-\frac{2B_2}{\varepsilon\lambda_1}\chi^2),\\
&\vartheta_3 = -\frac{\omega^*}{c_1}[\frac{1}{\lambda_1}+\frac{1}{8}\lambda_1(4B_1+\lambda_1^2\varepsilon^2)\chi^2].
\end{aligned} \tag{7.4.12}$$

It is easy to see that the roots $\pm\xi_1, \pm\xi_3$ correspond to quasi-elastic waves while $\pm\xi_2$ correspond to quasi-thermal waves. For given χ, the waves are attenuated and dispersed. The above results can be compared with those obtained in [235], [23].

Let us consider now the plane waves of assigned length. If we take ξ to be a real constant, then the equation (7.4.9) is an equation in χ. If the function $\chi = \chi(\xi)$ is known, then the group velocity is given by $v = c_1\chi'$, where $\chi' = d\chi/d\xi$.

Let $\chi_j = \chi_j(\xi), (j = 1, 2, 3, 4)$, be the roots of the equation (7.4.9). Let us consider the case of very long wave-length. If we take into account the relations

$$\begin{aligned}
&\chi_1(0) = 0, \chi_1'(0) = \sqrt{1+\varepsilon}, \quad \chi_1''(0) = -[\frac{2}{3}(1+\varepsilon) + i(1+\varepsilon)^{-1}],\\
&\chi_2(0) = 0, \chi_2'(0) = -\sqrt{1+\varepsilon}, \quad \chi_2'' = \chi_1''(0),\\
&\chi_3(0) = 0, \chi_3'(0) = 0, \quad \chi_3''(0) = -\frac{2i}{1+\varepsilon}, \quad \chi_3'''(0) = 0,\\
&\chi_4(0) = -\frac{i}{\tau_0}, \chi_4'(0) = 0, \quad \chi_4''(0) = 2i[1+\varepsilon(\tau_1-\tau_0)], \chi_4'''(0) = 0,
\end{aligned}$$

then we obtain

$$\chi_1 \approx \xi\sqrt{1+\varepsilon} - \frac{1}{2}\left[\frac{2}{3}(1+\varepsilon) + i(1+\varepsilon)^{-1}\right]\xi^2,$$

$$\chi_2 \approx -\xi\sqrt{1+\varepsilon} - \frac{1}{2}\left[\frac{2}{3}(1+\varepsilon) + i(1+\varepsilon)^{-1}\right]\xi^2,$$

$$\chi_3 \approx -\frac{i}{1+\varepsilon}\xi^2, \quad \chi_4 \approx -\frac{i}{\tau_0} + i\left[1+\varepsilon(\tau_1-\tau_0)\right]\xi^2,$$

when ξ is small. We may identity the roots χ_1, χ_2 with the quasi-elastic waves and χ_3, χ_4 with quasi-thermal waves. The quasi-thermal waves are standing waves. The roots $\chi_i, (i = 1, 2, 3)$, correspond to the harmonic plane waves from classical thermoelasticity.

c) **Equilibrium theory.** Let us consider the linear theory of thermoelastostatics for centrosymmetric, homogeneous and isotropic solids. As in classical theory we assume that the mechanical loadings are absent, the principal attention being devoted to the deformation due to the temperature field. The equations (7.3.8) and (7.3.33) reduce to

$$\tau_{ji,j} - \mu_{sji,sj} = 0 \tag{7.4.13}$$

and

$$q_{i,i} = -\rho_0 S, \tag{7.4.14}$$

respectively. Let us consider the boundary conditions

$$P_i = 0, R_i = 0, \theta = \widetilde{\theta} \text{ on } \partial B, \Pi_i = 0 \text{ on } C. \tag{7.4.15}$$

The field equations in terms of the displacement and temperature are given by

$$\begin{aligned}(\lambda + 2\mu)(1 - \ell_1^2\Delta) \text{ grad div } \mathbf{u} - \mu(1 - \ell_2^2\Delta) \text{ curl curl } \mathbf{u} - \\ -\beta \text{ grad } \theta = \mathbf{0},\end{aligned} \tag{7.4.16}$$

$$k\Delta\theta = -\rho_0 S. \tag{7.4.17}$$

The equation (7.4.17) with the corresponding boundary condition determines the temperature variation θ. In what follows we assume that θ is a known function. In the equation (7.4.16) the term $-\beta$ grad θ appears as a body force. If we introduce the notations

$$\begin{aligned}P_i^0 &= [\tau_{ji}^0 - \mu_{sji,s} + (\mu_{pji}n_p n_r - \mu_{pri}n_p n_j)_{,r}]n_j, \\ \tau_{ji}^0 &= \lambda e_{rr}\delta_{ij} + 2\mu e_{ij},\end{aligned} \tag{7.4.18}$$

then the boundary conditions (7.4.15) reduce to

$$P_i^0 = \beta\theta n_i, \quad R_i = 0 \text{ on } \partial B, \Pi_i = 0 \text{ on } C. \tag{7.4.19}$$

Thus, from (7.4.16) and (7.4.19) it follows that the problem of thermoelastostatics is reduced to the problem of elastostatics with the body force $-\beta$ grad θ and the surface traction $\beta\theta\mathbf{n}$.

d) **Exercises.**

1. *Elastic space with a spherical cavity.* Let us consider the equilibrium of an unbounded medium with a spherical cavity of radius a, with centre at the origin of coordinate. We assume that the heat source is absent and the boundary of the cavity is subjected to a constant temperature T^* and is free of tractions. In this case we obtain

$$\theta = \frac{a}{r}T^*, \tag{7.4.20}$$

where $r = (x_1^2 + x_2^2 + x_3^2)^{1/2}$. On the boundary of cavity we have the conditions

$$P_i = 0, R_i = 0 \text{ on } r = a. \tag{7.4.21}$$

We assume that

$$u_i = \varphi_{,i},$$

where $\varphi = \varphi(r)$ is an unknown function. The equilibrium equations (7.4.16) are satisfied if the function φ satisfies the equation

$$(1 - \ell_1^2\Delta)\Delta\varphi = \frac{\beta}{\lambda + 2\mu}\theta.$$

The solution of this equation which satisfies the conditions at infinity is

$$\varphi = \frac{1}{r}(A_1 + A_2 e^{-r/\ell_1}) + Ar, \tag{7.4.22}$$

where A_α are arbitrary constants and

$$A = \frac{aT^*\beta}{2(\lambda + 2\mu)}.$$

From (7.3.32) we obtain

$$\begin{aligned}
&\tau_{ij} = \lambda\delta_{ij}\Delta\varphi + 2\mu\varphi_{,ij} - \beta\theta\delta_{ij},\\
&\mu_{ijk} = \frac{1}{2}\alpha_1(\delta_{jk}\Delta\varphi_{,i} + 2\delta_{ij}\Delta\varphi_{,k} + \delta_{ik}\Delta\varphi_{,j}) +\\
&+\alpha_2(\delta_{jk}\Delta\varphi_{,i} + \delta_{ik}\Delta\varphi_{,j}) + 2\alpha_3\delta_{ij}\Delta\varphi_{,k} + 2(\alpha_4 + \alpha_5)\varphi_{,ijk}.
\end{aligned}$$

The conditions (7.4.21) on the boundary $r = a$ reduce to

$$\begin{aligned}
&\lambda\Delta\varphi + 2\mu\varphi'' - (\alpha_1 + 2\alpha_2)\Delta\Delta\varphi - (\alpha_1 + 2\alpha_3 + 4\alpha_4 + 4\alpha_5)(\Delta\varphi)'' -\\
&-\frac{2}{a}(2\alpha_1 + \alpha_2 + 3\alpha_3 + \alpha_4 + \alpha_5)(\Delta\varphi)' + 2(\alpha_4 + \alpha_5)\varphi^{(4)} = \beta\theta,\\
&(\alpha_1 + \alpha_2 + \alpha_3)(\Delta\varphi)' + (\alpha_4 + \alpha_5)\varphi''' = 0 \ \text{ on } r = a,
\end{aligned}$$

where $f' = df/dr$. Using (7.4.22), the above relations lead to the following system for the constants A_α

$$-4[\mu a^2 + 16(\alpha_4 + \alpha_5)]A_1 + e^{-p}(a^2 F + G) = 2A\zeta,$$
$$6(\alpha_4 + \alpha_5)A_1 - [(\alpha_1 + \alpha_2 + \alpha_3)d_1 + (\alpha_4 + \alpha_5)d_2]e^{-p}A_2 =$$
$$= 2(\alpha_1 + \alpha_2 + \alpha_3)Aa^2,$$

where

$$\zeta = 2\mu a^4 - \ell_1^2 a^2(\lambda + 2\mu) - 8a^2(\alpha_4 + \alpha_5), p = a/\ell_1,$$
$$F = \lambda p^2 + 2\mu(2 + 2p + p^2), G = 2(\alpha_1 + \alpha_2 + \alpha_3)p^2(1 + p - p^2) +$$
$$+2(\alpha_4 + \alpha_5)(24 + 24p + 9p^2 + p^3 - 3p^4),$$
$$d_1 = p^2(1 + p), d_2 = 6 + 6p + 3p^2 + p^3.$$

If $\alpha_s = 0, (s = 1, 2, \ldots, 5)$, then we obtain the solution in the classical theory.

2. *Determine the fundamental solutions in the theory of steady vibrations of homogeneous and isotropic bodies.*

Solution. We use the solution (7.4.7) in order to determine the fundamental solution of the equations (7.4.4) for the case of steady vibrations. We assume that

$$\mathbf{f} = \mathrm{Re}[\mathbf{f}^*(\mathbf{X})e^{-i\omega t}], \quad S = \mathrm{Re}[S^*(\mathbf{X})e^{-i\omega t}],$$
$$\mathbf{u} = \mathrm{Re}[\mathbf{u}^*(\mathbf{X})e^{-i\omega t}], \quad \theta = \mathrm{Re}[\theta^*(\mathbf{X})e^{-i\omega t}]. \tag{7.4.23}$$

If we introduce the notations

$$\mathbf{G} = \mathrm{Re}[\mathbf{G}^*(\mathbf{X})e^{-i\omega t}], \quad H = \mathrm{Re}[H^*(\mathbf{X})e^{-i\omega t}],$$
$$D_\alpha^* = (1 - \lambda_\alpha^2\Delta)\Delta + c_1^2\omega^2/c_\alpha^2, \quad D_3^* = \Delta + i\omega t + \tau_0\omega^2, \tag{7.4.24}$$
$$D_4^* = [c_1^2(1 - \lambda_1^2\Delta) - c_2^2(1 - \lambda_2^2\Delta)]D_3^* + c_1^2 i\varepsilon\omega(1 - i\tau_1\omega),$$

then from (7.4.7), (7.4.8) and (7.4.23) we obtain

$$\mathbf{u}^* = c_1^2[D_1^* D_3^* + i\omega\varepsilon(1 - i\omega\tau_1)\Delta]\mathbf{G}^* - D_4^* \text{ grad div } \mathbf{G}^* +$$
$$+\frac{\varepsilon c}{\beta}(1 - i\omega\tau_1) \text{ grad } H^*, \tag{7.4.25}$$
$$\theta^* = -\frac{1}{c}i\omega\beta c_2^2 D_2^* \text{ div } \mathbf{G}^* + D_1^* H^*$$

and

$$D_2^*[D_1^* D_3^* + i\omega\varepsilon(1 - i\omega\tau_1)\Delta]\mathbf{G}^* = -c_1^{-2}c_2^{-2}N\mathbf{f}^*,$$
$$[D_1^* D_3^* + i\omega\varepsilon(1 - i\omega\tau_1)\Delta]\, H^* = -KS^*. \tag{7.4.26}$$

Let us determine the fundamental solutions of the system (7.4.26). First we write

$$D_1^* D_3^* + i\omega\varepsilon(1 - i\omega\tau_1)\Delta = -\lambda_1^2(\Delta + \xi_1^2)(\Delta + \xi_2^2)(\Delta + \xi_3^2),$$
$$D_2^* = -\lambda_2^2(\Delta + k_1^2)(\Delta + k_2^2), \tag{7.4.27}$$

where ξ_i^2 are the rots of the equation (7.4.9) and

$$k_{1,2}^2 = [-c_2^2 \pm (c_2^4 + 4c_1^2 c_2^2 \lambda_2^2)^{1/2}]/2\lambda_2^2 c_2^2$$

Let us note that

$$k_1^2 = \kappa^2, \kappa > 0, k_2^2 = -\sigma^2, \sigma > 0. \tag{7.4.28}$$

If $f_i^* = \delta_{ik}\delta(X - Y), S^* = 0$, where $\delta(\cdot)$ is the Dirac delta, then from (7.4.26)-(7.4.28) we obtain

$$G_i^* = \delta_{ik} E(\mathbf{X}, \mathbf{Y}; \omega), \quad H^* = 0, \tag{7.4.29}$$

where

$$E(\mathbf{X}, \mathbf{Y}; \omega) = \sum_{i=1}^{5} a_i E_i(\mathbf{X}, \mathbf{Y}; \omega),$$
$$E_1 = \frac{A}{4\pi r} e^{i\kappa r}, E_2 = \frac{A}{4\pi r} e^{-\sigma r}, E_{2+s} = \frac{A}{4\pi r} e^{i\xi_s r}, s = 1, 2, 3,$$
$$r = [(X_i - Y_i)(X_i - Y_i)]^{1/2}, A = N/c_1^2 c_2^2 \lambda_1^2 \lambda_2^2,$$
$$a_p^{-1} = \prod_{j=1(j \neq p)}^{5} (h_p^2 - h_j^2), (p = 1, 2, \ldots, 5),$$
$$h_1^2 = k_1^2, h_2^2 = k_2^2, h_3^2 = \xi_1^2, h_4^2 = \xi_2^2, h_5^2 = \xi_3^2.$$

We denote by $u_i^{*(k)}$ the amplitude of the displacements and by $\theta^{*(k)}$ the amplitude of temperature caused by this concentrated force. From (7.4.25) and (7.4.29) we get

$$u_i^{*(k)} = c_1^2 \delta_{ik} [D_1^* D_3^* + i\omega\varepsilon(1 - i\omega\tau_1)\Delta] E - D_4^* E_{,ik}, \theta^{*(k)} = -\frac{1}{c} i\omega\beta c_2^2 D_2^* E_{,k}.$$

If $\mathbf{f}^* = \mathbf{0}, S^* = \delta(X - Y)$, then $\mathbf{G}^* = \mathbf{0}, H^* = \mathcal{E}$, where

$$\mathcal{E} = \sum_{i=1}^{3} p_i \mathcal{E}_i, \quad p_1^{-1} = (\xi_1^2 - \xi_2^2)(\xi_1^2 - \xi_3^2), p_2^{-1} = (\xi_2^2 - \xi_1^2)(\xi_2^2 - \xi_3^2),$$
$$p_3^{-1} = (\xi_3^2 - \xi_1^2)(\xi_3^2 - \xi_2^2), \quad \mathcal{E}_s = -\frac{K}{4\pi r} \lambda_1^{-2} e^{i\xi_s r}, s = 1, 2, 3.$$

The corresponding displacement vector and temperature are denoted by $u^{(4)}, \theta^{(4)}$. From (7.4.25) we obtain

$$u_i^{*(4)} = \frac{\varepsilon c}{\beta}(1 - i\omega\tau_1)\mathcal{E}_{,i}, \quad \theta^* = D_1^*\mathcal{E}.$$

The solution (7.4.7) can be used to solve other special problems (see [289]).

7.5 A Grade Consistent Theory of Micropolar Thermoelasticity

In [29], [30], Brulin and Hjalmars have developed a theory of micropolar thermoelastic media where the second-order displacement gradient is added to the classical set of independent constitutive variables. This theory was named the grade consistent micropolar thermoelasticity [30]. The fact that a dependence of second-order displacement gradient is necessary for the investigation of specific nonlocal phenomena was explained by Rymarz [322], Soós [347] and Hjalmars [151], p. 173. In this section we present the linear theory of grade consistent micropolar thermoelasticity.

a) **The field equations.** We assume that the deformation of a body is described by (5.1.1) where $\mathbf{u}$ is the displacement vector and $\boldsymbol{\varphi}$ is the microrotation vector. We postulate the energy balance in the form

$$\begin{aligned}\int_P (\rho_0 v_i \dot{v}_i + I_{ij}\nu_i\dot{\nu}_j + \rho_0\dot{e})dV &= \int_P \rho_0(f_i v_i + g_i\nu_i + S)dV + \\ &+ \int_{\partial P} (t_i v_i + m_i\nu_i + \mu_{ji}v_{i,j} + q)dA,\end{aligned} \tag{7.5.1}$$

for every part P of B and every time. Here, μ_{ji} is the hypertraction associated with the surface which were originally ∂P, measured per unit undeformed area, $v_i = \dot{u}_i$ and $\nu_i = \dot{\varphi}_i$.

If $\mu_{ji} = 0$, then (7.5.1) reduces to the balance law (5.9.1). Let us consider a motion of the body which differs from the given motion only by a constant rigid body translational velocity. As in Sections 1.1 and 5.9, from (7.5.1) we obtain (5.9.2) and the equations

$$t_{ji,j} + \rho_0 f_i = \rho_0\ddot{u}_i. \tag{7.5.2}$$

If we use (5.9.2) and (7.5.2), then the relation (7.5.1) reduces to

$$\begin{aligned}\int_P (I_{ij}\dot{\nu}_j\nu_i + \rho_0\dot{e})dV = \int_P [t_{ji}v_{i,j} + \rho_0(g_i\nu_i + S)]dV+ \\ + \int_{\partial P} (m_i\nu_i + \mu_{ji}v_{i,j} + q)dA.\end{aligned} \tag{7.5.3}$$

With an argument similar to that used in obtaining (1.1.7) from (1.1.6), we find that

$$(m_i - m_{ji}n_j)\nu_i + (\mu_{ji} - \mu_{kji}n_k)v_{i,j} + q - q_i n_i = 0, \tag{7.5.4}$$

where m_{ji} is the couple stress tensor, μ_{kji} is the hyperstress tensor and q_i is the heat flux vector. With the help of (7.5.4) and the divergence theorem, from (7.5.3) we obtain the local form of the energy balance

$$\begin{aligned}\rho_0\dot{e} = (m_{ji,j} + \rho_0 g_i - I_{ij}\dot{\nu}_j)\nu_i + (\mu_{kji,k} + t_{ji})v_{i,j} + m_{ji}\nu_{i,j}+ \\ +\mu_{kji}v_{i,jk} + q_{i,i} + \rho_0 S.\end{aligned} \tag{7.5.5}$$

Let us now consider a motion of the continuum which differs from the given motion only by a superposed uniform rigid body angular velocity, the body at time t being rotated back into original orientation. In this case ν_i are replaced by $\nu_i + b_i$, and v_i are replaced by $v_i + \varepsilon_{ijk}b_j x_k$, where b_i are arbitrary constants. We assume that $\rho_0\dot{e}, m_{ji}, \rho_0 g_i - I_{ij}\dot{\nu}_j, \mu_{kji}, t_{ij}, q_i$ and S are not affected by such motion. Thus, in the linear theory, we have

$$\begin{aligned}\rho_0\dot{e} = (m_{ji,j} + \rho_0 g_i - I_{ij}\dot{\nu}_j)(\nu_i + b_i) + (\mu_{kji,k} + t_{ji})(v_{i,j} + \varepsilon_{jik}b_k)+ \\ +m_{ji}\nu_{i,j} + \mu_{kji}v_{i,jk} + q_{i,i} + \rho_0 S.\end{aligned}$$

With the help of (7.5.5) we obtain

$$m_{ji,j} + \varepsilon_{irs}\tau_{rs} + \rho_0 g_i = I_{ij}\ddot{\varphi}_j, \tag{7.5.6}$$

where

$$\tau_{rs} = \mu_{krs,k} + t_{rs}. \tag{7.5.7}$$

In view of (7.5.6) the equation (7.5.5) reduces to

$$\rho_0\dot{e} = \tau_{ij}\dot{e}_{ij} + m_{ij}\dot{\kappa}_{ij} + \mu_{ijk}\dot{\kappa}_{ijk} + q_{i,i} + \rho_0 S, \tag{7.5.8}$$

where

$$e_{ij} = u_{j,i} + \varepsilon_{jik}\varphi_k, \kappa_{ij} = \varphi_{j,i}, \kappa_{ijk} = u_{k,ij}. \tag{7.5.9}$$

We postulate the entropy production inequality in the form

$$\int_P \rho_0 \dot{\eta} dV - \int_P \rho_0 \frac{S}{T} dV - \int_{\partial P} \frac{q}{T} dA \geq 0, \tag{7.5.10}$$

for every region P of B and every time, where $T(>0)$ is the absolute temperature. If we introduce the Helmholtz free energy ψ by (1.1.19) then from (7.5.8) we get

$$\rho_0(\dot{\psi} + \dot{T}\eta + T\dot{\eta}) = \tau_{ij}\dot{e}_{ij} + m_{ij}\dot{\kappa}_{ij} + \mu_{ijk}\dot{\kappa}_{ijk} + q_{i,i} + \rho_0 S. \tag{7.5.11}$$

Let θ be the temperature measured from the constant absolute temperature T_0 of the body in the reference configuration. We introduce the array $\Lambda = \{e_{ij}, \kappa_{ij}, \kappa_{ijk}, \theta, \theta_{,i}, X_p\}$ and assume that the material is characterized by the constitutive equations

$$\begin{aligned}
&\psi = \widehat{\psi}(\Lambda), \tau_{ij} = \widehat{\tau}_{ij}(\Lambda), m_{ij} = \widehat{m}_{ij}(\Lambda), \mu_{ijk} = \widehat{\mu}_{ijk}(\Lambda),\\
&\eta = \widehat{\eta}(\Lambda), q_i = \widehat{q}_i(\Lambda), m_i = \widehat{m}_i(\Lambda, n_k),\\
&\mu_{ji} = \widehat{\mu}_{ji}(\Lambda, n_k), q = \widehat{q}(\Lambda, n_k).
\end{aligned} \tag{7.5.12}$$

We assume that the response functions are of class C^2 on their domain.

For a given deformation, ν_i and $v_{r,s}$ in (7.5.4) may be chosen arbitrary so that, on the basis of the constitutive equations we obtain

$$m_i = m_{ji}n_j, \mu_{ji} = \mu_{kji}n_k, q = q_i n_i. \tag{7.5.13}$$

In view of (7.5.13), the inequality (7.5.10) implies the following local form

$$\rho_0 T\dot{\eta} - \rho_0 S - q_{i,i} + \frac{1}{T} q_i T_{,i} \geq 0. \tag{7.5.14}$$

It follows from (7.5.11) and (7.5.14) that

$$\tau_{ij}\dot{e}_{ij} + m_{ij}\dot{\kappa}_{ij} + \mu_{ijk}\dot{\kappa}_{ijk} - \rho_0\eta\dot{T} - \rho_0\dot{\psi} + \frac{1}{T} q_i T_{,i} \geq 0. \tag{7.5.15}$$

We introduce the notation $\sigma = \rho_0\psi$ and let $\Lambda^* = \{e_{ij}, \kappa_{ij}, \kappa_{ijk}, \theta, X_p\}$. In view of constitutive equations, from (7.5.15) we find that

$$\begin{aligned}
&\sigma = \widetilde{\sigma}(\Lambda^*), \tau_{ij} = \frac{\partial\sigma}{\partial e_{ij}}, m_{ij} = \frac{\partial\sigma}{\partial\kappa_{ij}},\\
&\mu_{ijk} = \frac{\partial\sigma}{\partial\kappa_{ijk}}, \rho_0\eta = -\frac{\partial\sigma}{\partial\theta}
\end{aligned} \tag{7.5.16}$$

and

$$q_i\theta_{,i} \geq 0. \tag{7.5.17}$$

It follows from (7.5.11) and (7.5.16) that the energy equation in the linear theory is

$$\rho_0 T_0 \dot{\eta} = q_{i,i} + \rho_0 S. \tag{7.5.18}$$

We note that from (7.5.17) we find

$$\widehat{q}_i(e_{rs}, \kappa_{rs}, \kappa_{pqr}, \theta, 0, X_p) = 0. \tag{7.5.19}$$

In the context of linear theory we assume that

$$\begin{aligned}\sigma = &\frac{1}{2}A_{ijrs}e_{ij}e_{rs} + B_{ijrs}e_{ij}\kappa_{rs} + \frac{1}{2}C_{ijrs}\kappa_{ij}\kappa_{rs} + D_{ijpqr}e_{ij}\kappa_{pqr} + \\ &+ E_{ijpqr}\kappa_{ij}\kappa_{pqr} + \frac{1}{2}F_{ijkpqr}\kappa_{ijk}\kappa_{pqr} - \beta_{ij}e_{ij}\theta - \\ &- \xi_{ij}\kappa_{ij}\theta - \zeta_{ijk}\kappa_{ijk}\theta - \frac{1}{2}a\theta^2.\end{aligned} \tag{7.5.20}$$

The constitutive coefficients have the following symmetries

$$\begin{aligned}&A_{ijrs} = A_{rsij}, C_{ijrs} = C_{rsij}, D_{ijpqr} = D_{ijqpr}, E_{ijpqr} = E_{ijqpr}, \\ &F_{ijkpqr} = F_{jikpqr} = F_{pqrijk}, \zeta_{ijk} = \zeta_{jik}.\end{aligned} \tag{7.5.21}$$

From (7.5.16), (7.5.20), (7.5.21) and (7.5.19) we obtain the following constitutive equations

$$\begin{aligned}\tau_{ij} &= A_{ijrs}e_{rs} + B_{ijrs}\kappa_{rs} + D_{ijpqr}\kappa_{pqr} - \beta_{ij}\theta, \\ m_{ij} &= B_{rsij}e_{rs} + C_{ijrs}\kappa_{rs} + E_{ijpqr}\kappa_{pqr} - \xi_{ij}\theta, \\ \mu_{ijk} &= D_{rsijk}e_{rs} + E_{rsijk}\kappa_{rs} + F_{ijkpqr}\kappa_{pqr} - \zeta_{ijk}\theta, \\ \rho_0\eta &= \beta_{ij}e_{ij} + \xi_{ij}\kappa_{ij} + \zeta_{ijk}\kappa_{ijk} + a\theta, \quad q_i = k_{ij}\theta_{,j}.\end{aligned} \tag{7.5.22}$$

In the case of an isotropic material with a centre of symmetry the constitutive equations are

$$\begin{aligned}\tau_{ij} &= \lambda e_{rr}\delta_{ij} + (\mu + \kappa)e_{ij} + \mu e_{ij} - b\theta\delta_{ij}, \\ m_{ij} &= \alpha\kappa_{rr}\delta_{ij} + \beta\kappa_{ji} + \gamma\kappa_{ij} + \beta_1\varepsilon_{jmn}\kappa_{min} + \\ &\quad + \beta_2\varepsilon_{rij}\kappa_{rss} + \beta_3\varepsilon_{rij}\kappa_{ssr},\end{aligned}$$

$$
\begin{aligned}
\mu_{ijk} = {} & \frac{1}{2}\alpha_1(\kappa_{rri}\delta_{jk} + 2\kappa_{krr}\delta_{ij} + \kappa_{rrj}\delta_{ik}) + \alpha_2(\kappa_{irr}\delta_{jk} + \\
& + \kappa_{jrr}\delta_{ik}) + 2\alpha_3\kappa_{rrk}\delta_{ij} + 2\alpha_4\kappa_{ijk} + \\
& + \alpha_5(\kappa_{kji} + \kappa_{kij}) + \frac{1}{2}\beta_1(\varepsilon_{rik}\kappa_{jr} + \varepsilon_{rjk}\kappa_{ir}) + \qquad (7.5.23) \\
& + \frac{1}{2}\beta_2(\varepsilon_{imn}\kappa_{mn}\delta_{jk} + \varepsilon_{jmn}\kappa_{mn}\delta_{ik}) + \beta_3\varepsilon_{kmn}\kappa_{mn}\delta_{ij}, \\
\rho_0\eta = {} & be_{rr} + a\theta, \quad q_i = k\theta_{,i}.
\end{aligned}
$$

We note that, in view of (7.5.7), the equations (7.5.2) can be written in the form

$$
\tau_{ji,j} - \mu_{kji,kj} + \rho_0 f_i = \rho_0 \ddot{u}_i. \qquad (7.5.24)
$$

Thus, the basic equations of the linear theory are: the equations of motion (7.5.24) and (7.5.6), the energy equation (7.5.18), the constitutive equations (7.5.22), and the geometrical equations (7.5.9).

In view of (7.5.13) and (7.3.13), the total rate of work over the boundary of the body becomes

$$
\begin{aligned}
\int_{\partial B}(t_i v_i + m_i \nu_i + \mu_{ji} v_{i,j})dA = {} & \int_{\partial B}(P_i v_i + m_i \nu_i + R_i D v_i)dA + \\
& + \int_C \Pi_i v_i ds,
\end{aligned}
\qquad (7.5.25)
$$

where P_i, R_i and Π_i are defined in (7.3.11).

To the system of field equations we adjoin the initial conditions (5.2.2) and the following boundary conditions

$$
\begin{aligned}
& u_i = \widetilde{u}_i, \ \varphi_i = \widetilde{\varphi}_i, \ Du_i = \widetilde{\nu}_i, \theta = \widetilde{\theta} \text{ on } S_1 \times (t_0, t_1), \\
& P_i = \widetilde{P}_i, \ m_{ji}n_j = \widetilde{m}_i, \ R_i = \widetilde{R}_i, \ q_j n_j = \widetilde{q} \text{ on } S_2 \times (t_0, t_1), \qquad (7.5.26) \\
& \Pi_i = \widetilde{\Pi}_i \text{ on } C \times (t_0, t_1),
\end{aligned}
$$

where $\widetilde{u}_i, \widetilde{\varphi}_i, \widetilde{\nu}_i, \widetilde{\theta}, \widetilde{P}_i, \widetilde{m}_i, \widetilde{R}_i, \widetilde{q}$ and $\widetilde{\Pi}_i$ are prescribed functions and S_1, S_2 and C are defined in Section 7.1.

Some results concerning existence, uniqueness and variational characterization of solution have been established in [185].

b) **Homogeneous and isotropic bodies.** From (7.5.9), (7.5.23), (7.5.24), (7.5.6) and (7.5.18) we obtain the field equations in terms of the displacement, microrotation and temperature

$$
\begin{aligned}
&(\mu+\kappa-\xi\Delta)\Delta\mathbf{u}+(\lambda+\mu-\nu\Delta)\ \text{grad div}\ \mathbf{u}+(\kappa-\zeta\Delta)\ \text{curl}\ \boldsymbol{\varphi}-\\
&\qquad -b\ \text{grad}\ \theta+\rho_0\mathbf{f}=\rho_0\ddot{\mathbf{u}},\\
&\gamma\Delta\boldsymbol{\varphi}+(\alpha+\beta)\text{grad div}\ \boldsymbol{\varphi}-2\kappa\boldsymbol{\varphi}+(\kappa-\zeta\Delta)\text{curl}\ \mathbf{u}+\rho_0\mathbf{g}=J\ddot{\boldsymbol{\varphi}},\\
&k\Delta\theta-c\dot{\theta}-bT_0\ \text{div}\ \dot{\mathbf{u}}=-\rho_0 S,
\end{aligned}
\tag{7.5.27}
$$

where

$$
\xi=2(\alpha_3+\alpha_4),\nu=2(\alpha_1+\alpha_2+\alpha_5),\zeta=\beta_3-\beta_1,c=aT_0.
$$

If we introduce the operators

$$
\begin{aligned}
&A_1=(\mu+\kappa-\xi\Delta)\Delta-\rho_0\frac{\partial^2}{\partial t^2},A_2=\gamma\Delta-J\frac{\partial^2}{\partial t^2}-2\kappa,\\
&A_3=k\Delta-c\frac{\partial}{\partial t},Q_1=\lambda+\mu-\nu\Delta,Q_2=\kappa-\zeta\Delta,\\
&Q_3=-bT_0\frac{\partial}{\partial t},
\end{aligned}
\tag{7.5.28}
$$

then the equations (7.5.27) can be written in the form

$$
\begin{aligned}
&A_1\mathbf{u}+Q_1\ \text{grad div}\ \mathbf{u}+Q_2\ \text{curl}\boldsymbol{\varphi}-b\ \text{grad}\ \theta=-\rho_0\mathbf{f},\\
&A_2\boldsymbol{\varphi}+(\alpha+\beta)\ \text{grad div}\ \boldsymbol{\varphi}+Q_2\text{curl}\ \mathbf{u}=-\rho_0\mathbf{g},\\
&A_3\theta+Q_3\ \text{div}\ \mathbf{u}=-\rho_0 S.
\end{aligned}
\tag{7.5.29}
$$

We denote

$$
\begin{aligned}
&\mathcal{L}_1=A_2A_3Q_1+bA_2Q_3-A_3Q_2^2,A_4=A_1+\Delta Q_1,\\
&\mathcal{M}=A_3A_4+b\Delta Q_3,\mathcal{L}_2=(\alpha+\beta)A_1-Q_2^2,\\
&\Omega=A_1A_2+\Delta Q_2^2,\mathcal{N}=A_2+(\alpha+\beta)\Delta.
\end{aligned}
\tag{7.5.30}
$$

Theorem 7.5.3. *Let*

$$
\begin{aligned}
&\mathbf{u}=A_2\mathcal{M}\mathbf{G}-\mathcal{L}_1\ \text{grad div}\ \mathbf{G}-Q_2\mathcal{N}\ \text{curl}\ \mathbf{H}+b\ \text{grad}\ X,\\
&\boldsymbol{\varphi}=A_1\mathcal{N}\mathbf{H}-\mathcal{L}_2\ \text{grad div}\ \mathbf{H}-Q_2\mathcal{M}\ \text{curl}\ \mathbf{G},\\
&\theta=A_4X-\Omega Q_3\ \text{div}\ \mathbf{G},
\end{aligned}
\tag{7.5.31}
$$

where $\mathbf{G},\mathbf{H}$ *and* X *satisfy the equations*

$$
\Omega\mathcal{M}\mathbf{G}=-\rho_0\mathbf{f},\quad \Omega\mathcal{N}\mathbf{H}=-\rho_0\mathbf{g},\quad \mathcal{M}X=-\rho_0 S.
\tag{7.5.32}
$$

Then $\mathbf{u}, \boldsymbol{\varphi}$ *and* θ *satisfy the equations (7.5.29).*

Proof. By (7.5.31),

$$\begin{aligned}
&A_1\mathbf{u} + Q_1 \operatorname{grad} \operatorname{div} \mathbf{u} + Q_2 \operatorname{curl} \boldsymbol{\varphi} - b \operatorname{grad} \theta = \\
&= (A_1 A_2 + Q_2^2)\mathcal{M}\mathbf{G} - (A_1\mathcal{L}_1 - Q_1 A_2 \mathcal{M} + Q_1 \mathcal{L}_1 \Delta + \\
&+ Q_2^2 \mathcal{M} - b\Omega Q_3) \operatorname{grad} \operatorname{div} \mathbf{G} + (A_1 + Q_1\Delta - A_4) b \operatorname{grad} X = \\
&= \Omega \mathcal{M} \mathbf{G}.
\end{aligned} \tag{7.5.33}$$

Similarly, we obtain

$$\begin{aligned}
&A_2\boldsymbol{\varphi} + (\alpha + \beta) \operatorname{grad} \operatorname{div} \boldsymbol{\varphi} + Q_2 \operatorname{curl} \mathbf{u} = \Omega \mathcal{N}, \\
&A_3\theta + Q_3 \operatorname{div} \mathbf{u} = \mathcal{M} X.
\end{aligned} \tag{7.5.34}$$

From (7.5.32)-(7.5.34) we obtain the desired result. □

The solution (7.5.31) has been used in [185] to study the effect of concentrated loads in the theory of steady vibrations.

The harmonic plane wave solutions for homogeneous and isotropic bodies have been investigated by Brulin [31].

References

[1] R.A.A. Adams, *Sobolev Spaces.* Academic Press, New York, 1975.

[2] G. Ahmadi and K. Firoozbakhsh, First strain-gradient theory of thermoelasticity. *Int. J. Solids Structures,* **11** (1975), 339–345.

[3] G. Ahmadi, Thermoelastic stability of first strain-gradient solids. *Int. J. Nonlinear Mech.*, **12** (1977), 23-32.

[4] M. Al-Hasan and J. Dyszlewicz, Radiation conditions and integral representations of a solution in coupled micropolar thermoelasticity. *J. Thermal Stresses*, **24** (2000), 1007-1018.

[5] K. Ames and B. Straughan, Continuous dependence results for initially prestressed thermoelastic bodies. *Int. J. Engng. Sci.*, **30** (1992), 7-13.

[6] G.L. Anderson, On forced vibration in the linear theory of micropolar elasticity. *Int. J. Engng. Sci.*, **11** (1973), 21-40.

[7] M.N. Anwar and H.H. Sherif, State space approach to generalized thermoelasticity. *J. Thermal Stresses*, **11** (1988), 353-365.

[8] R.J. Atkin and R.E. Craine, Continuum theories of mixtures: basic theory and historical development. *Quart. J. Mech. Appl. Math.*, **29** (1976), 209-243.

[9] L.Y. Bahar and R.B. Hetnarski, Coupled thermoelasticity of a layered medium. *J. Thermal Stresses*, **3** (1980), 141-152.

[10] J.R. Barber and R. Zhang, Transient behavior and stability for the thermoelastic contact of two rods of dissimilar materials. *Int. J. Mech. Sci.,* **30** (1988), 691-704.

[11] J.R. Barber, Thermoelasticity and contact. *J. Thermal Stresses*, **22** (1999), 513-525.

[12] H.W. Bargmann, On inertia effects in thermoelasticity and thermal shock parameters. *J. Thermal Stresses*, **15** (1992), 229-240.

[13] R.C. Batra, Thermodynamics of non-simple elastic materials, *J. Elasticity*, **6** (1976), 451-456.

[14] M.F. Beatty and K.J. Cheverton, The basic equations for a grade 2 materials viewed as an oriented continuum. *Arch. Mech.*, **28** (1976), 205-213.

[15] M.F. Beatty, Some theorems in the theory of small deformations superimposed on a finite deformation of a hyperelastic material of grade 2. *Int. J. Engng. Sci.*, **12** (1976), 339-351.

[16] A. Bedford and D.S. Drumheller, Theory of immiscible and structured mixtures. *Int. J. Engng. Sci.*, **21** (1983), 863-960.

[17] I. Beju, A uniqueness theorem in thermoelasticity with thermal relaxation. *Int. J. Engng. Sci.*, **30** (1992), 657-664.

[18] Z. Bem, Existence of a generalized solution in thermoelasticity with one relaxation time. *J. Thermal Stresses*, **5** (1982), 195-206.

[19] M. Birsan, A bending theory of porous thermoelastic plates. *J. Thermal Stresses*, **26** (2003), 67-90.

[20] R.D. Blevins, Elastic thermal stresses in bimetallic pipe welds. *J. Pressure Vessel Tech.*, **102** (1980), 430-432.

[21] F. Bofill and R. Quintanilla, Some qualitative results for the linear theory of thermo-microstretch elastic solids. *Int. J. Engng. Sci.*, **33** (1995), 2115-2125.

[22] F. Bofill, F. Martinez and R. Quintanilla, Existence of solutions of the equations of incremental thermoelasticity for unbounded domains. *J. Thermal Stresses*, **18** (1995), 1-12.

[23] B.A. Boley and J.H. Weiner, *Theory of Thermal Stresses.* John Wiley, New York, 1960.

[24] B.A. Boley and I.S. Tolins, Transient coupled thermoelastic boundary-value problems in the half space. *J. Appl. Mech.*, **29** (1962), 637-646.

[25] B.A. Boley and R.B. Hetnarski, Propagation of discontinuities in coupled thermoelastic problems. *J. Appl. Mech. ASME*, **35** (1968), 489-494.

[26] B.A. Boley, *Thermal stresses: A survey.* In "Thermal Stresses in Severe Environments" (Edited by D.P.H. Hasselman and R.A. Heller), pp. 1-11, Plenum Publ. Corp., New York, 1980.

[27] E. Boschi and D. Ieşan, A generalized theory of linear micropolar thermoelasticity. *Meccanica*, **7** (1973), 154-157.

[28] R.M. Bowen, *Theory of Mixtures.* In "Continuum Physics" (Edited by A.C. Eringen), vol. 3, Academic Press, New York, 1976.

[29] O. Brulin and S. Hjalmars, Linear grade consistent micropolar theory, *Int. J. Engng. Sci.*, **19** (1981), 1731-1738.

[30] O. Brulin, *Linear Micropolar Media.* In "Mechanics of Micropolar Media" (Edited by O. Brulin and R.K.T. Hsieh), pp. 87-146, World Scientific, Singapore, 1982.

[31] O. Brulin, Grade consistent micropolar theory of thermoelasticity. New Problems in Mechanics of Continua. Proc. 3rd Swed.-Pol. Symp. Mech., Jablonna 1982, *Stud. Ser.*, **17** (1983), 265-272.

[32] O. Brulin, I. Fischer-Hjalmars and S. Hjalmars, Continuum equations derived from an orthotropic polar model of the microstructure. *J. Techn. Phys.*, **29** (1988), 19-27.

[33] L. Brun, Méthodes énergétiques dans les systèmes évolutifs linéaires. Première Partie: Séparation des énergies. Deuxième Partie: Théorèmes d'unicité. *J. Mécanique*, **8** (1969), 125-166.

[34] L. Brun and M. Potier-Ferry, Mechanics of elastic solids. Thermodynamics and stability of the elastic solid. *C.R. Acad. Sci. Paris,* **296** (1983), 401-404.

[35] H.D. Bui, A. Ehralcher and Q.S. Nguyen, Propagation of a crack in dynamical thermoelasticity (in French), J. Mécanique, **19** (1980), 697-723.

[36] T.V. Burchuladze, Dynamic problems of nonclassical thermoelasticity solved in a closed form (Russian). *Tr. Tbilis. Mat. Inst. Razmadze*, **87** (1987), 29-44.

[37] G. Capriz and P. Podio-Guidugli, Structured continua from a Lagrangian point of view. *Ann. Mat. Pura Appl.*, **135** (1983), 1-25.

[38] G. Capriz, *Continua with Microstructure.* Springer Tracts in Natural Philosophy, vol. 35 (Edited by C. Truesdell), Springer-Verlag, New York, 1989.

[39] B. Carbonaro and J. Ignaczak, Some theorems in temperature-rate dependent thermoelasticity for unbounded domain. *J. Thermal Stresses*, **10** (1987), 193-220.

[40] D.E. Carlson, *Linear Thermoelasticity.* pp. 297-346 of Flügge's Handbuch der Physik, vol. VI a/2 (Edited by C. Truesdell), Springer-Verlag, Berlin-Heidelberg-New York, 1972.

[41] G. Caviglia, A. Morro and B. Straughan, Thermoelasticity at cryogenic temperatures. *Int. J. Non-linear Mech.*, **27** (1992), 251-264.

[42] P. Chadwick and B. Powdrill, Singular surfaces in linear thermoelasticity. *Int. J. Engng. Sci.*, **3** (1965), 561-595.

[43] D.S. Chandrasekharaiah, Heat-flux dependent micropolar thermoelasticity. *Int. J. Engng. Sci.*, **24** (1986), 1389-1395.

[44] D.S. Chandrasekharaiah, Thermoelasticity with second sound: A review. *Appl. Mech. Rev.*, **39** (1986), 355-376.

[45] D.S. Chandrasekharaiah, A uniqueness theorem in heat-flux dependent thermoelasticity. *J. Elasticity*, **18** (1987), 283-287.

[46] D.S. Chandrasekharaiah, Complete solutions in the theory of elastic materials with voids. *Q. Jl. Mech. Appl. Math.*, **40** (1987), 401-413.

[47] D.S. Chandrasekharaiah, Some theorems in generalized micropolar thermoelasticity. *Arch. Mech.*, **38** (1987), 319-333.

[48] D.S. Chandrasekharaiah, Hyperbolic Thermoelasticity: A Review of Recent Literature, *Appl. Mech. Rev.*, **51** (1998), 705-729.

[49] P.J. Chen, *Growth and Decay of Waves in Solids,* pp. 303-400 of Flügge's Handbuch der Physik, vol. VI a/3 (Edited by C. Truesdell), Springer-Verlag, Berlin-Heidelberg-New York, 1973.

[50] T.C. Chen and C.I. Weng, Coupled transient thermoelastic response in an axisymmetric circular cylinder by Laplace transform-finite element method. *Comput. Struct.*, **33** (1989), 533-542.

[51] K.J. Cheverton and M.F. Beatty, On the mathematical theory of the mechanical behavior of some non-simple materials. *Arch. Rational Mech. Anal.*, **60** (1976), 1-16.

[52] S. Chiriţă, Propagation of discontinuities in coupled micropolar thermoelasticity, *J. Thermal Stresses*, **3** (1980), 199-221.

[53] S. Chiriţă, Thermal stresses in a micropolar solid induced by a plane temperature field. *Rev. Roum. Sci. Techn., Méc. Appl.*, **26** (1981), 219-231.

[54] S. Chiriţă, Uniqueness and continuous dependence results for the incremental thermoelasticity. *J. Thermal Stresses*, **5** (1982), 161-172.

[55] S. Chiriţă, Uniqueness and continuous data dependence in dynamical problems of nonlinear thermoelasticity. *J. Thermal Stresses*, **5** (1982), 331-346.

[56] S. Chiriţă and Gh. Rusu, On existence and uniqueness in thermoelasticity of non-simple materials. *Rev. Roum. Sci. Techn., Méc. Appl.*, **30** (1985), 21-30.

[57] S. Chiriţă, Some applications of the Lagrange identity in thermoelasticity with one relaxation time. *J. Thermal Stresses*, **11** (1988), 207-231.

[58] S. Chiriţă and A. Scalia, On the spatial and temporal behaviour in linear thermoelasticity of materials with voids. *J. Thermal Stresses*, **24** (2001), 433-455.

[59] B.M. Chiu and J.D. Lee, On the plane problem in micropolar elasticity, *Int. J. Engng. Sci.*, **11** (1973), 997-1012.

[60] S.I. Chou and C.C. Wang, Estimates of error in finite element approximate solutions to problems in linear thermoelasticity. *Arch. Rational Mech. Anal.*, **77** (1981), 263-300.

[61] S.K.R. Choudhuri, Effect of rotation and relaxation times on plane waves in generalized thermoelasticity. *J. Elasticity*, **15** (1985), 59-68.

[62] A. Chrzeszczyk, Well-posedness of the equations of generalized thermoelasticity, *J. Elasticity*, **30** (1993), 69-80.

[63] Chuan Li and J.R. Barber, Stability of thermoelastic contact of two layers of dissimilar materials. *J. Thermal Stresses*, **20** (1997), 169-184.

[64] M. Ciarletta and D. Ieşan, On the nonlinear theory of nonsimple thermoelastic bodies. *J. Thermal Stresses*, **12** (1989), 545-558.

[65] M. Ciarletta and A. Scalia, Thermoelasticity of nonsimple materials with voids. *Rev. Roum. Sci. Techn., Méc. Appl.*, **35** (1990), 115-125.

[66] M. Ciarletta and A. Scalia, A solution of Galerkin type in the theory of thermoelastic materials with voids. *J. Thermal Stresses*, **14** (1991), 409-419.

[67] M. Ciarletta and A. Scalia, On uniqueness and reciprocity in linear thermoelasticity of materials with voids, *J. Elasticity*, **32** (1993), 1-17.

[68] M. Ciarletta and A. Scalia, On the nonlinear theory of nonsimple thermoelastic materials with voids. *ZAMM*, **73** (1993), 67-75.

[69] S.G. Ciumaşu, On the existence of solutions in the theory of linear thermoelasticity with microstructure. *Bull. Inst. Politeh. Iaşi,* Sect. I, **30** (**34**) (1984), 43-48.

[70] B.D. Coleman and W. Noll, The thermodynamics of elastic materials with heat conduction and viscosity. *Arch. Rational Mech. Anal.*, **13** (1963), 167-178.

[71] E. Cosserat e F. Cosserat, *Théorie des Corps Déformables*, Hermann, Paris, 1909.

[72] S.C. Cowin and F.M. Leslie, On kinetic energy and momenta in Cosserat continua. *ZAMP*, **31** (1980), 247-260.

[73] S.C. Cowin and J.W. Nunziato, Linear elastic materials with voids, *J. Elasticity*, **13** (1983), 125-147.

[74] S.C. Cowin and P. Puri, The classical pressure vessel problems for linear elastic materials with voids. *J. Elasticity*, **13** (1983), 157-163.

[75] S.C. Cowin, The stressed around a hole in a linear elastic material with voids. *Quart J. Mech. Appl. Math.*, **36** (1983), 247-260.

[76] I. Crăciun, Potential method in micropolar thermoelasticity. *Rev. Roum. Sci. Techn. Méc. Appl.*, **22** (1977), 19-36.

[77] I. Crăciun, Radiation conditions and uniqueness theorems, *Rev. Roum. Sci. Techn. Méc. Appl.*, **23** (1978), 165-180.

[78] C.M. Dafermos, On the existence and the asymptotic stability of solutions to the equations of linear thermoelasticity. *Arch. Rational Mech. Anal.*, **29** (1968), 241-271.

[79] C.M. Dafermos, Contraction semigroups and trend to equilibrium in continuum mechanics. IUTAM/IMU Symp. on Applications of "Methods of Functional Analysis to Problems in Mechanics" (Edited by P. Germain and B. Nayroles), Lect. in Mathematics 503, Springer-Verlag, Berlin 295-306, 1975.

[80] C.M. Dafermos, The second law of thermodynamics and stability. *Arch. Rational Mech. Anal.*, **70** (1979), 167-179.

[81] C.M. Dafermos and L. Hsiao, Development of singularities in solutions of the equations of nonlinear thermoelasticity. *Quart. Appl. Math.*, **44** (1986), 463-474.

[82] V.Y. Danilovskaya, Thermal stresses in an elastic half space due to a sudden heating of its boundary (in Russian). *Prikl. Mat. Mekh.*, **14** (1950), 316-318.

[83] W.A. Day, *Heat Conduction Within Linear Thermoelasticity.* Springer Tracts in Natural Philosophy, vol. 30, Springer-Verlag, New York, 1985.

[84] W.A. Day, An asymptotic formula for the maximum work done by a thermoelastic body. *Quart. J. Mech. Appl. Math.*, **44** (1991), 357-368.

[85] R. De Boer, Theory of Porous Media-Past and Present, *Z. Angew. Math. Mech.*, **78** (1998), 441-466.

[86] S. De Cicco and L. Nappa, On the theory of thermomicrostretch elastic solids, *J. Thermal Stresses*, **22** (1999), 565-580.

[87] S. De Cicco and L. Nappa, Some results in the linear theory of thermomicrostretch elastic solids. *Mathematics and Mechanics of Solids*, **5** (2000), 467-482.

[88] R.S. Dhaliwal and H.H. Sherief, Generalized thermoelasticity for anisotropic media. *Quart. Appl. Math.*, **38** (1980), 1-8.

[89] A.K. Dhar, Mechanical shock problem of coupled thermoelasticity in a nonsimple elastic material. *Indian J. Pure Appl. Math.*, **16** (1985), 174-178.

[90] S. Dost and B. Tabarrok, Generalized micropolar thermoelasticity. *Int. J. Engng. Sci.*, **16** (1978), 173-183.

[91] J.M. Doyle, A general solution for strain-gradient elasticity theory. *J. Math. Analysis Appl.*, **27** (1969), 171-180.

[92] L. Dragoş, *Principles of Mechanics of Continua* (in Romanian). Ed. Tehnică, Bucureşti, 1983.

[93] L. Dragoş, Fundamental solutions in micropolar elasticity. *Int. J. Engng. Sci.*, **22** (1984), 265-275.

[94] G. Duvaut et J.L. Lions, *Les inéquations en mécanique et physique*, Dunod, Paris, 1972.

[95] J. Dyszlewicz, Selected problems of linear asymmetrical thermoelasticity. *J. Thermal Stresses*, **19** (1996), 185-206.

[96] J. Dyszlewicz, *Boundary and Initial-Boundary Value Problems of Micropolar Theory of Elasticity,* WPW, Wroclaw, 1997.

[97] H.A. Erbay, An asymptotic theory of thin micropolar plates. *Int. J. Engng. Sci.*, **38** (2000), 1497-1516.

[98] J. Ericksen and C. Truesdell, Exact theory of stress and strain in rods and shells. *Arch. Rational Mech. Anal.*, **1** (1958), 295-323.

[99] J.L. Ericksen, *Introduction to the Thermodynamics of Solids.* Chapman & Hall, London, 1991.

[100] A.C. Eringen and E.S. Suhubi, Nonlinear theory of simple microelastic solids. *Int. J. Engng. Sci.*, **2** (1964), I: 189-203; II: 389-404.

[101] A.C. Eringen, *Mechanics of Continua*, John Wiley and Sons, New York, 1967.

[102] A.C. Eringen, *Foundations of Micropolar Thermoelasticity,* CISM Udine, Springer-Verlag, Wien, New York, 1970.

[103] A.C. Eringen and C.B. Kafadar, *Polar Field Theories.* In "Continuum Physics" (Edited by A.C. Eringen), vol. IV, Academic Press, New York, 1976.

[104] A.C. Eringen, Theory of thermo-microstretch elastic solids. *Int. J. Engng. Sci.*, **28** (1990), 1291-1301.

[105] A.C. Eringen, *Microcontinuum Field Theories.* I: *Foundations and Solids.* Springer-Verlag, New York, Berlin, Heidelberg, 1999.

[106] N.N. Eshel and G. Rosenfeld, Effects of strain-gradient on the stress-concentration at a cylindrical hole in a field of uniaxial tension. *J. Eng. Math.*, **4** (1970), 97-111.

[107] N.N. Eshel and G. Rosenfeld, Axisymmetric problems in elastic materials of grade two. *J. Franklin Inst.*, **299** (1975), 43-51.

[108] E. Feireisl, Forced vibrations in one-dimensional nonlinear thermoelasticity as a local coercive-like problem. *Comment. Math. Univ. Carolinae*, **31** (1990), 243-255.

[109] Fei-Yue Wang, On the solutions of Eringen's micropolar plate equations and of other approximate equations. *Int. J. Engng. Sci.*, **28** (1990), 919-926.

[110] G. Fichera, *Existence Theorems in Elasticity.* In vol. VI a/2 of the Handsbuch der Physik (Edited by C. Truesdell), Springer-Verlag, Berlin-Heidelberg-New York, 1972.

[111] G. Fichera, *Problemi Analitici Nuovi nella Fisica Matematica Classica.* Quaderni del Consiglio Nazionale delle Ricerche, Gruppo Nazionale di Fisica Matematica, Scuola Tipo-Lito "Istituto Anselmi", Napoli, 1985.

[112] W. Frydrychowicz and Z. Olesiak, Effect of a heated boundary on the distribution of normal stress in a micropolar semispace. *J. Thermal Stresses*, **2** (1979), 177-192.

[113] W. Frydrychowicz, S. Matysiak and Z. Olesiak, On thermal strsses in a micropolar layer, *J. Thermal Stresses*, **3** (1979), 437-447.

[114] A. Galka, Green's functions for the coupled problem of thermoelasticity, *Bull. Acad. Polon. Sci., Sér. Sci. Techn.*, **13** (1965), 369-376.

[115] A. Galka and R. Wojnar, One-dimensional dynamic thermal stresses generated in an elastic half-space by laser pulses, *J. Thermal Stresses*, **18** (1995), 113-140.

[116] R.D. Gauthier and E.W. Jahsman, A quest for micropolar elastic constants. II. *Arch. Mech.*, **33** (1981), 717-737.

[117] R.D. Gauthier and E.W. Jahsman, *Reexamination of micropolar wave experiment.* In "Continuum Models of Discrete Systems 4" (Edited by O. Brulin and R.K.T. Hsieh), pp. 225-231, North-Holland Publ., Amsterdam, 1981.

[118] T.G. Gegelia and R.K. Chichinadze, Basic static problems of elastic micropolar media. *Arch. Mech.*, **28** (1976), 89-104.

[119] J.A. Goldstein, *Semigroups of Linear Operators and Applications.* Oxford University Press, New York, 1985.

[120] M.A. Goodman and S.C. Cowin, A continuum theory for granular materials. *Arch. Rational Mech. Anal.*, **44** (1972), 249-266.

[121] M. Gotteland, M. Barbe, Etude comparative de la théorie des milieux micromorphique et micropolaires et de la théorie des directeurs appliquées aux surfaces, *J. Mécanique*, **20** (1981), 643-657.

[122] D. Graffi, Sui teoremi di reciprocità nei fenomeni non stazionari. *Atti Accad. Sc. Ist. Bologna*, **11** (1963), 33-40.

[123] A.E. Green and J.E. Adkins, *Large Elastic Deformations.* Oxford, Clarendon Press, 1960.

[124] A.E. Green, Thermoelastic stresses in initially stressed bodies. *Proc. Roy. Soc. London,* Ser. A, **266** (1962), 1-19.

[125] A.E. Green and R.S. Rivlin, Multipolar continuum mechanics. *Arch. Rational Mech. Anal.*, **17** (1964), 113-147.

[126] A.E. Green, Micro-materials and multipolar continuum mechanics. *Int. J. Engng. Sci.*, **3** (1965), 533-537.

[127] A.E. Green and W. Zerna, *Theoretical Elasticity*, 2nd Edition, Clarendon Press, Oxford, 1968.

[128] A.E. Green and P.M. Naghdi, On superposed small deformations on a large deformation of an elastic Cosserat surface, *J. Elasticity*, **1** (1971), 1-19.

[129] A.E. Green and N. Laws, On the entropy production inequality, *Arch. Rational Mech. Anal.*, **45** (1972), 47-59.

[130] A.E. Green and K.A. Lindsay, Thermoelasticity, *J. Elasticity*, **2** (1972), 1-7.

[131] A.E. Green and P.M. Naghdi, On undamped heat waves in an elastic solid. *J. Thermal Stresses*, **15** (1992), 253-264.

[132] A.E. Green and P.M. Naghdi, Thermoelasticity without energy dissipation. *J. Elasticity*, **31** (1993), 189-209.

[133] G. Grioli, Elasticità asimmetrica, *Ann. Mat. Pura Appl.* (IV), **50** (1960), 389-417.

[134] G. Grioli, Microstrutture. *Rend. Acc. Naz. dei Lincei*, Serie VIII, **49** (1970), I: 261-267, II: 379-385.

[135] G. Grioli, Extension of a theorem of Menabrea to the case of a microstructure in finite deformation. *Atti Accad. Naz. Lincei,* VIII. Ser., Rend., Cl. Sci. Fis. Mat. Nat., **75** (1983), 190-194.

[136] G. Grioli, Propagation of acceleration waves in structured continua. *Rend. Semin. Mat.,* Torino, **42** (1984), 11-24.

[137] R.A. Grot, Thermodynamics of a continuum with microstructure. *Int. J. Engng. Sci.*, **7** (1969), 801-814.

[138] M.E. Gurtin, Variational principles for linear elastodynamics. *Arch. Rational Mech. Anal.*, **16** (1964), 34-50.

[139] M.E. Gurtin, *The Linear Theory of Elasticity,* pp. 1-295 of Flügge's Handbuch der Physik, vol. VI a/2 (Edited by C. Truesdell), Springer-Verlag, Berlin-Heidelberg-New York, 1972.

[140] A. Hanyga and R. Teisseyre, The fundamental source solutions in the symmetric micromorphic continuum. *Rev. Italiana di Geofisica*, **22** (1973), 336-340.

[141] A. Hanyga and R. Teisseyre, Linear symmetric micromorphic thermoelasticity-source solutions and wave propagation. *Acta Geophys. Polonica*, **23** (1975), 147-157.

[142] D.P.H. Hasselman and R.A. Heller (Eds.), *Thermal Stresses in Severe Environments.* Plenum Press, New York, 1980.

[143] C.T. Herakovich and F. Mirzadeh, Discrete lawer analysis of orthotropic stiffened composite cylinders, *J. Thermal Stresses*, **20** (1997), 87-111.

[144] K.P. Hermann and V.V. Loboda, Contact zone models for an interface crack in a thermomechanical loaded anisotropic bimaterial, *J. Thermal Stresses*, **24** (2001), 479-506.

[145] R.B. Hetnarski, Coupled one-dimensional thermal shock problem for small times. *Arch. Mech. Stosow.*, **13** (1961), 295-306.

[146] R.B. Hetnarski, The fundamental solution of the coupled thermoelastic problem for small times. *Arch. Mech. Stosow.*, **16** (1964), 23-31.

[147] R.B. Hetnarski, An algorithm for generating some inverse Laplace transforms of exponential form. *ZAMP*, **26** (1975), 249-253.

[148] R.B. Hetnarski and J. Ignaczak, Generalized thermoelasticity: Closed form solutions. *J. Thermal Stresses*, **16** (1993), 473-498.

[149] R.B. Hetnarski and J. Ignaczak, Generalized thermoelasticity: response of semi-space to a short laser pulse. *J. Thermal Stresses*, **17** (1994), 377-396.

[150] R.B. Hetnarski and J. Ignaczak, Generalized thermoelasticity. *J. Thermal Stresses*, **22** (1999), 451-476.

[151] S. Hjalmars, *Non-linear Micropolar Theory.* In "Mechanics of Micropolar Media" (Edited by O. Brulin and R.K.T. Hsieh) pp. 147-189, World Scientific, Singapore, 1982.

[152] I. Hlaváček and M. Hlaváček, On the existence and uniqueness of solution and some variational principles in linear theories of elasticy with couple-stresses. *Aplikace Matematiky*, **14** (1969), I: 387-410; II: 411-426.

[153] I. Hlaváček and J. Nečas, On inequalities of Korn's type. *Arch. Rational Mech. Anal.*, **36** (1970), 305-311.

[154] M. Hlaváček, Uniqueness of the solution of the boundary-initial value problem for a linear elastic Cosserat continuum. *Aplikace Matematiky*, **16** (1971), 402-411.

[155] Y.S. Hsu, J.N. Reddy and C.W. Bert, Thermoelasticity of circular cylindrical shells laminated of bimodulus composite materials. *J. Thermal Stresses*, **4** (1981), 155-177.

[156] C.M. Ieşan, Existence theorem in the linear theory of multipolar elasticity. *Aplikace Matematiky*, **18** (1973), 385-390.

[157] D. Ieşan, Principes variationnels dans la théorie de la thermoélasticité couplée. *An. Şt. Univ. "Al.I. Cuza" Iaşi, Matematica*, **12** (1966), 439-456.

[158] D. Ieşan, Sur la théorie de la thermoélasticité micropolaire couplée. *C.R. Acad. Sci. Paris* , **265** A (1967), 271-275.

[159] D. Ieşan, On the plane coupled micropolar thermoelasticity. *Bull. Acad. Polon. Sci.*, Sér. IV, **16** (1968), I: 379-384; II: 385-390.

[160] D. Ieşan,On some theorems in the linear theory of micropolar thermoelasticity. *Rev. Roum. Math. Pures et Appl.*, **15** (1970), 1181-1195.

[161] D. Ieşan, On the linear coupled thermoelasticity with two temperatures. *ZAMP*, **21** (1970), 583-591.

[162] D. Ieşan, On thermal stresses in plane strain of isotropic micropolar elastic solids. *Acta Mechanica*, **11** (1971), 141-144.

[163] D. Ieşan, Torsion of micropolar elastic beams. *Int. J. Engng. Sci.*, **9** (1971), 1047-1060.

[164] D. Ieşan, Existence theorems in micropolar elastostatics. *Int. J. Engng. Sci.*, **9** (1971), 59-78.

[165] D. Ieşan, On the thermal stresses in beams. *J. Eng. Math.*, **6** (1972), 155-163.

[166] D. Ieşan, On some reciprocity theorems and variational theorems in linear dynamic theories of continuum mechanics. *Memorie dell'Accad. Sci. Torino, Cl. Sci. Fis. Mat. Nat.*, Serie 4, n.**17**, 1974.

[167] D. Ieşan, Thermal stresses in micropolar elastic cylinders, *Acta Mechanica*, **21** (1975), 261-272.

[168] D. Ieşan, Saint-Venant's problem for inhomogeneous and anisotropic elastic bodies. *J. Elasticity*, **6** (1976), 277-294.

[169] D. Ieşan, Saint-Venant's problem for heterogeneous anisotropic elastic solids. *Ann. Mat. Pura Appl.* (IV), **108** (1976), 149-159.

[170] D. Ieşan, Thermal stresses in composite Cosserat elastic cylinders, *J. Thermal Stresses*, **1** (1978), 149-162.

[171] D. Ieşan, Saint-Venant's problem for inhomogeneous and anisotropic Cosserat elastic solids. *ZAMM*, **58** (1978), 95-99.

[172] D. Ieşan, Deformation of composed Cosserat elastic cylinders. *Rev. Roum. Sci. Techn. Méc. Appl.*, **23** (1978), 823-845.

[173] D. Ieşan, Incremental equations in thermoelsticity. *J. Thermal Stresses*, **3** (1980), 41-56.

[174] D. Ieşan, *Saint-Venant's Problem in Micropolar Elasticity.* In "Mechanics of Micropolar Media" (Edited by O. Brulin and R.K.T. Hsieh) pp. 281-390, World Scientific, Singapore, 1982.

[175] D. Ieşan, Thermoelasticity of nonsimple materials, *J. Thermal Stresses*, **6** (1983), 167-188.

[176] D. Ieşan, Thermal stresses in heterogeneous anisotropic Cosserat elastic cylinders. *J. Thermal Stresses*, **8** (1985), 385-397.

[177] D. Ieşan, Some results in the theory of elastic materials with voids. *J. Elasticity*, **15** (1985), 215-224.

[178] D. Ieşan, The grade consistent theory of micropolar thermoelasticity. *Rev. Roum. Sci. Techn. Méc. Appl.*, **30** (1985), 3-19.

[179] D. Ieşan, On Saint-Venant's problem. *Arch. Rational Mech. Anal.*, **91** (1986), 363-373.

[180] D. Ieşan, A theory of thermoelastic materials with voids. *Acta Mechanica*, **60** (1986), 67-89.

[181] D. Ieşan, *Saint-Venant's Problem.* Lecture Notes in Mathematics. vol. 1279, Springer-Verlag, Berlin, Heidelberg, New York, London, Paris, Tokyo, 1987.

[182] D. Ieşan, Continuous dependence and uniqueness results in nonlinear theory of elastic materials with voids. *Rev. Roum. Sci. Techn. Méc. Appl.*, **32** (1987), 351-363.

[183] D. Ieşan, A theory of initially stressed thermoelastic materials with voids. *An. Şt. Univ. Iaşi, Matematica*, **33** (1987), 167-184.

[184] D. Ieşan, On some theorems in thermoelastodynamics. *Rev. Roum. Sci. Techn. Méc. Appl.*, **34** (1989), 101-111.

[185] D. Ieşan and R. Quintanilla, On the grade consistent theory of micropolar thermoelasticity. *J. Thermal Stresses*, **15** (1992), 393-318.

[186] D. Ieşan and A. Scalia, *Thermoelastic Deformations.* Kluwer Academic Publishers, Dordrecht, Boston, London, 1996.

[187] D. Ieşan, On the theory of thermoelasticity without energy dissipation. *J. Thermal Stresses*, **21** (1998), 295-307.

[188] D. Ieşan, On a theory of micromorphic elastic solids with microtemperatures. *J. Thermal Stresses*, **24** (2001), 737-752.

[189] D. Ieşan, On the micromorphic thermoelasticity. *Int. J. Engng. Sci.*, **40** (2002), 549-567.

[190] D. Ieşan and L. Nappa, Thermal stresses in plane strain of porous elastic bodies. *Meccanica* **39** (2004), 125-138.

[191] J. Ignaczak, A completeness problem for stress equations of motion in the linear elasticity theory. *Arch. Mech. Stosow*, **15** (1963), 225-235.

[192] J. Ignaczak and W. Nowacki, Singular integral equations of thermoelasticity. *Int. J. Engng. Sci.*, **4** (1966), 53-68.

[193] J. Ignaczak, *Tensorial equations of motion for elastic materials with microstructure.* In "Trends in Elasticity and Thermoelasticity" (Witold Nowacki Anniversary Volume), Wolters-Noordh. Publ., Groningen, 1971.

[194] J. Ignaczak, Uniqueness in generalized thermoelasticity. *J. Thermal Stresses*, **2** (1979), 171-175.

[195] J. Ignaczak, *Thermoelasticity with finite wave speeds. A survey.* In "Thermal Stresses in Severe Environments" (Edited by D.P.H. Hasselman and R. Heller), Plenum Press, pp. 15-30, 1980.

[196] J. Ignaczak and J. Bialy, Domain of influence in thermoelasticity with one relaxation time. *J. Thermal Stresses*, **3** (1980), 391-399.

[197] J. Ignaczak, A note on uniqueness in thermoelasticity with one relaxation time. *J. Thermal Stresses*, **5** (1982), 257-263.

[198] J. Ignaczak, A strong discontinuity wave in thermoelasticity with relaxation times. *J. Thermal Stresses*, **8** (1985), 25-40.

[199] J. Ignaczak, B. Carbonaro and R. Russo, Domain of influence theorem in thermoelasticity with one relaxation time. *J. Thermal Stresses*, **9** (1986), 79-92.

[200] J. Ignaczak, *Generalized thermoelasticity and its applications.* In "Thermal Stresses", vol. 3 (Edited by R.B. Hetnarski), Elsevier Sci. Publish., B.V., pp. 279-354, 1989.

[201] E. Inan and A.C. Eringen, Nonlocal theory of wave propagation in thermoelastic plates. *Int. J. Engng. Sci,*, **29** (1991), 831-844.

[202] V. Ionescu-Cazimir, Problem of linear coupled thermoelasticity. Theorems on reciprocity for the dynamic problem of coupled thermoelasticity. I. *Bull. Acad. Polon. Sci., Sér. Sci. Techn.*, **12** (1964), 473-480.

[203] V. Ionescu-Cazimir, Problem of linear coupled thermoelasticity. IV. Uniqueness theorem. *Bull. Acad. Polon. Sci., Sér. Sci. Techn.*, **12** (1964), 565-573.

[204] M. Jakubowska, Kirchhoff's type formula in thermoelasticity with finite wave speeds. *J. Thermal Stresses*, **7** (1984), 259-288.

[205] J. Jarić and Z. Golubović, Theory of thermoelasticity of granular materials. *Rev. Roum. Sci. Techn., Méc. Appl.*, **24** (1979), 793-805.

[206] J. T. Jenkins, Static equilibrium of granular materials. *J. Appl. Mech.*, **42** (1975), 603-606.

[207] L. Jentsch, Zur Thermoelastostatik stückweise homogener Körper. *Arch. Rational Mech. Anal.*, **24** (1967), 141-172.

[208] S. Jiang, Far field behavior of solutions to the equations of nonlinear 1-d-thermoelasticity. *Appl. Anal.*, **36** (1990), 25-35.

[209] S. Jiang, Rapidly decreasing behaviour of solutions in nonlinear 3-*d*-thermoelasticity. *Bull. Austral. Math. Soc.*, **43** (1991), 89-99.

[210] Z.H. Jin and R.C. Batra, Thermal fracture of ceramics with temperature-dependent properties. *J. Thermal Stresses*, **21** (1998), 157-176.

[211] P.M. Jordan and P. Puri, Thermal stresses in a spherical shell under three thermoelastic models. *J. Thermal Stresses*, **24** (2001), 47-70.

[212] D. Jou, J. Casas-Vazquez and G. Lebon, *Extended Irreversible Thermodynamics.* Springer-Verlag, Berlin, 1996.

[213] S. Kaliski and W. Nowacki, Wave equations of thermoelasticity. *Bull. Acad. Polon. Sci., Sér. Sci. Techn.*, **18** (1970), 209-218.

[214] M.A. Kattis, Thermoelastic plane problems with curvilinear boundaries. *Acta Mech.*, **87** (1991), 93-104.

[215] T. Katto, Linear evolution equations of hyperbolic type. II. *J. Math. Soc. Japan*, **25** (1973), 648-666.

[216] F.W. Keene and R.B. Hetnarski, Bibliography on thermal stresses in shells. *J. Thermal Stresses*, **13** (1990), 341-540.

[217] O.D. Kellogg, *Foundations of Potential Theory.* Springer-Verlag, New York 1976.

[218] Ken-Ichi Kanatani, A micropolar continuum theory for the flow of granular materials. *Int. J. Engng. Sci.*, **17** (1979), 419-432.

[219] J. Kiusalaas, W. Jaunzemis and J.C. Conway, A strain-gradient theory for prestrained laminates. *Int. J. Engng. Sci.*, **9** (1973), 1317-1330.

[220] R.J. Knops and L.E. Payne, On uniqueness and continuous dependence in dynamical problems of linear thermoelasticity. *Int. J. Solids Structures*, **6** (1970), 1173-1184.

[221] R.J. Knops and L.E. Payne, Uniqueness theorems in linear elasticity. *Springer Tracts in Natural Philosophy*, vol. 19. Berlin-Heidelberg-New York, Springer, 1971.

[222] R.J. Knops and E.W. Wilkes, Theory of Elastic Stability, pp. 125-302 of Flügge's Handbuch der Physik, vol. VI a/3 (Edited by C. Truesdell), Springer-Verlag, Berlin-Heidelberg-New York, 1973.

[223] R.J. Knops and B. Straughan, Continuous dependence theorems in the theory of linear elastic materials with microstructure. *Int. J. Engng. Sci.*, **14** (1976), 555-565.

[224] J.K. Knowles, On the spatial decay of solutions of the heat equations. *ZAMP*, **22** (1971), 1050-1056.

[225] Y.M. Kolyano, B.V. Protsyuk and V.M. Sinyuta, The axisymmetric static problem of thermoelasticity for a multilayered cylinder. PMM; *J. Appl. Math. Mech.*, **55** (1991), 920-926.

[226] G.V. Krishna Reddy and N.K. Venkatasubramanian, Saint-Venant's problem for a micropolar elastic circular cylinder. *Int. J. Engng. Sci.*, **14** (1976), 1047-1055.

[227] G.V. Krishna Reddy and N.K. Venkatasubramanian, Flexure of a micropolar elastic circular cylinder. *Proc. Indian Acad. Sci.*, **86A** (1977), 575-563.

[228] G.V. Krishna Reddy and N.K. Venkatasubramanian,On the flexural rigidity of a micropolar elastic cylinder. *J. Appl. Mech.*, **45** (1978), 429-431.

[229] R. Kumar and B. Singh, Reflection of plane waves from the flat boundary of micropolar generalized thermoelastic half space. *Int. J. Engng. Sci.*, **36** (1988).

[230] I.A. Kunin, *Elastic Media with Microstructure* II. Springer Series in Solid State Sciences, **44**, Springer-Verlag, Berlin 1983.

[231] V. Kupradze, *Dynamical Problems in Elasticity. Progress in Solid Mechanics.* vol. 3 (Edited by I.N. Sneddon), Amsterdam, 1963.

[232] V.D. Kupradze, T.G. Gegelia, M.O. Bashelishvili and T.V. Burchuladze, *Three-Dimensional Problems of Mathematical Theory of Elasticity and Thermoelasticity.* North-Holland Publ., 1979.

[233] J.E. Lagnese, The reachability problem for thermoelastic plates. *Arch. Rational Mech. Anal.*, **112** (1990), 223-267.

[234] J.E. Lagnese and J.L. Lions, *Modelling analysis of thin plates.* Collection Researches on Mathématiques Appliqueés 6, Masson, Paris, 1989.

[235] R.W. Lardner, The linear theory of second-grade elastic materials, *Quart. Appl. Math.*, **27** (1969), 323-334.

[236] R.W. Lardner, Singular stress concentrations in second-grade materials. *Quart. Appl. Math.*, **28** (1970), 111-121.

[237] R.W. Lardner, Dislocations in materials with couple stress. *J. Inst. Math. Appl.*, **7** (1971), 126-137.

[238] R.W. Lardner, The stress and couple-stress fields near the tip of an antiplane shear crack. *Arch. Mech.*, **23** (1971), 627-640.

[239] G. Lebon, *Variational Principles in Thermomechanics.* In "Recent Developments in Thermomechanics of Solids" (Edited by G. Lebon and P. Perzyna), Wien-New York, Springer Verlag, 1980.

[240] G. Lebon, A generalized theory of thermoelasticity. *Journal of Technical Physics* 23, **1** (1982), 37-46.

[241] J. Lenz, The action of an instantaneous heat source in the infinite hemitropic micropolar continuum, *Bull. Acad. Polon. Sci., Sér. Sci. Techn.*, **25** (1977), 209-215.

[242] L. Librescu and M. Stein, A geometrically nonlinear theory of shear deformable laminated composite plates and its use in postbuckling analysis. Thin-Walled Structures, *Special Aerospace Structures Issue*, **11** (1991), 177-201.

[243] L. Librescu, W. Lin, M.P. Nemeth and J.H. Starnes, Jr., Thermomechanical postbuckling of geometrically imperfect flat and curved panels taking into account tangential edge constraints. *J. Thermal Stresses*, **18** (1995), 465-482.

[244] L. Librescu, M.P. Nemeth, J.H. Starnes and W. Lin, Nonlinear response of flat and curved panels subjected to thermomechanical loads. *J. Thermal Stresses*, **23** (2000), 549-582.

[245] V.A. Lomakin, Problems of generalized theory of elasticity with couple-stresses. (in Russian) *Vestnik Mosk. Univ.* S. 1, **1** (1987), 82-88.

[246] H.W. Lord and Y. Shulman, A generalized dynamical theory of thermoelasticity. *J. Mech. Phys. Solids*, **15** (1967), 299-309.

[247] M.V. Lurie, Lamé problem in gradient elasticity theory (in Russian). *Dokl. Akad. N. SSSR*, **181** (1968), 1087-1089.

[248] C.-C. Ma, Plane solution of thermal stresses for anisotropic bimaterial elastic wedges, *J. Thermal Stresses*, **18** (1995), 219-246.

[249] K.Z. Markov, Green's tensor in micromorphic elasticity (in Russian), *Proc. of Nat. Congress Theor. Appl. Mech.* (Varna 1973), Sofia, 1975.

[250] K.Z. Markov, On the dilatational theory of elasticity. *ZAMP*, **61** (1981), 349-358.

[251] K.Z. Markov, Dilatation theories of elasticity. *New Problems in Mechanics of Continua.* Proc. 3rd Swed.-Pol. Symp. Mech., Jablonna, 1982, SM Stud. Ser., **17** (1983), 129-138.

[252] J.E. Marsden and T.J.R. Hugues, *Mathematical Foundations of Elasticity.* Prentice-Hall Inc. Englewood Cliffs, New Jersey, 1983.

[253] F. Martinez and R. Quintanilla, Existence and uniqueness of solutions to the equations of incremental thermoelasticity with voids, In *"Trends in Applications of Mathematics to Mechanics"*, Pitman Monographs 77, Longman Group Ltd., Harlow, 1995, 45-56.

[254] F. Martinez and R. Quintanilla, On the incremental problem in thermoelasticity of nonsimple materials. *Z. Angew. Math. Mech.*, **78** (1998), 703-710.

[255] M. Massoudi and G. Ahmadi, Rapid flow of granular materials with density and fluctuation energy gradients. *Int. J. Non-Linear Mechanics*, **29** (1994), 487-492.

[256] G.A. Maugin and A. Miled, Solitary waves in micropolar elastic crystals. *Int. J. Mech. Sci.*, **24** (1986), 1477-1499.

[257] G.A. Maugin and A. Berezovski, Material formulation of finite-strain thermoelasticity and applications. *J. Thermal Stresses*, **22** (1999), 421-449.

[258] M.F. McCarthy, *Singular surfaces and waves.* In vol. II of the Continuum Physics (Edited by A.C. Eringen), New York, Academic Press, 1975.

[259] A.J. Mc Connel, *Applications of Tensor Analysis.* Dover Publ. Inc., New York, 1957.

[260] M. Mihăilescu-Suliciu and I. Suliciu, An energetic restriction on a numerical scheme in one-dimensional dynamic thermoelasticity. *J. Thermal Stresses*, **22** (1999), 805-826.

[261] S.G. Mikhlin, *Multidimensional Singular Integrals and Integral Equations* (in Russian), Nauka, Moskow, 1962.

[262] R.D. Mindlin and H.F. Tiersten, Effects of couple stresses in linear elasticity. *Arch. Rational Mech. Anal.*, **11** (1962), 415-448.

[263] R.D. Mindlin, Microstructure in linear elasticity. *Arch. Rational Mech. Anal.*, **16** (1964), 51-77.

[264] R.D. Mindlin, Second gradient of strain and surface-tension in linear elasticity. *Int. J. Solids Structures*, **1** (1965), 417-500.

[265] R.D. Mindlin and N.N. Eshel, On first strain-gradient theories in linear elasticity. *Int. J. Solids Structures* , **4** (1968), 109-124.

[266] R.D. Mindlin, Galerkin stress functions for non-local theories of elasticity. *Int. J. Solids Structures*, **8** (1972), 1407-1411.

[267] I. Müller, The coldness, a universal function in thermoelastic bodies. *Arch. Rational Mech. Anal.*, **41** (1971), 319-332.

[268] I. Müller and T. Ruggeri, *Rational and Extended Thermodynamics*, Springer-Verlag, New York, 1998.

[269] M.A. Murad and J.H. Cushman, Thermomechanical theories for swelling porous media with microstructure, *Int. J. Engng. Sci.*, **38** (2000), 517-564.

[270] N.I. Muskhelishvili, *Some Basic Problems of the Mathematical Theory of Elasticity.* Noordhoff, Groningen, 1953.

[271] F.R.N. Nabarro, The calculation of thermal stresses in cylinders. *Int. J. Engng. Sci.*, **19** (1981), 1651-1656.

[272] P.M. Naghdi, *The Theory of Shells and Plates.* In vol. VI a/2 of the Handbuch der Physik, (Edited by C. Truesdell), Springer-Verlag, Berlin, Heidelberg, New York, 1972.

[273] L. Nappa, Thermal stresses in microstretch elastic cylinders. *J. Thermal Stresses*, **18** (1995), 537-550.

[274] L. Nappa, Variational principles in micromorphic thermoelasticity, *Mechanics Research Comm.*, **28** (2001), 405-412.

[275] C.B. Navarro and R. Quintanilla, On existence and uniqueness in incremental thermoelasticity. *ZAMP*, **35** (1984), 206-215.

[276] A. Nayfeh and S. Nemat-Nasser, Thermoelastic waves in solids with thermal relaxation. *Acta Mechanica*, **12** (1971), 53-69.

[277] J. Nečas, *Les méthodes directes en théorie des équations elliptiques.* Academia, Prague, 1967.

[278] J. Nečas and I. Hlavaček, *Mathematical Theory of Elastic and Elasto-Plastic Bodies: An Introduction.* Elsevier Scientific Publ. Comp. Amsterdam, New York, 1981.

[279] I. Nistor, Wave propagation in nonlinear thermoelastic Cosserat solids. *J. Thermal Stresses*, **3** (1980), 1-8.

[280] I. Nistor, Systematic methods of approximation in nonlinear theory of Cosserat media III. *Bul. Inst. Politeh. Iaşi,* Sect. I, **31** (1985), 19-25.

[281] I. Nistor, Variational principles for Cosserat body. *Int. J. Nonlinear Mech.*, **37** (2002), 565-569.

[282] N. Noda, Transient thermoelastic contact problem in a long, circular cylinder. *J. Thermal Stresses*, **7** (1984), 135-147.

[283] N. Noda, R.B. Hetnarski and Y. Tanigawa, *Thermal Stresses. Lastran Corp.* Partridge Hill, Honeoye Falls, 2002.

[284] J.P. Nowacki and W. Nowacki, Some problems of hemitropic micropolar thermoelasticity. *Proceedings of Tbilisi University*, **257** (1985), 131-145.

[285] W. Nowacki, *Thermoelasticity.* Reading, Mass., Addison-Wesley, 1962.

[286] W. Nowacki, Green functions for the thermoelastic medium. II. *Bull. Acad. Polon. Sci., Sér. Sci. Techn.*, **12** (1964), 465-472.

[287] W. Nowacki, Couple-stresses in the theory of thermoelasticity. *Bull. Acad. Polon. Sci., Sér. Sci. Techn.,* **14** (1966), I: 97-106; II: 203-212; III: 505-515.

[288] W. Nowacki, Green functions for micropolar thermoelasticity. *Bull. Acad. Polon. Sci. Sér. Sci. Techn.*, **16** (1968), 555-564.

[289] W. Nowacki, *Theory of Asymmetric Elasticity.* Polish Scientific Publishers, Warszawa and Pergamon Press, Oxford, New York, Paris, Frankfurt, 1986.

[290] J.L. Nowinski, *Theory of Thermoelasticity with Applications.* Sijthoff & Noordhoff Int. Publ. Alphen aan Den Rijn, Netherlands, 1978.

[291] J.W. Nunziato and E.K. Walsh, On the influence of void compaction and material non-uniformity on the propagation of one-dimensional acceleration waves in granular materials. *Arch. Rational Mech. Anal.*, **64** (1977), 299-316; **67** (1978), 395-398.

[292] J.W. Nunziato and S.C. Cowin, A nonlinear theory of elastic materials with voids. *Arch. Rational Mech. Anal.*, **72** (1979), 175-201.

[293] J.W. Nunziato and E.K. Walsh, On ideal multiphase mixtures with chemical reactions and diffusion. *Arch. Rational Mech. Anal.*, **73** (1980), 285-301.

[294] J.T. Oden, D.M. Rigsby and D. Cornett, On the numerical solution of a class of problems in a linear first strain-gradient theory of elasticity. *Int. J. Numer. Meth. Eng.*, **2** (1970), 159-174.

[295] Z. Olesiak, On a method of solution of mixed boundary-value problems of thermoelasticity. *J. Thermal Stresses*, **4** (1981), 501-508.

[296] W.C. Orthwein, Thermal buckling of a bimetallic strip. *J. Vib. Acoust. Stress Reliab.*, **106** (1984), 538-542.

[297] J. Padovan, *Anisotropic thermal stresses analysis.* In "Thermal Stresses" (Edited by R.B. Hetnarski), pp. 143-262, North-Holland, Amsterdam, 1986.

[298] J. Padovan, Thermoelasticity of unidirectionally micropolar cord-reinforced composites, *J. Thermal Stresses*, **18** (1995), 383-402.

[299] V.A. Palmov, Fundamental equations of the theory of asymmetric elasticity (in Russian). *Prikl. Mat. Mekh.*, **28** (1964), 401-412.

[300] S.L. Passman, Mixtures of granular materials. *Int. J. Engng. Sci.*, **15** (1977), 117-129.

[301] A. Pazy, *Semigroups of Linear Operators and Applications to Partial Differential Equations.* Springer, New York, Berlin, Heidelberg, Tokyo, 1983.

[302] M. Pitteri, On a statistical-kinetic model for generalized continua. *Arch. Rational Mech. Anal.*, **111** (1990), 99-120.

[303] A. Pompei and A. Scalia, On the steady vibrations of the thermoelastic porous materials. *Int. J. Solids Structures*, **31** (1994), 2819-2834.

[304] P. Puri, Steady-state thermoelastic problems for the half-space with couple stresses, *Arch. Mech. Stos.*, **22** (1970), 479-489.

[305] P. Puri and S.C. Cowin, Plane waves in linear elastic materials with voids. *J. Elasticity*, **15** (1985), 167-178.

[306] S.L. Pu and M.A. Sadowsky, Boundary layers in strain-gradient theory of linear elasticity. *Proc. IVth Southeastern Conf. Theor. Appl. Mech., New Orleans. Louisiana, 1968. Developments Theor. Appl. Mech.*, **4** (1970), 85-102.

[307] R. Quintanilla, A short note on incremental thermoelasticity. *Stochastica*, **7** (1983), 163-167.

[308] R. Quintanilla, Existence in thermoelasticity without energy dissipation. *J. Thermal Stresses*, **25** (2002), 195-202.

[309] R. Quintanilla, On the spatial decay for the dynamical problem of thermomicrostretch elastic solids. *Int. J. Engng. Sci.*, **40** (2002), 109-121.

[310] R. Racke and Y. Shibata, Global smooth solutions and asymptotic stability in one-dimensional nonlinear thermoelasticity. *Arch. Rational Mech. Anal.*, **116** (1991), 1-34.

[311] K.R. Rajagopal and L. Tao, *Mechanics of Mixtures.* World Scientific, Singapore, New Jersey, London, Hong Kong, 1995.

[312] S. Rionero and S. Chiriţă, The Lagrange identity method in linear thermoelasticity. *Int. J. Engng. Sci.*, **25** (1987), 935-947.

[313] R.S. Rivlin, The formulation of theories in generalized continuum mechaanics and their physical significance. Symposia Mathematica, Academic Press, London, **1** (1969), 357-373.

[314] D. Rogula, Moment stresses and symmetry of the stress tensor in bodies with nonlocal structure. *Bull. Acad. Polon. Sci., Sér. Sci. Techn.*, **18** (1970), 159-164.

[315] D. Rogula, Some basic solutions in strain gradient elasticity theory of an arbitrary order. *Arch. Mech.*, **25** (1973), 43-68.

[316] D. Rogula, Generalized interactions in nonlocal continua. Continuum models of discrete systems. Proc. 3rd Int. Symp., Freudenstadt 1979, *SM Stud. Ser.*, **15** (1980), 599-626.

[317] P.P. Roy, Wave propagation in a thinly two layered laminated medium with stress couples under initial stresses. *Acta Mech.*, **54** (1984), 1-21.

[318] R. Russo, Classical coupled thermoelasticity in unbounded domains. *J. Elasticity*, **22** (1989), 1-24.

[319] E. Rusu, Fundamental solution, for short time, in a generalized theory of linear micropolar thermoelasticity. *Bul. Inst. Politeh. Iaşi, Sect. I*, **33**(37) (1987), 65-72.

[320] G. Rusu, Existence theorems in elastostatics of materials with voids. *An. Şt. Univ. Iaşi, Matematica*, **30** (1985), 193-204.

[321] G. Rusu, On existence and uniqueness in thermoelasticity of materials with voids. *Bull. Acad. Polon. Sci., Sér. Sci. Techn.*, **35** (1987), 339-346.

[322] C. Rymarz, On the model of non-simple medium with rotational degrees of freedom. *Bull. Acad. Polon. Sci. , Sér. Sci. Techn.*, **16** (1968), 272-280.

[323] Cz. Rymarz, Gradient models of non-local media. *J. Techn. Phys.*, **20** (1979), 41-49.

[324] C. Rymarz, The models of material continuum. *Ann. Soc. Math. Pol., Ser. III, Mat. Stosow.*, **28** (1986), 5-32.

[325] N. Sandru, Le théorème de reciprocité dans l'élasticité asymétrique (cas dynamique), *Rend. Accad Lincei,* Ser. VIII, **38** (1965), 78-81.

[326] N. Sandru, On some problems of the linear theory of asymmetric elasticity. *Int. J. Engng. Sci.*, **4** (1966), 81-94.

[327] A. Scalia, On some theorems in the theory of micropolar thermoelasticity. *Int. J. Engng. Sci.*, **28** (1990), 181-189.

[328] A. Scalia, A grade consistent micropolar theory of thermoelastic materials with voids. *ZAMM*, **71** (1990), 2-14.

[329] E. Scarpetta, On the fundamental solutions in micropolar elasticity with voids. *Acta Mechanica*, **82** (1990), 151-158.

[330] N.H. Scott, A theorem in thermoelasticity and its application to linear stability. *Proc. R. Soc. Lond.* A, **424** (1989), 143-153.

[331] A.P.S. Selvadurai, *Mechanics of Poroelastic Media*, Kluwer Academic Publishers, Dordrecht, Boston, London, 1996.

[332] B.V. Semkin and A.V. Iushin, On Danilovskaia problem for the case of a boundary of an elastic half-space moving with constant velocity. *J. Appl. Math. Mech.*, **45** (1981), 291-292.

[333] M.U. Shanker and R.S. Dhaliwal, Singular integral equations in asymmetric thermoelasticity. *J. Elasticity*, **2** (1972), 59-71.

[334] M.U. Shanker and R.S. Dhalival, Dynamic plane strain problem of infinite micropolar thermoelastic body under the action of body forces and heat sources. *Utilitas Mathematica*, **8** (1975), 127-179.

[335] M.U. Shanker and R.S. Dhalival, Dynamic coupled thermoelastic problems in micropolar theory. *Int. J. Engng. Sci.*, **13** (1975), 121-148.

[336] J.N. Sharma, Some considerations on generalized thermoelasticity. *Int. J. Engng. Sci.*, **25** (1987), 1387-1392.

[337] J.N. Sharma and R.S. Chauhan, Mechanical and thermal sources in a generalized thermoelastic half-space, *J. Thermal Stresses*, **24** (2001), 651-676.

[338] H.H. Sherif and R.S. Dhalival, Generalized one-dimensional thermal shock problem for small times. *J. Thermal Stresses*, **4** (1981), 407-420.

[339] H.H. Sherif, Fundamental solution of the generalized thermoelastic problem for short times. *J. Thermal Stresses*, **9** (1986), 151-164.

[340] H.H. Sherif, On uniqueness and stability in generalized thermoelasticity. *Q. Appl. Math.*, **44** (1987), 773-778.

[341] D.I. Sherman, *On the problem of plane strain in nonhomogeneous media.* In "Non-Homogeneity in Elasticity and Plasticity". Pergamon Press, London, 1959.

[342] M. Šilhavý, An admissibility criterion for shocks and propagating phase boundaries via thermodynamics of non-simple materials. I, II. *J. Non-Equilibrium Thermodyn.*, **9** (1984), 177-186, 187-199.

[343] M. Slemrod, Global existence, uniqueness and asymptotic stability of classical smooth solutions in one-dimensional nonlinear thermoelasticity. *Arch. Rational Mech. Anal.*, **76** (1981), 97-134.

[344] I.N. Sneddon, *Fourier Transforms.* McGraw-Hill Book Co., New York, 1951.

[345] E. Soós, Reciprocity theorems for elastic and thermoelastic materials with microstructure. *Bull. Acad. Polon. Sci. , Sér. Sci. Techn.*, **16** (1968), 541-545.

[346] E. Soós, Uniqueness theorems for homogeneous, isotropic, simple elastic and thermoelastic materials having a microstructure. *Int. J. Engng. Sci.*, **7** (1969), 257-268.

[347] E. Soós, *Discrete and Continuous Models of Solids* (in Romanian), Editura Ştiinţifică, Bucureşti, 1974.

[348] A. Soyuçok, Free vibrations of a generalized thermoelastic sphere. *J. Thermal Stresses*, **14** (1991),173-192.

[349] J. Stefaniak, A generalization of Galerkin's functions for asymmetric thermoelasticity. *Bull. Acad. Polon. Sci., Sér. Sci. Techn.*, **16** (1968), 391-408.

[350] J. Stefaniak, The method of concentrated sources in solving thermal stress problems, *J. Thermal Stresses*, **21** (1998), 497-508.

[351] R. Stojanowić, *Mechanics of Polar Continua.* Lecture Notes. Int. Center Mech. Sci., Udine, Italy, 1969.

[352] B. Straughan, *Instability, nonexistence and weighted energy methods in fluid dynamics and related theories.* Research Notes in Mathematics 74, Pitman Advanced Publishing Program, Boston, 1982.

[353] B. Straughan, *Stability problems in electrodynamics, ferrohydrodynamics and thermoelectric magnetohydrodynamics.* In "Mathematical Topics in Fluids" (Edited by J.F. Rodrigues and A. Sequeira), pp. 163-192, Pitman, Longman, 1992.

[354] D.V. Strunin, R.V.N. Melnik and A.J. Roberts, Coupled thermomechanical waves in hyperbolic thermoelasticity. *J. Thermal Stresses*, **24** (2001), 121-140.

[355] E.S. Şuhubi, *Thermoelastic Solids.* In vol. II of the "Continuum Physics" (Edited by A.E. Eringen), New York, Academic Press, 1975.

[356] M.Zh. Svanadze, Existence theorems for solution of dynamic problems of thermoelasticity for piecewise homogeneous isotropic bodies. *Tr. Inst. Prikl. Mat. Im. I.N. Vekua*, **16** (1985), 216-255.

[357] R.T. Tauchert, *Thermal stresses in plates-statical problems.* In "Thermal Stresses" (Edited by R.B. Hetnarski), pp. 23-142, North-Holland, Amsterdam, 1986.

[358] R.T. Tauchert and R.B. Hetnarski, Bibliography on Thermal Stresses. *J. Thermal Stresses*, **9** (1986), 1-128.

[359] P.P. Teodorescu, On the action of concentrated loads in the case of a Cosserat continuum. IUTAM-Symposium 1967, Freudenstadt-Stuttgart, *Mechanics of generalized continua*, 1968, pp. 120-125.

[360] P.P. Teodorescu, *Dynamics of linear elastic bodies.* Abacus Press, Tunbridge Wells, Kent, England, 1975.

[361] C. Teodosiu, Nonlinear theory of the material of grade 2 with initial stresses and hyperstresses. *Bull. Acad. Polon. Sci., Sér. Sci. Techn.*, **15** (1967), I: 115-122; II: 123-130.

[362] T.Y. Thomas, *Plastic Flow and Fracture in Solids.* Academic Press, New York, London, 1961.

[363] T.Y. Thomas, *Concepts from Tensor Analysis and Differential Geometry.* Academic Press, New York, London, 1961.

[364] R.A. Toupin, Elastic materials with couple-stresses. *Arch. Rational Mech. Anal.*, **11** (1962), 385-414.

[365] R.A. Toupin, Theories of elasticity with couple-stress. *Arch. Rational Mech. Anal.*, **17** (1964), 85-112.

[366] C. Truesdell and R. Toupin, *The Classical Field Theories.* In vol. III/1 of the Handbuch der Physik (Edited by S. Flügge), Berlin-Heidelberg, New York, Springer-Verlag, 1960.

[367] C. Truesdell and W. Noll, *The Non-Linear Field Theories of Mechanics.* In vol. III/3 of the Handbuch der Physik (Edited by S. Flügge), Berlin-Heidelberg, New York, Springer-Verlag, 1965.

[368] C. Truesdell, *Rational Thermodynamics.* Springer-Verlag, New York, Berlin, Heidelberg, Tokyo, 1984.

[369] A. Tylikowski, Dynamic stability of rotating composite shells with thermoactive shape memory alloy fibers, *J. Thermal Stresses*, **21** (1998), 327-340.

[370] K.L. Verma, On the propagation of waves in layered anisotropic media in generalized thermoelasticity. *Int. J. Engng. Sci.*, **40** (2002), 2077-2096.

[371] D. Vieru, Some theorems in a grade consistent micropolar theory of thermoelasticity. *An. Şt. Univ. Al.I. Cuza Iaşi,* Sect. Ia, **31** (1985), 187-192.

[372] P. Villaggio, *Qualitative Methods in Elasticity*, Kluwer Academic Publishers, Dordrecht, Boston, London, 1996.

[373] D.K. Wagh, Propagation of Rayleigh waves in Cosserat continuum. *J. Indian Acad. Math.,* **5** (1983), 111-116.

[374] J. Wang and R.S. Dhaliwal, Fundamental solutions of the generalized thermoelastic equations, *J. Thermal Stresses*, **16** (1993), 135-162.

[375] J.H. Weiner, A uniqueness theorem for the coupled thermoelastic problem. *Quart. Appl. Math.*, **15** (1957), 102-105.

[376] N.S. Wilkes, Continuous dependence and instability in linear thermoelasticity. *SIAM J. Math. Anal.*, **11** (1980), 292-299.

[377] K. Wilmanski, Dynamics of bodies with microstructure. *Arch. Mech. Stos.*, **20** (1968), 705-717.

[378] K. Wilmanski, *Thermomechanics of Continua.* Springer, Heidelberg, 1998.

[379] K. Wilmanski, Mass exchange, diffusion and large deformations of poroelastic materials. In "Modeling and Mechanics of Granular and Porous Materials" (Edited by G. Capriz, V.N. Ghionna, P. Giovine), pp. 213-244, Birkhäuser, 2002.

[380] R. Wojnar, Uniqueness of the displacement-heat flux problem in thermoelasticity with two relaxation times. *Bull. Acad. Polon. Sci., Sér. Sci. Techn.*, **33** (1985), 5-16.

[381] C. Wozniak, Thermoelasticity of non-simple oriented materials. *Int. J. Engng. Sci.*, **5** (1967), 605-612.

[382] C. Wózniak, Nonstandard analysis and micromorphic effects in the multilayered bodies. *Bull. Acad. Polon. Sci., Sér. Sci. Techn.*, **34** (1986), 385-392.

[383] Cz. Wozniak, E. Wierzbicki and M. Wozniak, A macroscopic model for the heat propagation in the microperiodic composite solids, *J. Thermal Stresses*, **25** (2002), 283-294.

[384] K. Yosida, *Functional Analysis*, Springer-Verlag, Berlin-Heidelberg-New York, 1968.

[385] S. Zahorski, On motion and thermodynamics of non-simple continua with microstructure. *Arch. Mech. Stos.*, **19** (1967), 25-44.

Index

Mechanics

SOLID MECHANICS AND ITS APPLICATIONS

Series Editor: G.M.L. Gladwell

Aims and Scope of the Series

The fundamental questions arising in mechanics are: *Why?*, *How?*, and *How much?* The aim of this series is to provide lucid accounts written by authoritative researchers giving vision and insight in answering these questions on the subject of mechanics as it relates to solids. The scope of the series covers the entire spectrum of solid mechanics. Thus it includes the foundation of mechanics; variational formulations; computational mechanics; statics, kinematics and dynamics of rigid and elastic bodies; vibrations of solids and structures; dynamical systems and chaos; the theories of elasticity, plasticity and viscoelasticity; composite materials; rods, beams, shells and membranes; structural control and stability; soils, rocks and geomechanics; fracture; tribology; experimental mechanics; biomechanics and machine design.

1. R.T. Haftka, Z. Gürdal and M.P. Kamat: *Elements of Structural Optimization*. 2nd rev.ed., 1990 ISBN 0-7923-0608-2
2. J.J. Kalker: *Three-Dimensional Elastic Bodies in Rolling Contact.* 1990 ISBN 0-7923-0712-7
3. P. Karasudhi: *Foundations of Solid Mechanics.* 1991 ISBN 0-7923-0772-0
4. *Not published*
5. *Not published.*
6. J.F. Doyle: *Static and Dynamic Analysis of Structures.* With an Emphasis on Mechanics and Computer Matrix Methods. 1991 ISBN 0-7923-1124-8; Pb 0-7923-1208-2
7. O.O. Ochoa and J.N. Reddy: *Finite Element Analysis of Composite Laminates.* ISBN 0-7923-1125-6
8. M.H. Aliabadi and D.P. Rooke: *Numerical Fracture Mechanics.* ISBN 0-7923-1175-2
9. J. Angeles and C.S. López-Cajún: *Optimization of Cam Mechanisms.* 1991 ISBN 0-7923-1355-0
10. D.E. Grierson, A. Franchi and P. Riva (eds.): *Progress in Structural Engineering.* 1991 ISBN 0-7923-1396-8
11. R.T. Haftka and Z. Gürdal: *Elements of Structural Optimization.* 3rd rev. and exp. ed. 1992 ISBN 0-7923-1504-9; Pb 0-7923-1505-7
12. J.R. Barber: *Elasticity.* 1992 ISBN 0-7923-1609-6; Pb 0-7923-1610-X
13. H.S. Tzou and G.L. Anderson (eds.): *Intelligent Structural Systems.* 1992 ISBN 0-7923-1920-6
14. E.E. Gdoutos: *Fracture Mechanics.* An Introduction. 1993 ISBN 0-7923-1932-X
15. J.P. Ward: *Solid Mechanics.* An Introduction. 1992 ISBN 0-7923-1949-4
16. M. Farshad: *Design and Analysis of Shell Structures.* 1992 ISBN 0-7923-1950-8
17. H.S. Tzou and T. Fukuda (eds.): *Precision Sensors, Actuators and Systems.* 1992 ISBN 0-7923-2015-8
18. J.R. Vinson: *The Behavior of Shells Composed of Isotropic and Composite Materials.* 1993 ISBN 0-7923-2113-8
19. H.S. Tzou: *Piezoelectric Shells.* Distributed Sensing and Control of Continua. 1993 ISBN 0-7923-2186-3
20. W. Schiehlen (ed.): *Advanced Multibody System Dynamics.* Simulation and Software Tools. 1993 ISBN 0-7923-2192-8
21. C.-W. Lee: *Vibration Analysis of Rotors.* 1993 ISBN 0-7923-2300-9
22. D.R. Smith: *An Introduction to Continuum Mechanics.* 1993 ISBN 0-7923-2454-4
23. G.M.L. Gladwell: *Inverse Problems in Scattering.* An Introduction. 1993 ISBN 0-7923-2478-1

Mechanics

SOLID MECHANICS AND ITS APPLICATIONS

Series Editor: G.M.L. Gladwell

24. G. Prathap: *The Finite Element Method in Structural Mechanics*. 1993 ISBN 0-7923-2492-7
25. J. Herskovits (ed.): *Advances in Structural Optimization*. 1995 ISBN 0-7923-2510-9
26. M.A. González-Palacios and J. Angeles: *Cam Synthesis*. 1993 ISBN 0-7923-2536-2
27. W.S. Hall: *The Boundary Element Method*. 1993 ISBN 0-7923-2580-X
28. J. Angeles, G. Hommel and P. Kovács (eds.): *Computational Kinematics*. 1993 ISBN 0-7923-2585-0
29. A. Curnier: *Computational Methods in Solid Mechanics*. 1994 ISBN 0-7923-2761-6
30. D.A. Hills and D. Nowell: *Mechanics of Fretting Fatigue*. 1994 ISBN 0-7923-2866-3
31. B. Tabarrok and F.P.J. Rimrott: *Variational Methods and Complementary Formulations in Dynamics*. 1994 ISBN 0-7923-2923-6
32. E.H. Dowell (ed.), E.F. Crawley, H.C. Curtiss Jr., D.A. Peters, R. H. Scanlan and F. Sisto: *A Modern Course in Aeroelasticity*. Third Revised and Enlarged Edition. 1995 ISBN 0-7923-2788-8; Pb: 0-7923-2789-6
33. A. Preumont: *Random Vibration and Spectral Analysis*. 1994 ISBN 0-7923-3036-6
34. J.N. Reddy (ed.): *Mechanics of Composite Materials*. Selected works of Nicholas J. Pagano. 1994 ISBN 0-7923-3041-2
35. A.P.S. Selvadurai (ed.): *Mechanics of Poroelastic Media*. 1996 ISBN 0-7923-3329-2
36. Z. Mróz, D. Weichert, S. Dorosz (eds.): *Inelastic Behaviour of Structures under Variable Loads*. 1995 ISBN 0-7923-3397-7
37. R. Pyrz (ed.): *IUTAM Symposium on Microstructure-Property Interactions in Composite Materials*. Proceedings of the IUTAM Symposium held in Aalborg, Denmark. 1995 ISBN 0-7923-3427-2
38. M.I. Friswell and J.E. Mottershead: *Finite Element Model Updating in Structural Dynamics*. 1995 ISBN 0-7923-3431-0
39. D.F. Parker and A.H. England (eds.): *IUTAM Symposium on Anisotropy, Inhomogeneity and Nonlinearity in Solid Mechanics*. Proceedings of the IUTAM Symposium held in Nottingham, U.K. 1995 ISBN 0-7923-3594-5
40. J.-P. Merlet and B. Ravani (eds.): *Computational Kinematics '95*. 1995 ISBN 0-7923-3673-9
41. L.P. Lebedev, I.I. Vorovich and G.M.L. Gladwell: *Functional Analysis*. Applications in Mechanics and Inverse Problems. 1996 ISBN 0-7923-3849-9
42. J. Menčik: *Mechanics of Components with Treated or Coated Surfaces*. 1996 ISBN 0-7923-3700-X
43. D. Bestle and W. Schiehlen (eds.): *IUTAM Symposium on Optimization of Mechanical Systems*. Proceedings of the IUTAM Symposium held in Stuttgart, Germany. 1996 ISBN 0-7923-3830-8
44. D.A. Hills, P.A. Kelly, D.N. Dai and A.M. Korsunsky: *Solution of Crack Problems*. The Distributed Dislocation Technique. 1996 ISBN 0-7923-3848-0
45. V.A. Squire, R.J. Hosking, A.D. Kerr and P.J. Langhorne: *Moving Loads on Ice Plates*. 1996 ISBN 0-7923-3953-3
46. A. Pineau and A. Zaoui (eds.): *IUTAM Symposium on Micromechanics of Plasticity and Damage of Multiphase Materials*. Proceedings of the IUTAM Symposium held in Sèvres, Paris, France. 1996 ISBN 0-7923-4188-0
47. A. Naess and S. Krenk (eds.): *IUTAM Symposium on Advances in Nonlinear Stochastic Mechanics*. Proceedings of the IUTAM Symposium held in Trondheim, Norway. 1996 ISBN 0-7923-4193-7
48. D. Ieşan and A. Scalia: *Thermoelastic Deformations*. 1996 ISBN 0-7923-4230-5

Mechanics

SOLID MECHANICS AND ITS APPLICATIONS

Series Editor: G.M.L. Gladwell

49. J.R. Willis (ed.): *IUTAM Symposium on Nonlinear Analysis of Fracture*. Proceedings of the IUTAM Symposium held in Cambridge, U.K. 1997 ISBN 0-7923-4378-6
50. A. Preumont: *Vibration Control of Active Structures*. An Introduction. 1997 ISBN 0-7923-4392-1
51. G.P. Cherepanov: *Methods of Fracture Mechanics: Solid Matter Physics*. 1997 ISBN 0-7923-4408-1
52. D.H. van Campen (ed.): *IUTAM Symposium on Interaction between Dynamics and Control in Advanced Mechanical Systems*. Proceedings of the IUTAM Symposium held in Eindhoven, The Netherlands. 1997 ISBN 0-7923-4429-4
53. N.A. Fleck and A.C.F. Cocks (eds.): *IUTAM Symposium on Mechanics of Granular and Porous Materials*. Proceedings of the IUTAM Symposium held in Cambridge, U.K. 1997 ISBN 0-7923-4553-3
54. J. Roorda and N.K. Srivastava (eds.): *Trends in Structural Mechanics*. Theory, Practice, Education. 1997 ISBN 0-7923-4603-3
55. Yu.A. Mitropolskii and N. Van Dao: *Applied Asymptotic Methods in Nonlinear Oscillations*. 1997 ISBN 0-7923-4605-X
56. C. Guedes Soares (ed.): *Probabilistic Methods for Structural Design*. 1997 ISBN 0-7923-4670-X
57. D. François, A. Pineau and A. Zaoui: *Mechanical Behaviour of Materials*. Volume I: Elasticity and Plasticity. 1998 ISBN 0-7923-4894-X
58. D. François, A. Pineau and A. Zaoui: *Mechanical Behaviour of Materials*. Volume II: Viscoplasticity, Damage, Fracture and Contact Mechanics. 1998 ISBN 0-7923-4895-8
59. L.T. Tenek and J. Argyris: *Finite Element Analysis for Composite Structures*. 1998 ISBN 0-7923-4899-0
60. Y.A. Bahei-El-Din and G.J. Dvorak (eds.): *IUTAM Symposium on Transformation Problems in Composite and Active Materials*. Proceedings of the IUTAM Symposium held in Cairo, Egypt. 1998 ISBN 0-7923-5122-3
61. I.G. Goryacheva: *Contact Mechanics in Tribology*. 1998 ISBN 0-7923-5257-2
62. O.T. Bruhns and E. Stein (eds.): *IUTAM Symposium on Micro- and Macrostructural Aspects of Thermoplasticity*. Proceedings of the IUTAM Symposium held in Bochum, Germany. 1999 ISBN 0-7923-5265-3
63. F.C. Moon: *IUTAM Symposium on New Applications of Nonlinear and Chaotic Dynamics in Mechanics*. Proceedings of the IUTAM Symposium held in Ithaca, NY, USA. 1998 ISBN 0-7923-5276-9
64. R. Wang: *IUTAM Symposium on Rheology of Bodies with Defects*. Proceedings of the IUTAM Symposium held in Beijing, China. 1999 ISBN 0-7923-5297-1
65. Yu.I. Dimitrienko: *Thermomechanics of Composites under High Temperatures*. 1999 ISBN 0-7923-4899-0
66. P. Argoul, M. Frémond and Q.S. Nguyen (eds.): *IUTAM Symposium on Variations of Domains and Free-Boundary Problems in Solid Mechanics*. Proceedings of the IUTAM Symposium held in Paris, France. 1999 ISBN 0-7923-5450-8
67. F.J. Fahy and W.G. Price (eds.): *IUTAM Symposium on Statistical Energy Analysis*. Proceedings of the IUTAM Symposium held in Southampton, U.K. 1999 ISBN 0-7923-5457-5
68. H.A. Mang and F.G. Rammerstorfer (eds.): *IUTAM Symposium on Discretization Methods in Structural Mechanics*. Proceedings of the IUTAM Symposium held in Vienna, Austria. 1999 ISBN 0-7923-5591-1

Mechanics

SOLID MECHANICS AND ITS APPLICATIONS

Series Editor: G.M.L. Gladwell

69. P. Pedersen and M.P. Bendsøe (eds.): *IUTAM Symposium on Synthesis in Bio Solid Mechanics*. Proceedings of the IUTAM Symposium held in Copenhagen, Denmark. 1999 ISBN 0-7923-5615-2
70. S.K. Agrawal and B.C. Fabien: *Optimization of Dynamic Systems*. 1999 ISBN 0-7923-5681-0
71. A. Carpinteri: *Nonlinear Crack Models for Nonmetallic Materials*. 1999 ISBN 0-7923-5750-7
72. F. Pfeifer (ed.): *IUTAM Symposium on Unilateral Multibody Contacts*. Proceedings of the IUTAM Symposium held in Munich, Germany. 1999 ISBN 0-7923-6030-3
73. E. Lavendelis and M. Zakrzhevsky (eds.): *IUTAM/IFToMM Symposium on Synthesis of Nonlinear Dynamical Systems*. Proceedings of the IUTAM/IFToMM Symposium held in Riga, Latvia. 2000 ISBN 0-7923-6106-7
74. J.-P. Merlet: *Parallel Robots*. 2000 ISBN 0-7923-6308-6
75. J.T. Pindera: *Techniques of Tomographic Isodyne Stress Analysis*. 2000 ISBN 0-7923-6388-4
76. G.A. Maugin, R. Drouot and F. Sidoroff (eds.): *Continuum Thermomechanics*. The Art and Science of Modelling Material Behaviour. 2000 ISBN 0-7923-6407-4
77. N. Van Dao and E.J. Kreuzer (eds.): *IUTAM Symposium on Recent Developments in Non-linear Oscillations of Mechanical Systems*. 2000 ISBN 0-7923-6470-8
78. S.D. Akbarov and A.N. Guz: *Mechanics of Curved Composites*. 2000 ISBN 0-7923-6477-5
79. M.B. Rubin: *Cosserat Theories: Shells, Rods and Points*. 2000 ISBN 0-7923-6489-9
80. S. Pellegrino and S.D. Guest (eds.): *IUTAM-IASS Symposium on Deployable Structures: Theory and Applications*. Proceedings of the IUTAM-IASS Symposium held in Cambridge, U.K., 6–9 September 1998. 2000 ISBN 0-7923-6516-X
81. A.D. Rosato and D.L. Blackmore (eds.): *IUTAM Symposium on Segregation in Granular Flows*. Proceedings of the IUTAM Symposium held in Cape May, NJ, U.S.A., June 5–10, 1999. 2000 ISBN 0-7923-6547-X
82. A. Lagarde (ed.): *IUTAM Symposium on Advanced Optical Methods and Applications in Solid Mechanics*. Proceedings of the IUTAM Symposium held in Futuroscope, Poitiers, France, August 31–September 4, 1998. 2000 ISBN 0-7923-6604-2
83. D. Weichert and G. Maier (eds.): *Inelastic Analysis of Structures under Variable Loads*. Theory and Engineering Applications. 2000 ISBN 0-7923-6645-X
84. T.-J. Chuang and J.W. Rudnicki (eds.): *Multiscale Deformation and Fracture in Materials and Structures*. The James R. Rice 60th Anniversary Volume. 2001 ISBN 0-7923-6718-9
85. S. Narayanan and R.N. Iyengar (eds.): *IUTAM Symposium on Nonlinearity and Stochastic Structural Dynamics*. Proceedings of the IUTAM Symposium held in Madras, Chennai, India, 4–8 January 1999 ISBN 0-7923-6733-2
86. S. Murakami and N. Ohno (eds.): *IUTAM Symposium on Creep in Structures*. Proceedings of the IUTAM Symposium held in Nagoya, Japan, 3-7 April 2000. 2001 ISBN 0-7923-6737-5
87. W. Ehlers (ed.): *IUTAM Symposium on Theoretical and Numerical Methods in Continuum Mechanics of Porous Materials*. Proceedings of the IUTAM Symposium held at the University of Stuttgart, Germany, September 5-10, 1999. 2001 ISBN 0-7923-6766-9
88. D. Durban, D. Givoli and J.G. Simmonds (eds.): *Advances in the Mechanis of Plates and Shells The Avinoam Libai Anniversary Volume*. 2001 ISBN 0-7923-6785-5
89. U. Gabbert and H.-S. Tzou (eds.): *IUTAM Symposium on Smart Structures and Structonic Systems*. Proceedings of the IUTAM Symposium held in Magdeburg, Germany, 26–29 September 2000. 2001 ISBN 0-7923-6968-8

Mechanics

SOLID MECHANICS AND ITS APPLICATIONS

Series Editor: G.M.L. Gladwell

90. Y. Ivanov, V. Cheshkov and M. Natova: *Polymer Composite Materials – Interface Phenomena & Processes*. 2001 ISBN 0-7923-7008-2
91. R.C. McPhedran, L.C. Botten and N.A. Nicorovici (eds.): *IUTAM Symposium on Mechanical and Electromagnetic Waves in Structured Media*. Proceedings of the IUTAM Symposium held in Sydney, NSW, Australia, 18-22 Januari 1999. 2001 ISBN 0-7923-7038-4
92. D.A. Sotiropoulos (ed.): *IUTAM Symposium on Mechanical Waves for Composite Structures Characterization*. Proceedings of the IUTAM Symposium held in Chania, Crete, Greece, June 14-17, 2000. 2001 ISBN 0-7923-7164-X
93. V.M. Alexandrov and D.A. Pozharskii: *Three-Dimensional Contact Problems*. 2001 ISBN 0-7923-7165-8
94. J.P. Dempsey and H.H. Shen (eds.): *IUTAM Symposium on Scaling Laws in Ice Mechanics and Ice Dynamics*. Proceedings of the IUTAM Symposium held in Fairbanks, Alaska, U.S.A., 13-16 June 2000. 2001 ISBN 1-4020-0171-1
95. U. Kirsch: *Design-Oriented Analysis of Structures*. A Unified Approach. 2002 ISBN 1-4020-0443-5
96. A. Preumont: *Vibration Control of Active Structures*. An Introduction (2nd Edition). 2002 ISBN 1-4020-0496-6
97. B.L. Karihaloo (ed.): *IUTAM Symposium on Analytical and Computational Fracture Mechanics of Non-Homogeneous Materials*. Proceedings of the IUTAM Symposium held in Cardiff, U.K., 18-22 June 2001. 2002 ISBN 1-4020-0510-5
98. S.M. Han and H. Benaroya: *Nonlinear and Stochastic Dynamics of Compliant Offshore Structures*. 2002 ISBN 1-4020-0573-3
99. A.M. Linkov: *Boundary Integral Equations in Elasticity Theory*. 2002 ISBN 1-4020-0574-1
100. L.P. Lebedev, I.I. Vorovich and G.M.L. Gladwell: *Functional Analysis*. Applications in Mechanics and Inverse Problems (2nd Edition). 2002 ISBN 1-4020-0667-5; Pb: 1-4020-0756-6
101. Q.P. Sun (ed.): *IUTAM Symposium on Mechanics of Martensitic Phase Transformation in Solids*. Proceedings of the IUTAM Symposium held in Hong Kong, China, 11-15 June 2001. 2002 ISBN 1-4020-0741-8
102. M.L. Munjal (ed.): *IUTAM Symposium on Designing for Quietness*. Proceedings of the IUTAM Symposium held in Bangkok, India, 12-14 December 2000. 2002 ISBN 1-4020-0765-5
103. J.A.C. Martins and M.D.P. Monteiro Marques (eds.): *Contact Mechanics*. Proceedings of the 3rd Contact Mechanics International Symposium, Praia da Consolação, Peniche, Portugal, 17-21 June 2001. 2002 ISBN 1-4020-0811-2
104. H.R. Drew and S. Pellegrino (eds.): *New Approaches to Structural Mechanics, Shells and Biological Structures*. 2002 ISBN 1-4020-0862-7
105. J.R. Vinson and R.L. Sierakowski: *The Behavior of Structures Composed of Composite Materials*. Second Edition. 2002 ISBN 1-4020-0904-6
106. Not yet published.
107. J.R. Barber: *Elasticity*. Second Edition. 2002 ISBN Hb 1-4020-0964-X; Pb 1-4020-0966-6
108. C. Miehe (ed.): *IUTAM Symposium on Computational Mechanics of Solid Materials at Large Strains*. Proceedings of the IUTAM Symposium held in Stuttgart, Germany, 20-24 August 2001. 2003 ISBN 1-4020-1170-9

Mechanics

SOLID MECHANICS AND ITS APPLICATIONS

Series Editor: G.M.L. Gladwell

109. P. Ståhle and K.G. Sundin (eds.): *IUTAM Symposium on Field Analyses for Determination of Material Parameters – Experimental and Numerical Aspects.* Proceedings of the IUTAM Symposium held in Abisko National Park, Kiruna, Sweden, July 31 – August 4, 2000. 2003 ISBN 1-4020-1283-7
110. N. Sri Namachchivaya and Y.K. Lin (eds.): *IUTAM Symposium on Nonlnear Stochastic Dynamics.* Proceedings of the IUTAM Symposium held in Monticello, IL, USA, 26 – 30 August, 2000. 2003 ISBN 1-4020-1471-6
111. H. Sobieckzky (ed.): *IUTAM Symposium Transsonicum IV.* Proceedings of the IUTAM Symposium held in Göttingen, Germany, 2–6 September 2002, 2003 ISBN 1-4020-1608-5
112. J.-C. Samin and P. Fisette: *Symbolic Modeling of Multibody Systems.* 2003 ISBN 1-4020-1629-8
113. A.B. Movchan (ed.): *IUTAM Symposium on Asymptotics, Singularities and Homogenisation in Problems of Mechanics.* Proceedings of the IUTAM Symposium held in Liverpool, United Kingdom, 8-11 July 2002. 2003 ISBN 1-4020-1780-4
114. S. Ahzi, M. Cherkaoui, M.A. Khaleel, H.M. Zbib, M.A. Zikry and B. LaMatina (eds.): *IUTAM Symposium on Multiscale Modeling and Characterization of Elastic-Inelastic Behavior of Engineering Materials.* Proceedings of the IUTAM Symposium held in Marrakech, Morocco, 20-25 October 2002. 2004 ISBN 1-4020-1861-4
115. H. Kitagawa and Y. Shibutani (eds.): *IUTAM Symposium on Mesoscopic Dynamics of Fracture Process and Materials Strength.* Proceedings of the IUTAM Symposium held in Osaka, Japan, 6-11 July 2003. Volume in celebration of Professor Kitagawa's retirement. 2004 ISBN 1-4020-2037-6
116. E.H. Dowell, R.L. Clark, D. Cox, H.C. Curtiss, Jr., K.C. Hall, D.A. Peters, R.H. Scanlan, E. Simiu, F. Sisto and D. Tang: *A Modern Course in Aeroelasticity.* 4th Edition, 2004 ISBN 1-4020-2039-2
117. T. Burczyński and A. Osyczka (eds.): *IUTAM Symposium on Evolutionary Methods in Mechanics.* Proceedings of the IUTAM Symposium held in Cracow, Poland, 24-27 September 2002. 2004 ISBN 1-4020-2266-2
118. D. Ieşan: *Thermoelastic Models of Continua.* 2004 ISBN 1-4020-2309-X

Kluwer Academic Publishers – Dordrecht / Boston / London